The Evolution of Intelligence

By the same author

Crisis in Abundance
Unstated Assumptions in Education
Critical Issues in Polynesian Education in New Zealand
(editor and contributor)

The Evolution
of Intelligence

A general theory and some
of its implications

DAVID STENHOUSE

London: George Allen & Unwin Ltd
Ruskin House   Museum Street

First published in 1973

© George Allen & Unwin Ltd, 1974

ISBN 0 04 575017 3

Printed in Great Britain
in 10 pt. Times type
by Unwin Brothers Limited
Old Woking, Surrey

The Evolution of Intelligence

A general theory and some
of its implications

DAVID STENHOUSE

BOOKS
10 East 53d St. New York 10022
(a division of Harper & Row Publishers, Inc.)

ISBN–06–496518–x

Published in the U.S.A. 1974 by
HARPER & ROW PUBLISHERS, INC.
BARNES & NOBLE IMPORT DIVISION

Printed in Great Britain
in 10 pt. Times type
by Unwin Brothers Limited
Old Woking, Surrey

Acknowledgements

It is a pleasure to acknowledge the support and encouragement I have received in the long and often frustrating task of working out this 'general theory' to its present form. Several eminent and busy men took time to read an early draft of the basic argument (including Chapter III in virtually its present form), and gave encouragement to further work. The suggestion of book-length publication came from Dr Niko Tinbergen, FRS in 1964; Sir Julian Huxley, FRS and Dr Konrad Lorenz provided a number of searching and helpful comments in 1965; and in 1966 the late Dr W. C. Halstead, with whose neurologically based picture of intelligence my own theory had converged, remarked in a letter that my diagram of the operation of four-factor intelligence (published in *Nature* (1965), vol. 208, p. 815, see Fig. 8) was almost identical with what he had elaborated (but had not published) in the course of informal discussions with his own postgraduate students.

Many friends and past students, and colleagues in several universities, have abetted the theory-building in one way or another; and even more important, have provided the personal faith and support without which I should many times have abandoned the whole enterprise. It is impossible to name everyone who has supported the endeavour in this way – I am deeply grateful to all, and those named are to be understood as to some extent symbolic of the rest – but I must thank Desmond and Brenda Judge, Jack and Joan Greenwood, Malcolm and Sandra Thorne, Ian and Barbara Bock, Mrs Bev Priddle (*née* West), Stephen and Brenda Day, and Mrs Mary Ware; Professor Jack Campbell of the Department of Education, Dr George Naylor and Professor Don McElwain of the Department of Psychology, University of Queensland; Murray Elliot of Canberra; Lou Gurr and Professor W. C. Clarke of the Department of Zoology, Professor Jack Veale of the Department of Horticultural Science, and especially Professor Clem Hill and Dr Don McAlpine of the Department of Education, Massey University; and Jack Griffith, Duncan Neilson, and Gurmit Singh.

I am particularly grateful to Dr Ronald Fletcher, lately Professor of Sociology in the University of York. He has given invaluable support in the final preparation of this book and has read through the final manuscript. Ronald Fletcher has brought a great deal of sociological and older psychological literature to my attention—though it has proved impossible, in the circumstances, to incorporate discussion of it in the present volume. It is a disturbing symptom of our increasing professional compartmentalization that only in June 1971 did I become aware of Dr Fletcher's work. I would greatly have liked to attempt some discussion of this important work in the present volume – but this must be saved for a later book.

I would also like to thank the Clarendon Press, Oxford, for permission to reproduce illustrations from Niko Tinbergen's *The Study of Instinct*.

Naturally, none of the persons named above can be held responsible for the views expressed in this book.

I am conscious of the great debt of faith owed to my parents and my children and other relatives by blood and marriage, and above all to my wife who has supported despair and tempered the extravagances of hope, who typed many drafts of manuscript and coped with a ten year's deluge of books and papers within our home.

Contents

Illustrations

Introduction

Human behaviour achieves adaptiveness – and has become the major factor affecting the survival of life on earth – through being intelligent. The behaviour of all the other animals achieves adaptiveness by being instinctive. Yet man supposedly evolved from infra-human ancestors, if we accept the implications of the Darwinian revolution in the biological sciences. How, precisely, could the jump from an instinct-based to an intelligence-based system of behaviour have been accomplished? The fact that it *has* happened should not blind us to the theoretical difficulties that remain with regard to *how* it happened.

It is interesting and somewhat disturbing to note some of the different reactions to this problem within the social and biological sciences. Many biologists tacitly ignore or even deny the fact of human intelligence: there is a conspicuous paucity, for example in *Biological Abstracts*, of references to 'intelligence' (see Chapter 1). Many social scientists, on the other hand, deny the implications of man's membership of the animal kingdom. Explicit denials of the importance and even the presence of instinctive mechanisms in human behaviour are still to be found in many textbooks in sociology, education, and other fields. And many psychologists appear still tacitly to deny instinctive mechanisms in animals: 'learning theory' excludes everything else from consideration, all behaviour being assumed to be learned behaviour.

In contrast to this, some encouraging developments have recently occurred. David S. Palermo (1971) has just surveyed the evidence suggesting that a 'scientific revolution' (in the sense argued by Kuhn (1962) in his increasingly influential *The Structure of Scientific Revolutions*) is taking place in psychology. Ronald Fletcher's excellent book *Instinct in Man* has been re-issued in paperback form, presumably an indication that it is being extensively used – and Professor Fletcher, significantly enough, held the Foundation Chair at the University of York . . . in Sociology!

In a slightly different modality Desmond Morris and Robert Ardrey continue their theoretical explorations under the guise of 'popular' exposition. Their works have been called 'science fiction' by some reviewers – but while evidence may not yet be to hand to support some of their statements, and while other contentions may prove in fact to be unfounded, science fiction their work is not. Evidence and documentation, while often not given in detail, usually could be given to support many of their assertions; and where they go beyond the evidence at present available their suggestions are usually at least consistent with what is known, and could be regarded as merely pointing the way towards areas of research that might reveal decisive evidence. Both authors continue to figure in the 'bestseller' lists, bringing the viewpoint of evolutionary biology to an audience proportionately larger than at any time since Darwin himself. (Perhaps this is what makes some learned reviewers so annoyed with them.)

In general, then, it appears that the intellectual climate is becoming suitable for the investigation of a major problem in the 'interdisciplinary' area between the biological and the social sciences. The problem of accounting for the evolution of intelligence is perhaps the most important, intriguing, and difficult in this area.

One thing seems very clear: the problem is not going to be solved by pretending it does not exist. Denial of its existence involves denial of a substantial sector either of biological or of social science. Stated thus explicitly such denial is unthinkable. If, on the other hand, the existence of the problem, and its importance, are acknowledged, it may be possible to do something about it. An attempt at clarification may be made. Depending upon how the problem is formulated, it may be found that evidence is already (and may for long have been) available to indicate the course of past events in evolution. Such evidence may well have gone unnoticed in the absence of a theoretical framework which would have brought out its importance. Thus clarification in one area may lead to the discovery of evidence in another, and this in turn to further clarifications and discoveries. In particular, it seems obvious that at some stage or other in the evolution of man, a very substantial transition must have occurred, from a behavioural system dependent upon instinct to a behavioural system dependent upon intelligence. Since we know the beginning and the end states of the transition it should not be impossible to work out a likely sequence of intermediates. In attempting to do this two methodological essentials have to be kept in mind:

(1) Continuity of adaptiveness must have been preserved within

the behavioural systems in question, otherwise the phyletic line would have become extinct; and the account given must allow for this.

(2) Clarity at the conceptual level has to be maintained throughout the theoretical exploration despite possible loss of operational or evidential linkages. It is only *after* the theoretical re-orientation has been taken as far as possible that realignment with evidence becomes useful—if attempted prematurely it cannot but distort the argument.

This book is an attempt to work out just such a sequence for the transition from instinct to intelligence, in accordance with these axioms of methodology.

The starting-point of the present conceptual exploration lies in the body of ethological theory. This is outlined in Chapter II, various methodological issues having been covered and the general problem defined in Chapter I. The basic argument of the book, an abstract argument for the theoretical requirements for the evolution of intelligent behaviour, occupies Chapter III. Chapter IV is devoted to a short preliminary survey of some supporting evidence. In Chapter V an attempt is made to see if the new theory of evolutionary intelligence can provide illumination in the current major controversy between 'creativity researchers' and 'intelligence measurers': a means of resolving the 'IQ v. creativity' problem is suggested.

Evolutionary/phylogenetic issues are taken up again in Chapters VI and VII, where re-interpretation of existing evidence is undertaken to discover to what extent present views may have been implicit in already-established fact and theory.

Chapter VIII examines the 'recapitulationary' possibilities of similarity between the present theory and the findings of Piaget on the ontogeny of intelligence in the individual development of children.

Some indications are given in Chapter IX of the interaction of mechanisms of instinct with intelligence in social transactions; and the educational implications of the new picture of 'evolutionary intelligence' are explored in the relatively long Chapter X. (Here as elsewhere, the most salient revelation is of the great extent of further implications, at both theoretical and practical levels, that remain to be explored.)

Methodological issues are discussed in the last two chapters, those regarding learning theory in Chapter XI, and broader issues shading into philosophy of science and philosophy itself in Chapter XII.

Finally, in the Epilogue, the four-factor picture of evolutionary intelligence is transformed into five-factor 'intelligence' by the incorporation of 'instinct'; and in the Appendix some recently-published material relevant to the present theory is briefly discussed. (Since an attempt to incorporate discussion of this very recent work in the body of the text would lead to confusion at this stage I have compromised by inserting bare reference citations in the text and confining quotation and brief discussion to the Appendix.)

I

The Need for an Evolutionary Account of Intelligence; Some Preliminary Clarifications

Evolutionary necessity . . . also applies to the origins of human thinking (RENSCH 1966 : 348).

Controversies over the nature of intelligence, what intelligent behaviour *is*, have distracted attention from its evolution, and have obscured the possibility that enhanced understanding of its evolution might lead to an improved understanding of its nature. Hebb (1949) remarked that 'psychology has an intimate relation with the other biological sciences, and may . . . look for help there'. And as Huxley (1964) has said of the evolution of 'mind': 'The only satisfactory approach . . . is an evolutionary one.' This book comprises an attempt to work from evolution to intelligence.

1. Animal behaviour is among the most recent parts of animate objective reality to open up its causal background to the human understanding. The barriers against the admission of animal behaviour into the subject-matter of science were for long not technical but mental. For centuries we asked the wrong questions, we looked at the doings of animals through the wrong eyes. Then, about the beginning of the present century, a few people began to dispense with the human-coloured spectacles they had been wearing; they began to watch animals, without anthropomorphic preconceptions, and this allowed them to ask the right sorts of questions, questions to which answers could, by the use of appropriate observational and experimental techniques, be obtained. In those early days the only people who studied animal behaviour were the individualists, the explorers and originators

who were independent of, and who therefore in the long run led and moulded, scientific public opinion. Besides the sustained patience, the subtlety, breadth, and pertinacity of mind which are indispensible for real increases in the understanding of animal behaviour at any time, the pioneers in the subject had to possess, or to develop, various attributes of personality which even now cannot be described except in seeming paradox:[1] hardihood with sensitivity; self-confidence and self-doubt; a disregard for convention, along with the ability to use the conventions as a protective garment; an immense humility, and an immense . . . pride? No, not pride – awareness of capability, faith in their own abilities might describe it best; and it is accompanied usually by a strong sense of responsibility for the use of this intellectual power. Through these qualities an understanding of animal behaviour has been gained; and the methods of this understanding have been brought into acceptance within the world of science. The study of animal behaviour has in the past fifteen years become unexceptionable. Ethology has been incorporated into the scientific orthodoxy. Many people are now following the tracks cut by the pioneers, and broadening and smoothing them and straightening the corners. The special qualities of the pioneers have become redundant for most contemporary work on the behaviour of animals – and these qualities have, in fact, largely disappeared. Nowadays we can get to know about the lives and reactions of insects and fish, migrating birds and socialistic monkeys simply by reading a few books on the theory of instinct and going out and practising the research methods that have been found appropriate. No need, nowadays, to indulge in the risky sport of imaginative speculation – that phase is past for most parts of the subject. Yet there are, still, a few problems for which we have not found the right mental formulae: there are still parts of the frontier to be broken in. This book is an attempt to explore one small section of the frontier, a region that is not unimportant though few of the pioneers have so far tried to enter it. Journeys have been made all round, and the 'blank patch' has been noted and remarked upon – in tantalizing and sometimes forbidding terms! So this journey is not of Columban type – it is perhaps more like a trip into the Matto Grosso. Yet though the Matto Grosso is not the continents of the Americas, it still needs

[1] cf. below, Chapter V. Comfort (1970) writes of Sir Julian Huxley as exhibiting a 'mixture of great self-confidence with great antithetical vulnerability'.

to be explored. The concept of 'intelligence' is one which, like it or not, is widely used – and though difficulties with regard to its detailed elucidation are numerous, our accepted linguistic usages (apart from anything else) do indicate that we know what we are talking about, generally, when we use the word. This is, of course, on the general or commonsense level. As we move from the popular to the learned, however, we find a tendency for 'intelligence' to drop out of the vocabulary. Educationalists still use the word, so do many psychologists (see Guilford 1967), especially those concerned with clinical, educational and other 'applied' work; whereas the more academic psychologists, along with most experimental workers in the behavioural sciences generally, tend to avoid it – if they can.

2. For a number of years there has been this curious situation in the literature of behaviour and of biology generally. On the one hand the evolutionary biologist apparently takes it as a commonplace that the mammals, for example, are characterized and distinguished by their high intelligence (Colbert 1945; Simpson 1953, 1967), and that man is the most intelligent of the mammals (Rensch 1966). People actively and directly concerned with experimental behavioural work (refs. in Hinde 1966), on the other hand, have been chary of using the term 'intelligence' at all (or even 'intelligent behaviour' which avoids, one would have thought, even the appearance of the fallacy of reification). Perusal of the subject index of the *Zoological Record* or of *Biological Abstracts* will reveal many entries on a multiplicity of behavioural topics, but usually *nothing* on intelligence or intelligent behaviour. What, one may reasonably ask, is the reason for this peculiar state of affairs? Is it that the general evolutionary biologists rush in where the 'experts', knowing better, disdain to tread? Or is it that the 'experts' have got lost among the trees, and only the evolutionary biologists can perceive the real outlines of the wood? In this regard perhaps educationalists also have retained a greater degree of biological realism than workers in the more particularized and experimental behavioural sciences. It seems not unlikely that they have. People involved in the hurly-burly of the practice of education are subject to 'selection pressures' towards pragmatical soundness – if one's theories and practice are too far out of touch with reality, one suffers for it, personally and directly – and the achievement of 'workability' does necessitate a concordance with reality, even though the

realities may not be understood, consciously and explicitly, in detail.

Of one thing there can be no doubt. *Any* question relating to intelligence is bound, by the very nature of the subject-matter, to be difficult, hence likely to be controversial (c.f. Harlow 1958; Hebb 1949; Introduction). Few writers (except those who have denied the existence of or the possibility of investigating intelligence) have cared to make general and categorical statements about it. Yet because difficulties have been encountered does not mean that we should abandon our attempts to come to grips with intelligence – even less do they entail the non-existence of the presumptive 'object' of inquiry. On this last point I am in full agreement, in terms both of scientific and philosophical methodology and of factual belief, with the words of Konrad Lorenz: 'Somehow, I put great faith in common parlance and I think there *is* something very real that corresponds to the word "intelligence"!' (Pers. comm.)

It appears likely that the reluctance of biologists, anthropologists, psychologists *et al.* to attempt an investigation of the evolution of intelligence is due, like the earlier reluctance to investigate animal behaviour in general, to mental rather than technical barriers. The 'avoidance behaviour' of the learned world seems likely to have been a learned reaction! It is probably activated by the covert precept, held widely indeed, that science can and should deal only with facts. These 'facts' seem usually to be understood in a narrow physical and concrete sense. If science were constituted exclusively by the direct study of such facts the evolution of intelligent behaviour would of necessity be excluded from its subject-matter. Behaviour is not very concrete, the behaviour of long-dead animals if possible even less so. 'Behaviour does not fossilize,' as Mayr (1958) remarks. So there might appear to be no 'facts' on which a theory of the evolution of intelligence could be based. This is, of course, quite mistaken. There are just as many facts, and facts of the same sort, relevant to the phylogeny of intelligence as there are to the phylogeny of any other behavioural feature. The facts are: *present* features of behaviour, plus structures, etc., relating to behaviour, which can be found in fossilized form. The reconstruction of behavioural phylogenies (e.g. Delacour and Mayr 1945; see also Mayr 1958) need involve only slightly more inferential treatment than does the reconstruction of a structural phylogeny (e.g. of any of the structures of horses mentioned by Simpson 1955).

Facts would appear to provide 'constitutive' rather than 'constituted' rules (Piaget 1932) for scientific activity. One does not decry the factual. Bascially, all science is disciplined by the objective, the concrete, the factual. It is nevertheless an erroneous view that on the one hand what we know as 'facts' are simple and in some sense primary or primitive; and, on the other, that science is composed of nothing but such 'facts'. 'Theory' is not something undesirable which unfortunately does come into science, as a sort of unavoidable accident, but which should be minimized. As Thorpe (1963: 470) very pertinently says (and one could wish to make his words compulsory reading for all students of science): 'Science in order to progress must also philosophize and must enter upon a thorough criticism of its own foundations.' Philosophers have known for a long time that 'facts' are not simple (this is a main implication of the work of Locke (1632–1704) to cite only one example); and argument which, though ultimately it must rest on a factual basis, is itself highly abstract, is the very essence not only of philosophy but also of science (see Kuhn 1962; Toulmin 1961; Thornton 1968).

The attempt to elucidate the evolution of intelligence must, then, be abstract and generalized, at least in the initial stages. Along with the abstract argument in this book, however, it will be shown that a number of factual items, otherwise rather isolated and disjointed and of problematical significance, can be brought into meaningful relationship. This does not establish the truth of the theory here advanced – to assume that it did would involve the logical fallacy of affirming a consequent – but it does tend to substantiate the claim that this theory can be a fruitful means of working towards the truth. So long as no contradiction of established fact is embodied in this general theory for the evolution of intelligent behaviour, any re-orientation of outlook and attitude which may result from its promulgation can be heuristically useful. Mayr wrote in 1958: 'The backward state of the field of animal behaviour is, to a considerable extent, due to the former absence of working hypotheses and heuristic schemes. . . . It is the particular merit of Lorenz to have provided a set of hypotheses and theories that have tremendously stimulated research in this area. This is a merit which is not decreased even if some of these hypotheses were over-simplifications, or even entirely wrong.' And it must be remembered that no scientific theory has so far proved absolute: the power of a theory is judged by *increase in understanding*, not by a transition from total ignorance to perfect knowledge (see Toulmin 1961).

3. It is needful to attempt the formulation of a definition of intelligence at the highest level of generality. Highly generalized definitions have in the past been rejected as lacking in immediate 'operational' utility. Such utility is not to be decried. It is a necessity, for example, for the practising psychologist. But the initial formulation of a general definition does not rule out the possibility of subsequent 'operational' deductions; and the practical implications may be made very much more revealing, more useful, if a sufficiently wide sweep of reality has been encompassed by the original work at the generalized level. So with regard to intelligence it may be that an evolutionary approach can lead eventually to clarifications at the 'operational' or practical level which could have been attained in no other way – indeed, I believe (and, naturally, hope) that this will prove to be the case with the theory presented here.

Precise definitions of 'intelligence' or 'intelligent behaviour' are notoriously difficult to make. The *reductio ad absurdum*, so far as verbal definitions are concerned, is 'Intelligence is what is measured by an intelligence test' (cf. Thorndike 1921). Apparently as a result of this inability to pinpoint in words the subject of their discussions, most writers have been rather vague as to the manner in which it has come into existence. As Lashley (1949) points out, the evolution of intelligence usually comes to be 'described in terms of performance in situations requiring learning. This view is based on a confusion of definitions.' He goes on to give an excellent discussion of the relevance of the investigation of perceptual mechanisms, rather than learning experiments, to the understanding of intelligence – reference will be made later to various points arising from his work – but despite this promising start he deals with sensory and neurological mechanisms, not with intelligent behaviour as such. Harlow (1958) uses 'intelligence' as being apparently synonymous with 'learning ability', in several places (e.g. p. 271). On the question of the evolution of 'learning-cum-intelligence', his position seems a fairly typical one: 'As long as increasingly complex receptor systems provide the organism with slight survival advantages, one can be assured that increasingly complex nervous systems will develop; and as long as increasingly complex nervous systems develop, the organisms will be endowed with greater potentialities which lead made) and differentiated as much as possible (thus facilitating the sorting out of distinguishable elements from often intricately compounded complexes of behaviour).[1] These requirements are elaborated in (3) and (4) below

(3) As a particularization from (2) above the theory must be able to show clear relationships between intelligent behaviour, however defined and characterized, and the other types of behaviour found in the animal kingdom. The major desideratum is that a clear logical differentiation should be made between intelligent behaviour and instinctive behaviour. It must be emphasized right from the start that the differentiation sought is a logical or conceptual one, and that even the clearest and best-established conceptual distinction does not necessarily enable every real-life incident of behaviour to be pigeonholed decisively as being of one category or the other. In terms of the currently orthodox distinction between 'instinctive' and 'learned', for example, it is now generally recognized that, even apart from the virtual impossibility of making a categorical all-or-nothing assignment of a particular activity to one side or the other, it is usually a matter of considerable difficulty to disentangle even the elements contributed from the side of learning from those contributed by instinct. Lorenz (1965, esp. Chapter V) reviews evidence for rejecting simplistic and reductionist interpretations, which have often been accepted as showing that 'all behaviour, down to its smallest elements, owes its adaptiveness to both processes', i.e. both learning and instinct. Specifically, he shows (Chapter VII) the limitations in the conclusions that can legitimately be drawn from deprivation experiments. 'The deprivation experiment, taken by itself, can justify direct assertions only about what is not learned' (p. 85). 'If . . . our subject would have failed to show a specific response [as a result of deprivation of relevant "information"], we should *not* be justified in asserting that this response is normally dependent on learning' (my emphasis added). Since much of the 'evidence' purporting to show a preponderance of learned elements in behaviour is from deprivation experiments of one sort or another, it follows that the status of learning with

---

[1] Shouksmith (1970), in a lucid and adaptively orientated discussion, remarks that 'creativity, intelligence, and problem-solving may be treated as facets of one process, yet still be effectively analysed'; and the basic part of such analysis is to work out the necessary prerequisites for each type of functioning, as is attempted in Chapter III.

regard to a great deal of natural adaptive behaviour must be regarded at the present time as questionable. The position was summed up succinctly in an incident at the 1955 Macy Conference. An experiment was described in which a young monkey had been kept from birth with boxes fastened over its hands and feet. When released it failed to show the normal scratching and grooming activity of the species; and this was asserted as proof that grooming was a learned not an instinctive activity. Tinbergen asked: 'Did it *not learn to*; or did it *learn not to*?'

Intelligent behaviour and instinctive behaviour must, then, be characterized in the clearest possible terms at the conceptual level, so that they can stand in maximal contrast and thus facilitate the identification of the essential features of each within the confusions and interpenetrations of actual behaviour. In connection with this it must be reiterated that the establishment of a clear-cut *logical* distinction does not automatically ensure clear-cut separation of different types of behaviour in practice. Conversely, difficulties of practical identification and differentiation must not be taken as vitiating sound theoretical distinctions.

(4) As a further particularization from (2) and (3) above, the contrasting definitions of intelligent and instinctive behaviour should allow for the derivation of operational consequences susceptible to incorporation in an evolutionary account in which selection pressures could plausibly be assigned for the production of the changes postulated as necessary. It should be possible to assign selection pressures, appropriate to particular habitats and modes of life, which could plausibly be expected, in terms of known mechanisms in analogous cases, to produce the features observable in intelligent behaviour. These features should be shown to be produced in a plausible temporal sequence, a reasonable order of phylogenetic appearance. It would be unreasonable, for example, if the bulk of postulated changes were to be dependent on the very early development of true speech. In such a case, not only would the postulated sequence of events be inconsistent with our present inferences as to what actually happened, but 'higher' functions of restricted distribution would be involved to explain the appearance of 'lower' and more widely distributed functions.

Finally the evolutionary or phylogenetic account should provide points of detail for comparison with actual characteristics of living animals selected as being equivalent, in relevant

respects, to the postulated ancestral types in the phylogeny of intelligent behaviour.

I claim that all of these requirements are met, substantially even if not exhaustively, in the theory elaborated in the subsequent chapters of this book.

GENERAL DEFINITIONS OF INSTINCT AND INTELLIGENCE

Since the requirement is to set up initial definitions of instinct and intelligence that are at the same time of the highest possible degree of generality and in the greatest possible contrast, it is clear that they must abstract from a great deal that is important at the operational and descriptive level with regard to both types of behaviour. The consequence of this is that the general definitions are unlikely to provide, in themselves, recognizable descriptions of the behavioural categories involved.

With regard to the concept of instinctive behaviour, Thorpe (1963: 50) suggests that: 'The most essential and fundamental part of the concept is that of "internal drive", and that this drive cannot be considered solely as the expression of some primary physiological need in the ordinary sense of that term.' To this statement may be added: 'Lorenz's chief contribution to the concept of drive was that of action specific "energy".'

Besides the postulate of 'drive resulting from the accumulation of action specific energy', derived as above, Thorpe suggests that 'other important characteristics of instinct are three: (a) It is an inherited system of co-ordination; (b) it involves more or less rigid inherited action patterns; and (c) more or less rigid inherited releasing mechanisms.'

Further explanations of the evidence for and significance of the concepts of action specific energy-releasing mechanisms, and so on, are given in Chapter II. At present the only concern is to formulate a general definition of instinctive behaviour; and on the basis of the salient features of Thorpe's suggestions, it appears that this can be done, in the following terms: The instinctive behaviour of an animal forms a rigid system, inherited as a whole, which is adaptive within the normal circumstances of life of the species in question.

It needs to be pointed out that many components in such a system may involve learning; but that this changes neither the systematic nature nor the overall nor detailed rigidity of an animal's behaviour. 'What rules ontogeny, in bodily as well as in behavioural development', says Lorenz (1965: 42), 'is obviously the hereditary blueprint contained in the genome and not the environmental circumstances indispensable to its realization. It is not the bricks and mortar that rule the building of a cathedral but a plan that has been conceived by an architect and which, of course, also depends on the solid causality of bricks and mortar for its realization. The plan must allow for a certain amount of adaptation that may become necessary during building; the soil may be looser on one side, necessitating compensatory strengthening of the fundaments. The phylogenetically adapted blueprint of the whole may rely on subordinate parts modifying each other. . . . The adaptive modification effected by one part on the other may even take the form of true learning. The gosling "knows innately" that it should copulate with a fellow member of the species stretching out low in water, but it has to learn what a fellow member of the species is.'

In short, various provisos are admitted regarding instinctive behaviour as dealt with *at the descriptive level*. Relative to reflexes in the strict sense, it is more flexible, as Hebb (1949) among others has pointed out. Individual variation within a population can be considerable, due to genetic and epigenetic differences especially in learning. But within the lifetime of the individual, once each particular instinctive (plus learned) pattern has formed, it is largely, even if not absolutely, irrevocably set. Descriptively, of course, a particular species' behaviour may involve intelligence, hence may be variable – but this is due to the interaction of instinct with intelligence.

Relationships between instinct, learning, and intelligence are discussed more fully in Chapter XI. For the moment the only point to be made – even if only provisionally – is that of the rigidity of the instinct system as a whole.

Turning now to the question of intelligence, the most extreme divergence from the concept of instinctive behaviour as being fixed and rigid is of course the logical contrary, namely, behaviour that is variable: non-fixed and non-rigid. Since we assume throughout a context of evolutionary causation through the mechanisms of natural selection, it is clear that *all* the behaviour in question, whether instinctive, 'learned', or intelligent, whether fixed or variable, must *in general* be adaptive. The behavioural repertoire of a species is adaptive only in general, because specific activities on particular occasions may well be non-adaptive. The 'rolling up in a ball' defensive reaction of the hedgehog (*Erinaceus*) is non-adaptive with regard to a 'predator' in the form of a car or truck – nevertheless up until the twentieth century this reaction has served the species well, as no doubt it still does against 'natural' predators away from the highways. As with any characteristic or trait of a living organism it is the overall balance of effects that is decisive, and an adaptive (or positively adaptive) trait is not necessarily totally lacking in maladaptive features (see Dobzhansky, 1956). Thus the adaptiveness in question with regard to intelligent behaviour must be the adaptiveness of the particular action, not a general adaptiveness of whole classes of actions. This is not to say that every attempt to achieve adaptiveness must be successful: even human genius makes stupid mistakes! And at the other extreme it is possible for some chance-achieved innovation to be highly beneficial. So at the level of judging actual activities in real-life situations it is unwise to make categorical assertions on the basis of limited observation. What is being attempted here is not the judgement of actual incidents, however, but the determination of what judgements *would* be made if conditions were known in complete detail. The situational logic of judgement is being explored. All statements are in a sense operationally hypothetical, in that what is asserted would be dependent, in an ordinary descriptive statement, upon extensive series of observations and exhaustive knowledge of circumstances – and these conditions, in practice, are often not met. But the conditionals are valid provided that the conditions are sometimes met, or that they could be met; and this does sometimes happen. Sometimes we are *not* mistaken when we ascribe the adaptiveness of an (human) action to intelligent intent. (For an exemplification of this type of operational logic see Ziedins 1956 following Wittgenstein 1953.)

A further proviso must be incorporated in the definition of intelligent behaviour, namely that the variability in question must occur within the lifetime of the individual. It is necessary that this be specified, since phylogenetic variability is present in all behaviour even of the most rigid instinctive type, just as it is within the structural rigidities upon which our recognition and classification of species depends; and it is not this phylogenetic variability that is in question.

Thus the general definition of intelligent behaviour at which we arrive is: Intelligent behaviour is behaviour that is adaptively variable within the lifetime of the individual.

Before leaving this chapter it will be convenient to make several provisos as to terminology.

'Intelligence' and 'intelligent behaviour' are often used interchangeably in what follows. Where behaviour *per se* is in question I have usually felt it desirable to use the phrase 'intelligent behaviour'. Where the *capacity* for intelligent behaviour is in question I talk of 'intelligence'; but I have on occasion used the latter term as shorthand for the phrase, especially where reiteration of 'intelligent behaviour' would seem pedantic. In the context of an evolutionary discussion there seems little danger of generating confusion. 'Intelligence' in the 'capacity' sense can be regarded as deriving directly from part of the genotype, namely that part of it that is required for the development of 'intelligent behaviour' at the phenotypic level. These two usages can roughly equate with Hebb's (1949: 294–6) 'intelligence A' and 'intelligence B': 'One is (A), an *innate potential*, the capacity for development, a fully innate property that amounts to the possession of a good brain and good neural metabolism. The second is (B) the functioning of a brain in which development has gone on, determining an *average level of performance or comprehension* by the . . . person.'

Note that Hebb is mistaken in suggesting that 'a good brain and good neural metabolism' are 'fully innate'. Only the genome, the information coded in the chromosomes and cytoplasm of the zygote, is innate. As Lorenz says (1965: 37), 'No biologist . . . will forget that the blueprint contained in the genome requires innumerable environmental factors in order to be realized in the phenogeny of structures and functions.' Thus the brain and its functioning can be but a product of the interaction of genome with factors of the environment. Nevertheless the brain itself is a prerequisite for behaviour, and particular capacities of the brain must be regarded as prerequisite for intelligent behaviour. It is unnecessary that these capacities should be specified at this stage. Neither is it necessary that we should be committed to envisaging them as structures or as functions: an amorphous structure and/or function is sufficient for the present. What is important is that the relationship of 'capacity for intelligent behaviour' and 'intelligent behaviour itself' should be appreciated.

The specific features or capacities of the brain that constitute the 'capacity for intelligent behaviour' are subjected to evolutionary selection pressures only through their effect on behaviour, notably by making it adaptively variable, intelligent. Selection to increase 'capacity for intelligent behaviour' thus operates directly through 'intelligent behaviour itself' – the former can increase only if the latter increases (and vice versa). Behind the 'capacity for intelligent behaviour', understood as structural and/or functional and thus itself phenotypic, is, finally, the genome; and this in turn can be selected for or against only through the series of 'layers' arranged concentrically outward, as it were, to the phenome. The outermost 'layer' could well be regarded as purely behavioural: it is through its behaviour that an animal makes its most immediate and most intimate adjustment to the environment, hence through behaviour that selection pressures operate most strongly.

Thus, since the whole intention of the following arguments is to elucidate evolutionary possibilities and probabilities, it should make no difference whatever if, on occasion, there should be a terminological conflation between intelligent behaviour *per se* with the capacity which makes it possible. The two are necessarily interrelated, and while it is recognized that a capacity or potentiality need not always be actualized, there is no way within the evolutionary mechanisms as now understood in which a capacity can be progressively and massively increased except by positive selection of its actualization in behaviour. Provided this is kept in mind there seems no need for pedantic punctiliousness in the general argument to separate our usages of 'intelligent behaviour' and 'the capacity for behaving intelligently' within the more general concept of 'intelligence'.

Further discussion of various logical and methodological issues will be offered after the main arguments have been elaborated into the general theory (Chapter Three), and after some of the implications of this theory in a variety of fields have been explored. But before moving on to the central arguments regarding the evolution of intelligence it is necessary to give a brief outline of our theoretical understanding of the type of behaviour that was the precursor of intelligence and that forms the matrix within which intelligence has developed. Instinctive behaviour must have temporal priority over intelligence in both phylogeny and ontogeny for reasons that will become sufficiently obvious later, so Chapter Two presents some of the essentials of contemporary ethological instinct theory.

B

II

Modern Behaviour Theory: Ethology

Our understanding of the behaviour of animals advanced into a new dimension with the formulation and refinement of the body of theory associated with the ethological school of which the pre-eminent names are those of Lorenz and Tinbergen, with whom are associated Thorpe, Hinde, Marler, Morris and others. (A historical appreciation, in brilliant miniature, is to be found in Huxley 1964). The emergence of the new methodology may be dated at 1935 with the publication of Lorenz's *Der Kumpan in der Umwelt des Vögels*; its classical exposition in English is Tinbergen's *The Study of Instinct* (1951). Since then there has been a rapidly increasing volume of experimental and observational research within the ethological framework. In the 1950s ethology had to fight for recognition as a legitimate way of studying the behaviour of infra-human animals. The 1960s have seen the still-continuing theoretical controversies largely ignored by growing battalions of empirical researchers, working in every continent on representatives from all the major groups of animals. As we enter the 1970s there are signs that ethological assumptions and methods are about to be applied on an increasing scale in the study of human behaviour. This may be expected to precipitate a crisis at the theoretical level between the learning theorists and the instinctualists . . .

In order to appreciate the significance of the ethological advance it is necessary to know something of the historical context of attempts to achieve a general scientific explanation of animal behaviour; and within this to understand the methodological limitations of the early theories. Thus what is specific to the ethological approach will be seen in relation to the problems for the explication of which ethological theory was generated.

HISTORICAL AND METHODOLOGICAL BACKGROUND

The first properly scientific explanations of animal behaviour with which we need concern outselves are those that appealed to experimental findings on the 'reflex', or reflex-arc reactions. These reactions were significant as being the first that allowed the demonstration of a clear-cut relationship between a structural pattern at the neuronal level running into and within the central nervous system, and circumscribed and specifiable behavioural activities. The central nervous system had been opaque to detailed investigation prior to this, hence the great significance that attached to the early demonstrations of specific reflexes. It was soon shown (Sherrington 1906) that the reflex was often not so functionally isolated as had originally been thought: the reaction to a standard stimulus was found to vary with the general physiological state of the animal and also as a result of the mediation of 'higher' centres of the CNS. Great advances in the understanding of behaviour did result, nevertheless, from 'reflexology', even though many of the early assumptions and techniques were revealed as being over-simplified, and a great deal of productive research is still being carried on at this level.

Both the strengths and the limitations of reflexology lie in its reductionist presuppositions. That many simple (and not so simple) movements of obvious adaptive value could be shown to result from a particular and often quantifiable stimulus 'firing' a device whose structure could be demonstrated, and whose fixity of structure guaranteed fixity of response, allowed explanations of 'adaptation' and 'purposiveness' in objective mechanistic and hence scientifically respectable terms. Descriptions of specific and unvarying causal mechanisms could replace the anthropomorphic and teleological explanations to which behavioural 'science' had previously been restricted. But while the reduction of various activities, 'behaviours', to reflexes does permit 'objective' explanations to be given, it is impossible to use reflexes as the basis for a general account of behaviour without *either* doing violence to the plausibility of the descriptions of the behaviour to be explained *or*, on the other hand, so extending the notion of the reflex that its structural and mechanistic simplicity *and demonstrability* is lost.

There are several features of 'behaviour in general' which must be covered by any explanatory theory aiming at comprehensiveness and generality. At this stage in the discussion these may simply be stated to be:

(a) complexity;
(b) variability;
(c) spontaneity.

They will be further discussed and justified as we proceed.

It is theoretically possible to give an account of complex behaviour in terms of simple behavioural 'units' if the number of such units is multiplied and if a satisfactory account can be given of their manner of interrelationship and interaction. This last requirement, it should be noted, implies that the 'units' cannot be all of the same sort: reflexes must be connected by units which themselves are not reflexes as strictly defined. In practice, those workers who attempt to build a general behaviour theory exclusively upon 'reflexes' have usually been guilty of extending the concept of the reflex to the point where its empirical demonstrability and its operational simplicity have been lost. From the denotation of a particular structural arrangement (illustrated in the well-known 'reflex arc' which channels a specific quantity and type of imprest energy (the 'stimulus') into a specific response, 'reflex' is broadened in connotation until it means 'whatever mechanisms cause the behaviour in question'. This is a 'blanket' meaning; and in this limiting case the term has no denotation over and above the observed behaviour it is supposed to explain.

Reflexes cannot be made the exclusive basis of variable behaviour, or of spontaneous behaviour, if the term 'reflex' is to retain its original denotation. Variability and spontaneity cannot be generated out of units that are themselves neither variable nor spontaneous. Possibly this logical difficulty was obscured, for some people, by the psychological difficulties of visualizing the great complexity of the neurological and other causal interactions behind much behaviour – but the psychological barriers raised by the sheer complexity of a phenomenon cannot be allowed to distract attention from the logical status of the conceptual relationships, because these determine the logical feasibility or otherwise of a particular line of explanation. Thus the conflation of the problems of complexity with those of variability, etc. may have fostered the illusion that they could all be surmounted by a purely 'reflex' theory, albeit one of great complexity, whereas this is in fact not possible. Thus a general theory based on reflexes must either deny variability and spontaneity, or must change the concept of the reflex to include within it the possibility of variation and spontaneity, or must allow the inclusion of other mechanisms besides reflexes.

This last possibility has been of greatest importance historically. Pavlov (see Pavlov 1927) added the 'conditioned' to the 'unconditioned' reflex to increase the repertoire of behavioural units available as 'explanatory raw material'. But while the flexibility of different possibilities of conditioning was added to the range of a species behaviour by this means, and while in a sense the range of potentiality open to the individual *before his potentialities were actualized* was increased, it must be recognized that conditioned reflexes once established are just as unvarying as structurally or physiologically fixed reflexes. The dog can be conditioned to salivate in response to the ringing of a bell or the flashing of a light or the lab attendant putting on his overalls or to combinations of these stimuli – but once it has been conditioned it cannot not respond to the conditioned stimulus, neither can it respond to a different stimulus to which it has not yet been conditioned. The actualization of a potential for conditioning automatically closes off further actualization. Except that de-conditioning and re-conditioning may be possible, the end result of conditioning is still fixed and unvarying behaviour. The series of possibilities is closed by the conditioning, whereas a general explanatory theory must account also for behaviour that generates open-ended series of possibilities. Conditioned reflexes are open to variation, in fact, only *before* they have been established, only before the conditioning has taken place.

Similar logical and methodological difficulties to those that arise in connection with reflexes arise also with regard to other 'behavioural units' that have been discovered and that have been used as the basis for attempts to give a single general explanation of all behaviour. Tropisms (Loeb 1918), taxes and kineses all contribute to the behaviour and hence to the adaptation of various animals, and the investigation of these phenomena has illuminated fields of study as diverse as ecology, physiology, comparative anatomy and palaeontology, as well as adding very substantially to our understanding of behaviour in itself. But tropisms, taxes and kineses are discoverable precisely because they involve fixed responses to determinable stimuli; and again there arises the dilemma for a general theory, of either denying that variability occurs in the behaviour to be explained, or of altering the explanatory concepts of taxes, etc. to allow them to be variable; which would vitiate them. To illustrate, consider an example of tropotaxis (a type of locomotory movement originally included by Loeb along with what are now termed 'tropisms' in a narrower sense than his original one). The crustacean *Armadillium* (Muller 1925, cited in Hinde 1966, pp. 111–14) is positively tropotactic to light, in that if placed equidistant from two equal light sources and off the axis between them, it will usually move towards a point roughly midway between the two lights. In doing so it apparently equalizes by simultaneous comparison the light impinging upon its two sides. After reaching the midway point the animal will usually move directly to one or other of the lights; or it may remain static for a time; but as Hinde points out, there may on occasion be direct movement from starting-point to one of the light sources. The causation and control of the behaviour patterns 'available' to the animal in the experimental situation, simple as it is, are thus more complex than the control circuitry of gadgets like the 'phototropotactic dog' (see Rensch 1966: 346; Walter 1953), and it might appear that the concept of the taxis must therefore be expanded to allow variability of reaction. This is false, however. A more satisfactory account is given by saying that each tactic response is fixed, but that the animal has several alternative taxes available for use at a given time. Thus the variability lies between the several taxes, not within each one. The total control system itself is also fixed, but the alternatives available within it allow variation in the actual behaviour exhibited. The complexity of even a minimally-complex tactic control system is well brought out by Hassenstein *et al.*, 1959 (see discussion in Hinde 1966: 116–32). But while a great deal even of human behaviour does appear to be based on fantastically intricate concatenations of reflexes, and even though much of the behaviour of the infra-human animals does appear rigid and invariable, there is a great deal even in the most rigid systems that cannot plausibly be explained in terms of taxes, etc. which are unvarying and essentially reactive, i.e. non-spontaneous. The female spider, for example, which at one time allows the male to court her, will at another time kill and eat him irrespective of his courtship signals (cf. Bristowe 1958). An explanation is needed for the non-operation of the mating response at one time and of the feeding response at the other.

Variability and spontaneity of behaviour may sometimes be difficult to distinguish by observation and in practice until the behaviour of the particular species in question is intimately known – but together they form an obstacle course which reflexive and tropistic theories have great difficulty in passing, and which they may appear to pass only at the cost of fatal imprecision in their own basic terms. The strength of the concepts of 'reflex', 'taxis', and so on, is that they can be applied in a way that is precisely determinable, and often indeed measurable in quantified terms. The stimulus which causes a reflex or tactic reaction can be discovered, described and quantified. The mediating structures and reactions within the organism, in receptor, nerve-pathways, and effector, even if in some cases not in practice able to be specified in detail, can at least be simulated in model in terms of known components with known functions. The model, 'built' out of known types of neurone, for example, with known reactive properties, can explain some behaviour *with nothing left over*. But not all types of behaviour can be explained in this way; and, in particular, variation of response, and the spontaneous initiation of action, cannot be explained. These features must remain as 'bits left over' after reflexes, taxes, etc. have been used to explain as much as they are capable of doing. It might be said that a bird seeking nesting material is actuated by nothing but reflexes – but if this were the case the whole notion of a 'reflex' would have to have changed. The otherwise useful specifiability of the reflex would have been lost. What, for example, is the stimulus? The absence of a piece of nest-material? But lots of other things are absent – food, a copulation-partner, water, and so on – and the absence of these things does not lead to feeding, copulation, drinking, etc. The absence of something cannot *by itself* be a stimulus. . . .

Historically, the proximate stimulus for the genesis of ethological theory seems to have been Lorenz's (1935) observation of 'vacuum activity' hunting and feeding behaviour in a starling. The starling was confined, and provided with an abundance of static and 'dead' food. Lorenz noticed, on the basis of his acquaintance with the behaviour of the species under natural conditions, that the captive starling 'hunted' and 'ate' non-existent insects: it went through the motions of hunting, capturing, and swallowing active insect prey even when it could clearly be seen that no insects were present. Lehrman's (1953) objection, that it is difficult to be categorically sure about the complete absence of an external stimulus, is beside the point, since even if some miniscule stimulus such as a speck of dust or some excessively minute insect were proved to have been present there would still be the substantial disparity to be explained between this and the usual type of stimulus. Putting it the other way round, minute objects do not normally function as stimuli at all in this context, and whether the 'hunting' behaviour is regarded as being completely or as only relatively spontaneous makes little difference to the causal explanation which is demanded. The causal explanation must point to some internal factor as the main or sole immediate stimulus.

In brief, the solution to this problem of spontaneity which was offered by ethological theory (the major exposition being Tinbergen's (1951) *The Study of Instinct*) is that there accumulates within the individual animal a store of action-specific energy (A.S.E.) which is normally directed into expression as adaptive behaviour through the operation of 'releaser' stimuli in actuating an hierarchically arranged series of innate releasing mechanisms (I.R.M.s). The 'vacuum' hunting behaviour of the starling would thus be explained roughly as follows.

The make-up of the bird is such that general environmental influences cause the (usually slow) accumulation of energy specific to foraging and hunting. Normally this energy is released and directed in the form of overt observable hunting behaviour when the animal encounters suitable prey. Perceiving the prey probably also stimulates greater and more rapid generation of A.S.E. for hunting. But the A.S.E. still accumulates even when no prey is available; and assuming that 'storage' is limited the specificity of the accumulated energy means that once it reaches a certain level it is liable to burst through the normal releasing mechanisms, even though these have not been triggered off by the usual external stimulus, i.e. the prey. Thus the hunting behaviour occurs with no (or very little) external stimulus: it occurs spontaneously, or *in vacuo* (as far as the normal external stimulus is concerned).

This quick sketch glosses over the difficulties and weak points – but also the further strong points, of the theory. It is necessary to elaborate upon both sides – but it may as well be said that, despite twenty-odd years of experimentation and theoretical disputation, no substitute has been found for the general outlines of this theory. As with evolutionary theory since Darwin, the story has become more complicated, but the same story still stands.

The 'key concepts' of ethological theory are those of:

(1) Endogenous Action Specific Energy.

(2) The 'releaser' stimulus,
(3) A hierarchical model of the neural mechanisms (understood in a functional rather than a structural sense) of the central nervous system, which mediate the causality of behaviour and thus achieve control and directiveness. Included here are the Innate Releasing Mechanisms.

THE HIERARCHICAL MODEL

The third strategic master-stroke of ethology was to take the logical model of an animal's behaviour, in the form of a tabulated classification of the various activities which make up the species' repertoire, and to convert this into a functional model of the controlling neural mechanisms.

To illustrate the simplicity and effectiveness of this it will be convenient to refer to the behaviour-patterns of an animal whose repertoire has been intensively studied by ethological methods. Such an animal is the Three-spined Stickleback, *Gasterosteus aculeatus*. An extensive descriptive and critical literature has grown up in relation to the ethological investigations centred upon it, so that, although the behaviour of this species will be known directly by only a limited number of readers, most will be familiar with some of the literature. (The general accessibility of the literature is an advantage, since controversial issues, when of marginal relevance to the evolutionary relationships between instinct and intelligence, can be dealt with by means of brief reference to published works.)

First, a brief outline description of the behaviour of the male Stickleback over the breeding season.

In the winter the little fish are gregarious, and swim around their pond (or aquarium, for those under laboratory observation) in shoals. As spring comes on, the appearance of both males and females changes, the females assuming an over-all silvery colour and a swollen 'abdomen' as the eggs ripen, the males becoming conspicuous with the ventral side of the body bright red, a blue ring around the eyes, etc. The behaviour changes also. Male fish become aggressive to each other, and they separate out from the shoal and each settles down in a particular area of the pond or aquarium. This combination of site attachment plus aggressiveness adds up to the phenomenon of territoriality which is found in one form or another in most of the 'higher' animals and which has a variety of functions (spacing out and sharing

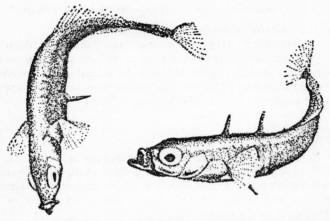

Figure 1. Male Sticklebacks fighting (*after Ter Pelkwijk and Tinbergen, 1937*).

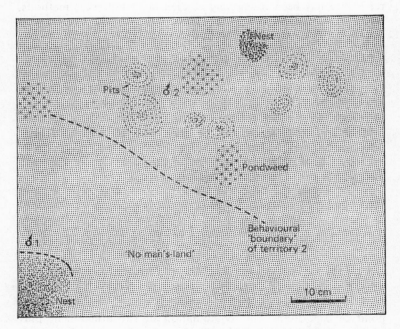

Figure 2. Territories of two male Sticklebacks showing pits due to displacement/threat digging in territory of ♂2 (*after Tinbergen and Van Iersel, 1947*).

the suitable breeding sites, food supplies, predator attacks, and so on) which have been discussed by Tinbergen (1957, 1964); Hinde (1956). Each male Stickleback lives in his territory and spends his time feeding, fighting his neighbours if they intrude, nest-building, courting passing females, and just swimming about ready to swing into any of these sorts of behaviour as circumstances dictate. Later there will be actual mating with females who will lay eggs in the nest, after which the male will spend a great deal of his time in care of the eggs and, later, of the young.

There are several highly-general headings under which Stickleback activities, as with those of most other animals, can be classified. In the broadest terms, each particular activity can be classified as having to do with either feeding, sleeping, reproduction, care of the body surface and so on. Each of these general categories can be subdivided. Since reproductive behaviour is susceptible to the most clear-cut subdivision, attention will be concentrated on it – on the understanding that findings with regard to reproductive behaviour will apply, *mutatis mutandis* and unless otherwise stated, to other types of activity.

A large number of activities can be seen, once the full repertoire of the species is understood, to relate to reproduction. The taking up and defence of territories by the males, for example, appears orientated mainly towards the subsequent reproductive patterns – even though territoriality probably confers selective advantage in other ways as well (cf. reviews by Tinbergen 1957, Hinde 1956). Of more obvious and direct reproductive significance is male nest-building activity; then there are the courtship displays, the actual fertilization of eggs, and the several activities involved in the care of eggs and young. Thus under the general heading 'Reproduction' can be subsumed a number of diverse types of activity: fighting, nest-building, courting, caring for eggs and young. The logical relationship between these classes of behaviour may be represented as follows:

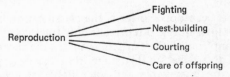

(It is not necessary to attempt, here, an exhaustive classification. The intention is simply to show the stages in the construction of the hierarchical model.)

Within each of the subordinate classes of behaviour, 'Fighting', 'Nest-building' and so on, there must be distinguished further subdivisions. Fighting between male Sticklebacks may take any of several forms. There are agonistic displays, such as the Vertical Threat Posture, and the Threatening Approach with dorsal spines raised, as well as Biting and Chasing. Similarly, under the heading 'Nest-building' must be subsumed a number of different actions, such as Digging, Bringing Pondweed, Cementing, Tunnelling. The tabular representation of the behaviour of the male Stickleback must therefore be extended as in Fig. 3.

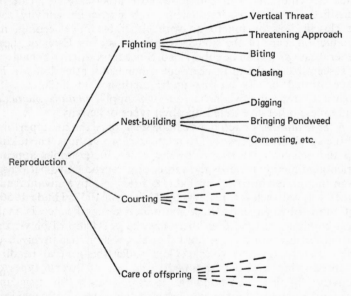

Figure 3. Hierarchical classification of instinctive activities.

All that has been done so far is the classification of behaviour patterns. The lines in Fig. 3 represent only the logical relationship of the subsumption of the particular under the more general. It is important to notice, at this stage, that the only parts of the schema that are observable and identifiable *as actual activities* are those in the right-hand column, i.e. only those at the most particular level. 'Fighting' cannot be observed as anything distinct from, or in addition to, *either* Biting *or* Vertical Threat *or* Chasing . . . and so on. 'Fighting' must always be understood as a logical and purely abstract category. (An apparent exception to this will be discussed below, in relation to the distinction between the appetitive and the consummatory phases of an instinctive activity.)

The hierarchical classification of behaviour illustrated in Fig. 3 can be turned into an hierarchical model of the mechanisms controlling behaviour, by taking the lines (which in the classification indicate the logical relationship of subsumption of particular under general) and assuming that these are neural pathways within the CNS; and by taking the 'boxes' and assuming that these now stand for the functional centres which control the observable activities.

In the simplest terms the hierarchical 'control system' model works by the flowing of energy from high-level centres (at the left of the diagram) down to lower ones and finally out of the CNS (to the right of the diagram) in the form of actual and overt motor behaviour. The 'energy' may be specific to varying degrees. It is generated in the top-level centres under the influence of immediate external stimuli and/or the influence of hormonal and other humoral changes which are affected by the environment in only a general and long-term way. Guided and directed by combinations of internal and external causal factors, the energy flows down and accumulates in the lower centres. From these it is released usually by highly specific combinations of external stimuli, the specificity of the 'releaser situation' normally ensuring the adaptive appropriateness of the resulting activity.

The 'flow' of energy from higher to lower centres, and the transition in control level from general to specific releasing and orientating stimuli, is marked in descriptive categorization by the distinction between appetitive and consummatory behaviour. Discussion of a particular analysis may make clear several points of significance.

The stimulus for a territory-holding male to attack another male is not a simple one. It is not a question merely of proximity. A nearby male, but which happens to be on the far side of the territory-boundary, is not attacked; whereas a distant one who is actually trespassing within the territory is attacked, if he stays long enough for the territory-holder to get to him! Thus the 'stimulus' is in fact a highly complex total situation, rather than any particular isolatable element within it; and though some sets of features can be demonstrated to be of greater significance

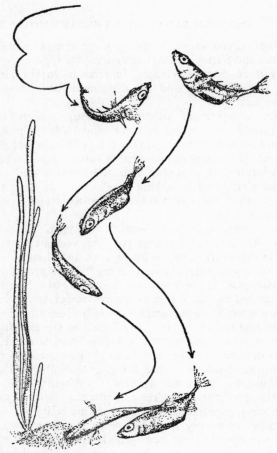

Pictorial schema.

SEQUENCE OF REACTIONS

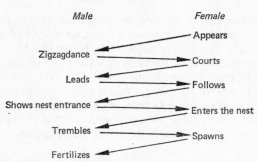

Male	Female
	Appears
Zigzagdance	Courts
Leads	Follows
Shows nest entrance	Enters the nest
Trembles	Spawns
Fertilizes	

Figure 4. Courtship of Stickleback (*Gasterosteus aculeatus*).

than others within the total complex, these sub-sets (the actual 'releasers' in the original usage of Lorenz 1935) are themselves complex and essentially configurational or relational (see Tinbergen 1951, Hinde 1966: 44–54 and refs. therein). But the stimuli eliciting a particular activity are not presented simultaneously in array, rather serially in a sequence of sub-sets. The early sub-sets may be ambiguous or generalized, and the activity they elicit may be determinable or directable in any of several different ways – it may be regarded as the common part of several different possible sequences. Thus the male Stickleback which observes an intruder in a distant part of his territory swims towards it. On closer approach the intruder is identified by its appearance and behaviour either as a male or as a female. If it is a female, and if both animals are in appropriate condition, courtship and possibly mating will follow. If the intruder is a male, the territory-holder will assume a threat approach, which builds up into higher-intensity displays or direct-contact fighting or the chasing-off of the intruder, depending upon the behaviour of the latter. It is useful, therefore, to distinguish between the *appetitive* phase of a behaviour pattern – the approach from a distance towards the intruder, in the case being considered – and the *consummatory* phase (or phases), in which an increasingly specific series of actions finally completes the function appropriate to the situation: in this case, the driving away of the intruder. In the appetitive phase a considerable range of possibilities for further action is left open – namely, all the various patterns of attack (in the event that the intruder is a male) *or* of courtship (in the event that it is a female) – and the appetitive behaviour itself is variable, depending on the particular circumstances, the whereabouts of the territory-holder in relation to the intruder, the physical features of the territory, and so on. The appetitive phase tends to bring the individual into a position where the further 'releasers' for the consummatory phases will be encountered; if they are encountered, and depending on precisely *which* ones are encountered, a particular sequence of consummatory activity will be elicited. If appetitive behaviour is regarded as occurring to the left of the centre column in Fig. 3, p. 44, where it can be common to both Fighting and Courting in the situation in question, the transition from appetitive to consummatory phases will involve a shift towards the right; the alternative distinguishable consummatory activities constituting the right-hand column. Thus perception of the 'male signals' ('red

belly underneath', Hinde 1966: 45) of the intruder may elicit the
Vertical Threat Posture; if this draws similar display from the
intruder, the territory-holder may resort to Biting; and finally,
if the intruder retreats, to Chasing. Considerable complexity of

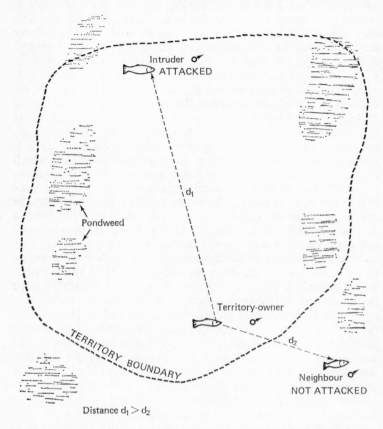

Figure 5. Total releaser-situation for territorial attack in Stickleback

interactions is possible, indeed it is from the laborious analysis of
the relative frequencies of particular sequences that an understanding of the causation, both external and internal, of such
behaviour has been gained (see Hinde 1966: 295–311, for references
and discussion of a number of analyses). But the possibility of

such analysis depends upon the different component activities, the different activities subsumed under 'Fighting' for example, being each relatively stereotyped and therefore recognizable. That each activity should appear in an all-or-nothing fashion at 'typical intensity' (Morris 1957) has adaptive value in many instances, notably when the activity forms part of or is associated with a social signal (see Hinde 1966: 275–6). The consummatory phases of instinctive behaviour sequences are thus made up of sets of behavioural units which are themselves relatively invariant; and though variation is possible within the set or sequence, strict limits are imposed on variability by the number of 'units' available in a particular contest, under particular conditions of motivation, and so on. Appetitive behaviour is much more variable, involving as it often does little in the way or recognizable 'units' beyond those of locomotion. It is of course inferred that appetitive behaviour involves substantial internal changes in 'set' or 'expectancy', substantial lowering of specific sensory thresholds, but this is not directly observable. Appetitive behaviour is often elicited, too, by purely internal stimuli, changes in blood sugar level or in hormone concentration, etc. (see Hinde 1966, Chapters Eight to Fifteen and references therein). Appetitive behaviour is thus spontaneous in the sense of being the result of internal causal factors often long-term in their operation and actuated only remotely by environmental changes; and it is variable not only in involving variation in, say, locomotor activity conformable to the terrain, but also in allowing the releaser stimuli which are encountered to direct energy into any of several specific channels. This allows the possibility of switching channels (so long as the consummatory phase of an activity is not too far advanced) with obvious adaptive gain in many circumstances.

Besides the mechanisms which set off appetitive behaviour of one sort or another and which direct the behavioural outflow towards increasing specificity culminating in appropriate consummatory sequences, the hierarchical model also provides a basis for the explanation of various other phenomena. In fighting between male Sticklebacks, for example, a single performance of, say, Vertical Threat does not usually suffice to end the fight. Normally a sequence of activities would be involved on the part of both males. The 'chain reaction' in fighting is relatively variable, so it will be convenient to illustrate the sequential and rigid nature of a typical instinctive pattern by reference to the

Model   No. of responses

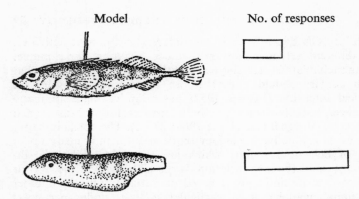

Figure 6. Models of female Stickleback used to test responses of males (*after Tinbergen 1951*).

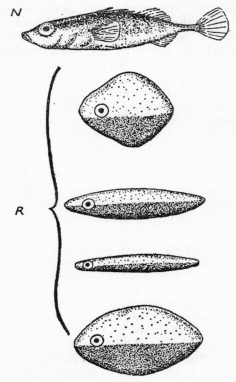

Figure 7. Models of male Stickleback used to test fighting responses of males. Top model, lacking red underside, is less efficacious (*after Tinbergen 1951*.)

courting and mating behaviour of the Stickleback. Fig. 4 shows in schematized form the chain reaction between male and female, from Zigzag Dance of male to Egg Laying by female and Fertilization of eggs in nest by male. The performance of any one action is itself the stimulus for the performance by the other partner of the next action on the chain. Within the individual animal each activity may be assumed to be 'switched off' as it is completed by a feedback loop which actuates an inhibitory 'block' on the neural channel leading into the 'centre' for that activity.

Many displays, sometimes elaborately ritualized and with specially-developed structures associated with them, can be given an evolutionary explanation in terms of the hierarchical model. If in relation to Fig. 3 we imagine motivational levels for attacking and fleeing to be high and equally balanced, so that 'fighting' energy is not in fact discharged through any of the activities in the right-hand column, this energy could be visualized as 'backing up' and finally spilling over, through the next intermediate-level centre, into Nest-building. Thus a bout of Digging might sometimes be observed among bouts of fighting. Such 'displacement' Digging does in fact occur (see Tinbergen 1951), and it does have a threat function. It is distinguished from 'true' or autochthonous Digging by the spines in the dorsal fin being raised; and it may be presumed to be in process of evolutionary elaboration into a new and distinct agonistic display, additional to those already used. Two types of change are involved (Blest 1961) in the evolutionary development of such a display:

(a) Ritualization: increasing the effectiveness of the signalling function of the display, usually by making it more conspicuous through the development of colours, movements of greater amplitude, etc.

(b) Emancipation: altering the motivational basis of the display so that it becomes more closely associated with its newly-acquired function. In the case of displacement Digging in the Stickleback, this would entail being shifted up into the group of Fighting activities, neural connections being rearranged accordingly. That is to say, it would gradually change from being *displacement* Digging to being true Threat Digging.

Ethological theory thus makes considerable provision for both spontaneity and variability in behaviour. Spontaneity: in that the appetitive, and even on occasion the typical consummatory sequences of the species, may appear in the absence of any immediate external stimulus, due to internal changes in, say, blood chemistry and/or the accumulation of 'action-specific energy' (which at some time, no doubt, will itself be specifiable in terms of particular physiological processes). Variability is in a sense the distinguishing feature of the appetitive phases of instinctive behaviour, as contrasted with the 'fixed action pattern' of the consummatory phases. Even in the latter, however, the recognition of the importance and the subtlety and complexity of the external and internal causal factors allows for some variation. That a male Stickleback fails to court one gravid female then a little later does perform courtship to another, can be explained, not in terms of 'caprice' or of pathology, but simply by reference to changing hormone levels or to the fact that one was outside the territory of the particular male while the other was inside. Variability also occurs through the re-sorting and re-ordering of the 'units' of fixed action patterns, as has been indicated; and considerable variation between the individuals of a species is possible, due to the fact that learning plays an increasing part in the ontogeny of behaviour especially in 'higher' forms (see Thorpe 1963, Hinde 1966 especially Chapter Six). Thorpe says: 'Learning is manifested particularly within appetitive behaviour, and is regarded biologically as the process of adjusting more or less fixed automatisms and releasing mechanisms to a perpetually changing environment' (p. 73).

Thus different individuals of the same species may learn to 'key' their consummatory behaviour sequences to different external stimuli, different releasers. Adaptability of the species to variations in the environment is thus made possible. From the point of view of the behaviour of the individual however, it must be emphasized that under natural conditions the fact that learning has entered even quite largely into the development of particular behaviour patterns does not mean that these patterns are variable. As Ewer (1957) has pointed out, the weight of evidence indicates that learned patterns, or patterns of activity which have involved learning to some extent or other, are just as rigid, stereotyped and inflexible, once learned, as are the 'machine-like' instinctive sequences of insects. The observations of Howard (1952) show that, while a considerable diversity is to be found in the fine details of the behaviour of a group of individuals of one species – the differences clearly being due partly to learning – each individual has its own peculiar patterns of activity, which are constant and which are in fact used as individual identifying characteristics. Thus while the *capacity* for learning is adaptive and does contribute to the total range of variation of species behaviour, in that when an individual is confronted with an unusual situation it may be able to 'learn' an adaptive response, if 'what is learned' (i.e. the new response) is repeated to the extent that it becomes a habit, the potential for variability has disappeared. Some issues involved in the relationships between learning, instinct and intelligence are discussed elsewhere (see Chapter Eleven); for the moment it may simply be asserted, as an admittedly rather sweeping generalization, that an individual's behaviour-system, whether of 'pure' instinct or of instinct-modified-by-learning, is variable but not *adaptively* variable. In other words, the individual is not able to vary his behaviour to follow moment-by-moment demands of novelty in environmental situations. Such adaptive variability is the characteristic of intelligence. The implications of its evolution will be explored in Chapter Three and subsequently. The argument up to the present may be summed up, to conclude this chapter, as follows.

Phylogenetically, the behavioural advance, in terms of instinctive behaviour plus learning, from the platyhelminths to the birds and mammals, is to a great extent merely quantitative. A greater number of behavioural 'units' becomes available to each species and to the individual members of a species; and concomitantly with this there is a greater range of 'flexibility of assortment' within each 'family' of 'behavioural units' (e.g. the alternative forms of fighting in male Sticklebacks). Along with this increase in complexity go increases in variability and in spontaneity – but these are limited in ways that have been indicated, notably in general in being confined to the appetitive phases of the instinctive patterns. Such variations as do occur in the consummatory phases are the product of, and are subject to, evolutionary processes. They change only in the time-scale of evolution. Within the life-span of the individual they are virtually invariant. Given the standard releaser situation and appropriate motivational background and build-up through appetitive phases, the standard consummatory sequence must follow.

It is against this setting that the problem of the evolution of intelligence is to be posed.

III

Evolution of Intelligence From and Within the Framework of, an Instinct System

For the purpose of the present argument a view of instinctive behaviour broadly following that of the Lorenz-Tinbergen ethological school, as exemplified most comprehensively in Tinbergen's *The Study of Instinct* (1951), will be taken. A brief account of ethological theory is given in Chapter II of the present work. It should be noted that, although the main concepts peculiar to ethological theory, namely those of action-specific energy and of a hierarchical model of neural functioning, have been explained in Chapter II, the argument to be given in Chapter III does not depend in any way upon these concepts. My argument is in fact highly abstract, and its basis is the systemic and relatively fixed nature of instinctive sequences and learned habits, rather than the particular details of the mechanisms involved.

The argument of this chapter may appear over-terse, indeed it was my own intention up until final-draft stage that it should be expanded and amplified. Now, however, I have decided that it is best to leave the bones of the argument as bare as is consonant with intelligibility *at this stage*, and to attempt in the subsequent chapters to insert various needful provisos, qualifications and amplifications. This results in an apparent lack of balance between the several chapters, certainly in terms of their mere size – but the logic of communication sometimes demands that the most important message should be short, and should attain impact through simplicity. So I have attempted here to present a skeleton round which in the rest of the book some at least of the requisite muscles, nerves, skin, etc. will be moulded.

1. INSTINCTIVE BEHAVIOUR

Apart from simple mechanistic kineses and taxes, the behaviour of the vast majority of animals is of the sort which is by general consensus called 'instinctive' (cf. Hebb 1949; Lashley 1938). The behavioural repertoire of an instinctive individual consists of a number of inbuilt responses to the stock situations likely to be encountered (cf. Ewer 1957) in the course of the normal way of life in the typical habitat of the species. These responses occur because they are biologically useful to the species; or, more accurately, because they have in the past been useful to the ancestors of the present populations – they have therefore been selected for in the course of evolution. Without going into the minutiae of the controversies over innate *v.* learned, genetic control *v.* environmental effects, we can generalize that instinctive behaviour is adapted to the normal, average, or standard conditions under which the particular species lives: instinctive behaviour patterns therefore tend to be stereotyped and automatic (Ewer 1957). The essence of instinctive as contrasted with intelligent behaviour is its conservatism, whether the species or the single individual be considered. (The place of learning in instinct is discussed elsewhere, especially Chapter Eleven. At this point it may be remarked merely that learned components of instinctive patterns, *once learned*, are as fixed and unvarying as extreme 'instinctive' actions. Again, see Ewer 1957, 1961).

2. DYNAMIC SYSTEM OF INSTINCTS

The behavioural repertoire of an individual can be visualized as an instinctive *system*. One is forced to recognise that:

(1) there are a number of activities necessary for the life of the individual and the propagation of the species; and
(2) efficiency generally demands concentration on one activity at a time.

It follows that there must be principles and mechanisms for the allocation of priorities from time to time, otherwise the continual presence of mating partners, for example, would result in continuous reproductive activity and no time for eating, sleeping, etc. In other words, the instincts must form, collectively, an integrated and dynamic system (cf. Ewer 1961: 259). We can neglect discussion of the mechanisms by which alternation of activities is effected (see Chapter Two). The main points are: that alternation is known to occur, and that it is adaptive.

3. PROTO-INTELLIGENCE DYSGENIC

Since intelligent behaviour is essentially *variable* within the lifetime of the individual (see Chapter One), it looks as though any intrusion of intelligence into the delicately-balanced instinctive system would be likely to lead to deleterious effects, and would therefore be selected against. Once intelligence has been developed to the stage that it can 'take over' from the instinct system at least some of the normal everyday activities of an animal, a net gain has been achieved: for the unusual can be dealt with. But intelligence is useless in its early developmental stages, phylogenetic or ontogenetic. The instinctive system must therefore be left functional and the embryonic 'intelligence' must be allowed to develop within, and as it were supported by, an instinctive framework.

Rensch (1966) enunciates the same general principle in a slightly different context: 'After a harmoniously working structural type had developed in phylogeny, it was easier that variants survived in which a new development was added to the *final* stages, because all earlier alterations would more disturb or destroy the established harmony.' (My italics. The essential point is that variant individuals in whom the pre-existing integration was too much disturbed would not, in fact, survive.)

It is interesting that Darwin (1873) remarked: 'It is not improbable that there is a certain amount of interference between the development of free intelligence and of instinct.' For the development of intelligence 'the various parts of the brain must be connected by very intricate channels of the freest intercommunication; and as a consequence each separate part would perhaps tend to be less well fitted to answer to particular sensations or associations in a definite and inherited – that is instinctive – manner' (p. 103).

The general problem thus turns into two particular problems, those of:

(1) Allowing for the requisite degree of variability to be incorporated within the instinctive system without destroying or much diminishing its adaptiveness.

(2) Proposing some means whereby this intelligent variability could have been selected for and thus progressively increased.

4. INADEQUACY OF SELECTION MERELY FOR SENSORY IMPROVEMENT

It is inadequate merely to remark with Harlow (1958) that selection must favour improvement in sensory equipment and that this must entail improvement in the central nervous system. He himself points out that there are 'fish . . . reptiles, birds, and mammals that have the receptor potentialities . . . for human-type thought'. It is precisely this discrepancy which constitutes the problem of 'the evolution of learning' as Harlow puts it, and *a fortiori* of the evolution of intelligence.

5. NECESSITY FOR SELECTION AFFECTING CONSUMMATORY PHASE

Consider the life of an 'instinctive' animal. Most of its time is spent doing 'nothing in particular' (Tinbergen 1953), i.e. when not strongly motivated towards one particular type of activity the animal waits passively or (more commonly) moves randomly within its habitat. As time passes either internal changes or chance encounters with particular external stimuli incline the animal towards a specific activity. Either way, the appetitive phase of a specific instinctive activity ensues. This (gradually) brings the animal into a situation appropriate for the performance of the consummatory phase, which achieves some biologically necessary end (either for the individual or for the species) and leaves the animal to resume 'nothing in particular' (see Hinde 1953 on the appetitive-consummatory distinction, discussed in Chapter Two – which remains a methodologically useful one.)

Note that the division of each instinctive activity into 'appetitive' and 'consummatory' phases must be subject to two opposing evolutionary tendencies:

(a) to prolong the appetitive phase so that the animal can 'change its mind', in response to environmental changes or emergencies, right to the last possible moment (selection pressure in this direction helps to account for the very existence of an appetitive phase); and

(b) to reduce both appetitive and consummatory phases, so that each particular activity may be 'got over with' as quickly as possible.

It seems fairly obvious that, in the early evolutionary stages, the variability characteristic of intelligent behaviour can come only into the appetitive or the 'nothing in particular' periods: the crescendo of particularity of a consummatory phase cannot be diverted. But precisely *what* can come into the behaviour? Variability *per se* is of no value (except possibly in human aesthetic experience), and would tend to be eliminated. Variability must be biologically useful if it is to be selected for, and while non-adaptive features can sometimes survive, apparently for long periods (Dobzhansky, 1956), much more is involved in the evolution of intelligence than the mere survival of a certain amount of harmless variability. Variability has become a basic feature of the behaviour of some animals and must have been actively, indeed one might say massively, under positive selection pressure. Etkin (1954) rightly emphasizes the need for strong selection pressures (though the types of pressure he postulates to explain human evolution seem rather inadequate in detail).

6. VARIABILITY IN CONSUMMATORY PHASE

An animal's instincts have been evolved specifically to meet its life requirements, satisfy the biological needs of the individual and the species; and the consummatory acts are the parts of its total activity which most immediately and most importantly do this.

No notable change in behaviour can come about except through alteration in the consummatory phases of a sequence. This is true of all changes in instinctive patterns, e.g. changes in agonistic and reproductive displays (see Hinde and Tinbergen 1958 for discussion and references.)

Since appetitive behaviour is itself variable, a change is unlikely even to be noticed unless it affects the more fixed and recognizible sequences towards the consummatory climax; and it is difficult to see that variations in appetitive behaviour *without* variation in consummatory sequences could ever make much difference to an individual or a species (except in a negative sense), hence be much affected by selection pressures. It becomes apparent that we must allow for variability in the consummatory phases of instinctive behaviour, if this variability is to be under selection pressures of magnitude sufficient for intelligence eventually to result.

Variability in the consummatory phase can be either of two kinds (Tinbergen, 1951):

(a) The activity may be released and orientated by stimuli ('releasers') different from the normal.

(b) The activity itself may differ, in motor pattern or orientation or both, from the normal.

Variability as to releasing and orientating stimuli is possible within the lifetime of the individual, if the stimuli are learned rather than totally inbuilt and automatic. That many releasers have to be learned, has been known for some time. The phenomena first studied were the imprinting responses of young Anatiform birds (Lorenz, 1935; Thorpe 1951) and imprinting was originally thought to be a type of learning unique in its rapidity and irreversibility, and in occurring very early in the life of the individual. Later work has shown, however, that imprinting is an end of a spectrum of types of learning, rather than an isolated type of process: varying degrees of rapidity and irreversibility have been shown to occur (Thorpe 1963). One of the most interesting results of this line of work has been the revelation of *sensitive periods* (see Thorpe, 1961) in the life of the individual, when specific learning can and must take place – failure to learn during the sensitive period means that a particular response can never be incorporated into the individual's repertoire. Sensitive periods are adapted to the nature of the response to be learned, e.g. many male Passerines learn their species' song at the only appropriate time, i.e. in their first breeding season at the age of 9–11 months.

That a particular activity can be learned only during a circumscribed and often short sensitive period, and that this learning is relatively irreversible, are parts of the mechanism for ensuring specificity, which in this context is but one remove from adaptiveness. The shortness of the imprinting period for the 'following' response of the newly-hatched gosling is a good example: the mother is under natural conditions almost certainly the first 'imprintable' object the goslings see – and later, when they are likely to encounter others, they have lost the tendency to respond. Activities which do not have to be specific, on the other hand, have less need of a circumscribed learning period, i.e. they may be learned more or less as and if opportunity offers. Also, and importantly, the relative irreversibility of the 'imprinting' type of learning is unnecessary. Such activities, in relatively small, short-lived, rapidly-reproducing species, are likely to be those that are less important to individual and species welfare, since the important functions tend to be dealt with on an instinctive basis, with a fairly high degree of specificity in the releasing mechanisms

60 THE EVOLUTION OF INTELLIGENCE

(Ewer 1957; Thorpe 1965, esp. p. 511). (The exception is exploratory behaviour which will be discussed below.) In a species that
is long-lived and where there is a sufficiently long association
between parents and offspring for learning by sustained repetition
or imitation to take place, the rapidity, irreversibility, and
limited sensitive period characteristic of the 'imprinting' end of the
spectrum become unnecessary, indeed undesirable; and the
possibility of adaptive variation from the previous normal
behaviour is introduced.

The adaptiveness of learned variations can come about, it is
important to note, only if reinforcement is associated with the
attaining of the biological 'goal' or 'purpose' of the behaviour. If
reinforcement were consequent only on the mere performance of a
particular activity-pattern, then response to inappropriate
releasers would be learned just as easily (in the first instance,
anyway) as to appropriate ones. Presumably internal reinforcement mechanisms have to be evolved, or adapted, to achieve this
link between behaviour and 'goal' or 'purpose'. Barnett (1958b)
defines 'consummatory state' as 'an afferent pattern, impinging on
the central nervous system, which reduces or ends an activity'. It
would seem that such a consummatory state would allow for the
termination of an activity to be associated directly with the
achieving of its biological 'purpose', since the consummatory
state achieved through an original and instinctive behaviour
sequence would presumably be the same as that achieved through
a derived and intelligently-modified sequence. Barnett remarks
further: '. . . an important feature of a consummatory state is that
it has reward value: an animal tends to learn to achieve it'. This
is, of course, even better for the present argument!

The learning of new responses could on the basis so far established be of a trial-and-error type.

Reinforcement (reward) must be even more directly linked with
adaptiveness (the achieving of the biologically useful goal or
purpose of the behaviour – we talk, of course purely from the
spectator's point of view, with no suggestion that the 'behaving
animal' knows anything of purpose or goal) if the motor patterns
of the consummatory activity are to be varied (see (b) above.)
The connection must be close and immediate otherwise the
introduced variations will tend towards the non-performance of
the vital function of the consummatory activity, and this will
lead to selection *against* the variation and against variability in
general.

7. NEED FOR RAPID ADAPTIVENESS

'Trial-and-error' (Thorpe 1963: 88) learning might thus be incorporated into the consummatory phases of an instinctive pattern in order to increase adaptively its variability. It is apparent that such learning would have to be accomplished quickly – consummatory phases *are* such because they perform functions which are biologically necessary to the individual and/or the species, and these functions cannot generally be omitted or much delayed – hence reward or reinforcement would have to be associated directly and immediately with the successful performance of a new variant of the function, if the variation were to have any hope of perpetuation.

The chief difficulty about trial-and-error learning, in relation to the performance of vital functions, is that the making of one trial will in many situations prevent the essay of a second, i.e. some vital functions can be performed only if *no* errors are made, if the animal 'gets it right first time'. As Ashby (1964: 230) remarks: 'The common name of "trial and error" is about as misleading as it can be. "Trial" is in the singular, whereas the essence of the method is that the attempts go on and on. "Error" is also ill-chosen, for the important element is the success at the end.' Certainly it can be said that there would be selection in favour of reducing the number of unsuccessful trials, i.e. against 'pure' trial-and-error learning.

If adaptive variation is to be brought with the minimum of trial-and-error into the consummatory phases of its activities an animal must have built up, prior to an opportunity for consummatory action, some knowledge or 'understanding' if its situation. A rat, or even a cockroach (Barnett 1958), will adaptively vary its escape reactions in proportion to the length of time it has had to explore a particular area. Thus some sort of 'store' of knowledge, or memory, must be involved. There must also be involved, in principle at least, power of abstraction, of 'seeing the common quality', of generalizing. It must be recognized that *this* situation is effectively similar to *that* past situation in which action X was performed with non-adaptive results, if the animal is to avoid or modify X *this* time (and *vice versa*). In terms of orthodox learning theory, it looks as though it would be advantageous if 'latent' learning were to have occurred prior to the consummatory situation, if even 'insight learning' were to have occurred. The term 'insight learning' would not normally, of course, be applied except as a last resort, i.e. if an experimental animal appears to make even the vestiges of a 'trial' this is regarded, on the principle of parsimony, as ground for rejecting 'insight'. I feel that total and unquestioning adherence to Ockham's Razor may be mistaken, particularly during the phase of speculative exploration of a new field (as in the present instance). The simplest or lowest-level explanation is to be accepted, as such, only *when all else is equal;* otherwise, the most illuminating or most suggestive explanation may be preferable, certainly in a provisional sense. So in this case I suggest: that a modicum of 'insight learning' probably occurs in many learning activities and reduces the number of unsuccessful trials; and that this may occur even in quite lowly animals (see Thorpe 1963: 116 ff.) Is there a real difference in kind between 'latent' and 'insight' learning? There is a conventional usage, certainly, which differentiates them, but we are far from being able to assert the existence, far less the nature, of any difference in the basic neurological processes involved. The cockroach which *after exploration* can increase adaptiveness of its escape reaction, without any 'trials', is said to show 'latent' learning, but this might equally be described as 'insight'. (My position on these issues I find to be similar in many ways to that of Hebb 1949, see especially pp. 158 ff.)

Ewer (1961) regards the genesis of a new adaptive variation in behaviour as always occurring in a 'problem situation'. The new behaviour is synthesized out of 'parts of patterns which are appropriate to other situations'; presumably these 'parts of patterns' are susceptible to at least some degree of change during their incorporation into the new activity. Ewer suggests a similarity between the motor patterns of instinctive behaviour and those of formed (i.e. already learned) motor habits. The difference between instinctive and learned behaviour would thus be only in mode of origin, instinctive patterns being 'endogenous' in her (1957) sense. She notes the importance of *play* as a means of ensuring what might be called 'flexibility of translocation': that motor patterns should not be irrevocably fixed in one particular part of the individual's repertoire, but should be capable of use (perhaps in modified form) in other parts according to need. This flexibility is important indeed. Ewer's acount, however, does not seem to visualize an animal's retaining, let alone increasing, its adaptive flexibility of action; nor does she emphasize the need for the 'unlearning' of no-longer-adaptive patterns – though what she does say is not inconsistent with the inclusion of this important feature of intelligent behaviour within the theoretical corpus. Increasing *adaptability* as a way of raising the general adaptiveness of behaviour to a new level has been of immense importance in human development. (Considerations of this sort lie behind Huxley's (1961) proposal of a new Grade, Psychozoa, to indicate Man's unique place in the Animal Kingdom. For an accessible general account, see Huxley 1953.) Barnett (1958b) has emphasized the importance of what he calls 'deutero-learning', learning-to-learn, as a basis for subsequent adaptability: the aspect of adaptability of behaviour which has, so far, been understressed is that of 'unlearning', of *not* performing actions which are not, or are no longer, adaptive, (or, often the most significant in the human context, actions which are not the *most* adaptive of those possible). This issue is further developed in Chapter Nine.

The situation where the individual animal changes a less adaptive for a more adaptive action can be regarded as a problem-situation in Ewer's (1961) sense if either:

(1) the individual is in some way aware of the disparity between what *was* being done and what *might* be done; or
(2) the individual is able to refrain from the performance of the less adaptive action, and then (possibly by chance) performs one that is more adaptive.

(1) presupposes insight, indeed a fairly advanced level of abstract understanding and imagination; (2) involves a notably loose connection between the action and the stimulus-situation which elicits it. That there would have to be a causal factor responsible for such 'loose connection' may be regarded as highly significant in the light of what follows (see section 9 below, also Chapters Four and Six on the functioning of the P-factor of intelligence). These are two extremes of a range of possible cases, and most actual instances would presumably be of an intermediate kind. Note, however, that in all cases an improvement in performance, increased adaptiveness of a behaviour pattern, is dependent on at least some degree of abstention from, or delay in, the overt motor activity previously usual in the 'problem situation'. Something similar to (2) above can also occur in a *play* situation, i.e. when no particular activity is required of the individual. A new type of response can be built up in play and then used 'in earnest'.

A brief diversion into some of the implications of exploratory behaviour may be useful at this stage.

8. EXPLORATORY BEHAVIOUR

Exploratory behaviour is a type of instinctive activity whose consummation is learning (see Shillito 1963). It apparently has the function (among others) of building up knowledge of the environment, i.e. a sort of mental map or topographical model (McReynolds 1962). It is linked, as Barnett (1958a) points out in a review of the subject, with the possession by the individuals of a species of 'specific homes nests, or other centres' (see also Barnett 1958b). As the home range or territory becomes familiar, the individual becomes better at finding food or shelter, escaping from predators, etc. (*v.* reviews of territoriality by Hinde, 1956, and Tinbergen, 1957). If the home range is relatively stable in configuration, the knowledge of it may be particular, the 'items' in the memory store may be of *this* rock, *this* fallen branch, and so on. Generalization and abstraction are necessary insofar as the *relationships* between items must be known. Generalization and abstraction must play an increasing part, however, in proportion as the home range itself is variable. 'This', 'that' become functionally inadequate; it is necessary to progress to 'this sort', 'that sort'. It is a commonplace that the variable arboreal environment gave rise to the increased intelligence of the high primates – and it seems probable that the demands of a wide-ranging life in the changeable sea have produced the apparently high intelligence of some Cetaceans. (It is only recently that Cetacean intelligence has been scientifically investigated, and not much is known of it – methodological difficulties are considerable – but similarities between porpoise and human brains were remarked upon by Ray in 1671, and a number of recent workers have shown that some Odontocetes are nearer to man, on degree of cerebralization than are the great apes! (Wirz 1950) There can be little doubt of the direct evidence for Cetacean intelligence even if some of this evidence is not absolutely unequivocal (cf. Kellogg 1961; Lilley 1960).

In accepting that the end-result of exploratory behaviour *is* the construction of a 'mental map', it is important not to overlook or understress one of the most significant features of the behaviour: this is that the 'map' in some sense incorporates 'evaluative judgements' relevant to the welfare of the individual or the species of animal concerned. In human usage, the very idea of a model presupposes that the original, the 'thing modelled', is not going to be reproduced exactly and in totality: some simplification, and therefore distortion and falsification is accepted as inescapable. This is one of the basic assumptions in cartography. The skill of the cartographer consists in choosing the most suitable projection *for a given purpose.* The question: 'Which is *the* best projection?' has no answer. Similarly, in exploratory activity an animal cannot memorize *everything:* it will notice and store items of information roughly in proportion to their usefulness to it. This limitation would seem to be imposed by the capacity of the central nervous system (cf. Thorpe 1963: 145 ff.); and various mechanisms contribute to the sorting and compression (or codification) of information which is the necessary concomitant. The overall sensory capacities of the species are of obvious relevance; so are the diverse types of filtering mechanisms (see Marler 1961). Note that any sorting which is not due to in-built mechanisms working purely on the physical nature of the sensory inflow must necessarily include or presuppose a capacity to generalize. 'Useful', 'non-useful', 'dangerous', 'safe' – the use of such categories logically entails generalization and abstraction – and a very important feature of novel adaptation is the extension of these and similar categories to include the subsumption of new objects (cf. the use of novel tools by primates: Köhler 1925; Bolwig 1961).

9. NECESSITY FOR INHIBITION OR DELAY OF RESPONSE

It seems apparent that animals generally do, by overt 'exploratory behaviour' or by less obvious 'latent learning' processes, build up 'latent knowledge' or 'memory stores'. With regard to the adaptiveness of behaviour little has been gained, however, if despite the build-up of 'latent learning' the animal still responds in its original automatic but no-longer-optimally-adaptive manner when it encounters the releasers for consummatory sequences. It is absolutely necessary that the original rigid and automatic connection between releasers and particular activity sequence should be broken. This sounds like an exceedingly radical change. In fact it is not. Non-response to a releaser situation is commonplace. It is one of the features for which ethological theory set out to account (Lorenz 1935; Tinbergen 1951). The ethological explanation is in terms of various internal factors. Very striking examples of non-response are due to an inappropriate hormonal balance (Tinbergen 1951; Beach 1948). It would appear at present, however, that in order to change from the slow

C

humoral mechanisms which are appropriate for non-reaction to releasers for reproduction, feeding, etc. in 'instinctive' animals, to the faster-acting (and, particularly, the *faster removed*) inhibiting mechanisms necessary for intelligent behaviour, a change to neural and possibly neo-cortical mechanisms must be necessary. At the human level, the looseness of the connection between 'releasers' (understood in a highly general sense) and activity would equate with 'flexibility of mind', 'originality', 'creativity' – and several workers have recently emphasized the importance of 'personality factors', 'emotional factors', in this flexibility (Hudson 1962; Getzels and Jackson 1962). This 'emotional' side suggests that humoral components may be still present and important in human 'flexibility mechanisms'; and this in turn might explain the effectiveness of certain drugs, e.g. mescalin (Huxley 1954) in inducing 'novel types of experience'. It seems wise, however, considering the present state of knowledge in this field, to avoid further discussion for the moment. (For an excellent and suggestive discussion of this and related topics, see Huxley 1964.)

The mechanisms for the sort of non-response to a releaser situation which could be significant in terms of adaptive variability (i.e. intelligent behaviour) would be likely, then, to be different in detail from the mechanisms appropriate for alternation between the major instincts. They would themselves necessarily incorporate relatively quick flexibility: they must needs operate, and cease to operate, very rapidly. This point can be appreciated by considering its opposite. A mechanism causing complete and long-term *non*-response to the releaser situation militates against adaptive variability just as effectively, just as completely, as does fixity in the automatic and immediate instinctive, previously normal response to it. In neither case is there any possibility whatsoever of the past experience of the individual being used towards the elaboration of a response that is new and different and hence with the possibility of greater adaptiveness. The only possibility of novelty depends on the withholding of the previously normal instinctive response *for a time* – namely, while a new and different response is elaborated and attempted – but the withholding mechanism must be such as to allow this new response, this new variation on the old theme (as it may be), to be put into operation, to be actually tried. This ability to withhold the 'standard' response must be operable like the brake of a car, able to be applied and released more or less instantaneously. And it appears that this requirement for quick and immediate action necessitates that the main mechanisms involved be central and neurological, rather than peripheral and humoral. This is not to say that mechanisms of the latter sort have no part to play, but the main conclusion of this section of the argument must be that adaptive variability and innovation have been dependent in their evolution upon a behavioural factor (and its underlying potentiality in the form of brain mechanisms, see Chapter One, pp. 32–3) for the withholding of the previously normal consummatory sequence. In other words intelligent behaviour depends upon and necessarily incorporates this withholding or 'procrastination' factor, without which the abstracting and generalizing factor and the 'memory store' would be quite unavailing.

10. A THREE-COMPONENT INTELLIGENCE

From the several foregoing lines of discussion it is apparent, I suggest, that three basic factors must be developed in the individual if the evolution of intelligent behaviour is to occur:

(i) The most important factor is that which gives the individual animal the power *not to respond* in the usual way to the stimulus-situation which previously initiated an instinctive sequence culminating in a consummatory act. This power not to respond may be absolute, or may be merely the ability to delay the response – withhold it provisionally, as it were – but its absence would negate the very possibility of adaptive variability in behaviour.

(ii) If a new adaptive response is to be achieved by anything other than the merest trial-and-error, some latent and/or insight learning must have occurred, and this presupposes, logically, some sort of 'memory store'. (This is to be understood only in a functional sense at present; and the term 'memory store' gives, of course, only a crude and static picture of what will be elaborated later, see Chapter Four.)

(iii) Both the 'stocking up' and the 'using' of the 'memory store' presuppose abstraction and generalization. This is necessary, both at what might be termed the 'factual' level (the abstraction of common qualities in objects, e.g. 'hard', 'heavy', 'red', etc.) and at the level of 'evaluation', where the common qualities are obviously relational, e.g. 'dangerous', 'useful'. It may be emphasized at this point that no sharp distinction can be made between these two levels – neither can one be made between this 'generalizing' factor in intelligence and the 'generalizing' factors in perception. As Harlow (1958) remarks, 'to be efficient, reception involves both differentiation and generalization. Generalization merges into transposition and transfer . . .' It is unnecessary to assume that the generalizing function should be associated exclusively either with intelligence or with perception, i.e. that they are separable. It is clear that the generalizing processes of perception are largely central rather than peripheral, as Harlow indicates, hence they can contribute both to functions which might be regarded as perceptual and to those which might be regarded as pertaining to intelligence. Also, they can contribute both to instinctive and to intelligent behaviour, and to learning – so here again it is necessary to keep in mind that we are establishing and tracing the implications of logical distinctions: we are not suggesting that the factors are separable, only that they are distinguishable.

Having now completed a preliminary elucidation of the factors which would appear to be necessitated within intelligence by the requirements for its evolution from the basis of an instinct system, it is time to attempt both an amplification and particularization of the theoretical implications, and also to find empirical evidence – if any can be found – which might support the purely abstract inferences made so far. As is usual with any fairly high-level theoretical re-orientation (see Chapter Twelve), the discovery of empirical evidence is likely to be a question rather of what is to count *as* evidence, it is a question of criteria, of seeing new significancies in already-known phenomena, rather than straightforward fact-finding by observation and experiment. That, of course, can come later. But before a programme of empirical research can be planned, it is necessary to determine what fields offer prospects for fruitful work, what types of phenomena are likely to be worth investigating and in what ways. To do this it is necessary to survey some of the established fields, to see if clues can be picked up as to the profitability of reinterpretation and re-assessment in terms of the new theory. The fields initially considered are likely to be selected on the basis of 'historical accident', the idiosyncracies of personal experience – but this need not detract from their methodological relevance, since this could be assumed to be the attention-attractant in the first place.

POSTCRIPT TO CHAPTER III

Two issues of great importance have been raised by Ronald Fletcher (pers. comm. 9/1/72) since the foregoing was written:

(a) That consideration of what earlier workers have written on the instinct-intelligence-learning-emotion complex would have enriched the present arguments. This is undoubtedly true. In extenuation I can make only two pleas:

(i) That my own education, in zoology and philosophy, and the pressure of day-to-day teaching and research, has entailed that the names of Lloyd Morgan, Hobhouse, Spencer and many others are indeed little more than names to me. I am delighted to find that others have forayed before me in this territory, and I look forward to an opportunity to become familiar with their writings.

(ii) It occurs to me that many others are likely to be in a similar position, as regards background, to myself; and that my rashness in exploring, not having examined the maps of my predecessors, while it causes some parts of the argument to be left barer than they need be, at least may have the merit of focusing attention on what I believe to be new, in my own theory, namely the Postponement or Withholding Factor of Intelligence. I would hope that readers of this book may be led, as I myself have been, into awareness of new realms of the literature.

(b) That by rigorous concentration on adaptation and the adaptiveness of behaviour I have understressed the role of 'transformation' in the evolutionary processes. It is of course true and profoundly important that man has not only adapted himself to the environment, he has also transformed it. But if I have obscured the distinction between 'transformation' and 'adaptation' understood in a narrow sense, it is only because I have throughout been using 'adaptation' in a broad sense such that 'transformation' is implicitly included. I think this is clear from the opening sentence of the Introduction. The matter of the distinction is discussed explicitly, albeit briefly and without using 'transformation', in Chapter Nine, section 4. That man transforms his environment, and is in turn transformed by it, the interplay of individual and society – these are great and fascinating themes indeed. That I have touched on some aspects less than fully I can only hope to excuse – with thanks to Fletcher, who has been in this position himself, for offering the excuse to me (Fletcher 1968: xvi) – by saying that one cannot do everything in one book.

IV

Clarification and Amplification: preliminary survey of supporting evidence

1. INTRODUCTION

Before commencing the task of exploring various realms of literature in a survey of already extant support for the theory of intelligence just propounded, it will be well to attempt at this point some amplification and clarification of a few salient issues. (I should perhaps reiterate that the more detailed elaborations to be presented in this chapter are still not intended to be, and cannot be taken as, definitive in the sense of giving a full and final picture. The only reasonable methodology, with regard to a subject so excessively complex as intelligent behaviour, is to work from the general and relatively abstract toward increasing particularity and concreteness—see Chapters XI and XII.)

2. THE CENTRAL 'MEMORY STORE'

I have deliberately left my picture of the 'memory store' as simple as possible up to this point so as not to encumber the flow of the argument. It is now desirable to amplify this picture, arguing as before on the abstract basis of what is necessitated if adaptively variable behaviour is to be generated out of the rigidity of an instinct system.

Let us try to infer some of the particular features of the 'memory store', working on the basis of the adaptive needs of the organism.

'Size' or 'capacity', in the sense of 'information storage capacity', must obviously be significant. Other things being equal the larger the memory store, the greater the number of 'items' that can be stored in it, the more useful it must be to its possessor. 'Size' and 'capacity' must of course be understood purely in a functional sense: there need be little correlation, at the level of detail and considering organisms closely related to one another or forming adjacent 'links' in a single phylogeny, between functional capacity and the gross size of the whole brain or even of its parts. (Differences in 'capacity' and retentiveness, such as have been demonstrated by Rensch and his associates (see Rensch, 1956, 1966) to exist between 'related' species of differing body- and brain-size, are in fact significant in a rather different frame of reference from that here envisaged. Rats and mice are not to be regarded as closely related, within the probably narrow phylogeny of intelligence. The rate/mouse comparison is a useful one because besides being related in a fairly general way, there is also a broad similarity in habitat and way of life – it is the conjunction of this with phylogenetic relationship which allowed Rensch to make the working assumption that all variables save size had been controlled.) Besides the capacity to 'store' a large number of items of information it is also desirable that items should be able to remain in store for as long as possible; and that they should be able to resist deterioration or distortion through the passage of time. All the work in progress at the present time on the mechanisms of 'memory' (see Gerard 1961 for a review) is enlarging our knowledge of the ways in which the 'memory store' operates.

To some extent in antagonism to the 'large capacity' and 'indefinite preservation' requirements, however, is another requirement of optimum adaptibility: that the store should not easily become overloaded, clogged up with so many items of information that the sorting to find items relevant to a particular situation would be maladaptively time-consuming. This requirement is largely a demand for system and order in the memory-store. To a great extent it is met, at the functional level, by the continuous resorting involving interaction of 'memory' and 'abstraction' factors which is discussed below. But the desirability from some aspects of limiting at least the intake of information, if not the absolute capacity of the store, is worth stressing at this point in order that the adaptiveness of having two different sets of mechanisms of memory, the 'short-term' and the 'long-term', can be appreciated (Gerard 1961). 'Long-term memory' can apparently be regarded as virtually permanent. 'Short-term memory', on the other hand, seems to involve mechanisms in which the items of information 'decay' fairly rapidly, and disappear if not transferred into the 'long-term' system. In terms of adaptiveness at the behavioural level, the importance of having the two systems might be indicated in this way, that experiences which are both unusual (in the sense of being of infrequent occurrence) and unimportant to the organism, will not be recorded for long, the 'records' will be discarded and the 'storage capacity' kept uncluttered. Thus the memory store operates as a two-stage system, the 'short-term memory' functioning as a sort of filter to ensure that only those items which have a reasonable likelihood of future significance go into permanent storage in the 'long term memory'. (It should be noted that even short-term memory is more central than, and operates on items that have passed through, the 'filtering mechanisms' discussed by Marler 1961a.)

Another general requirement of a 'memory store' is that items should be accessible to recall when needed; and, further on the same line of argument, that items should be 'grouped', functionally if not structurally, in sets relating to the needs of particular behaviour patterns. This could be effected to some extent by what might be termed 'contiguity of input': the fact that items functionally related to each other would normally tend to be acquired at the same time and/or in similar circumstances. This 'contiguity of input' might be expected to result in 'contiguity of storage' (the metaphorical nature of this suggestion must be stressed), in that items could be expected to have 'inbuilt cross-references' probably only of a partial nature, which would result in their being associated or recalled in 'families' or part-families rather than singly.

The desirability of 'functional grouping' of memory items leads to a further elaboration, that items in the memory store should not as it were be left static, like information-cards stacked in the drawers of a filing system. Enhanced adaptiveness of behaviour could come about as the result of the perception of hitherto unsuspected relevancies, new connections of one sort or another between items. For this to occur it would appear necessary that a process of resorting and reappraisal should have taken place, so that items could be added to the 'families of items' already established by the mechanisms mentioned above. This 'perceiving of new similarities and differences, new relevancies' among the memory-items must involve the operation of the 'abstracting and generalizing' factor isolated in the argument of Chapter III. In terms of the 'card index system' analogy, a fundamental desideratum can now clearly be distinguished: that the cards must not be confined rigidly to the drawers in which originally they happened to be placed; they must be capable of translocation from one drawer to another, so that the system becomes not only flexible but also dynamic. The optimum of adaptiveness in adaptability at the behavioural level must depend upon optimal flexibility of reassortment of items within the memory store.

This requirement is sharply accentuated when it is considered that requirements for action, for behaviour, are often of a more or less emergency nature. As mentioned in Chapter III, behavioural responses often have to be urgent. Optimal adaptiveness of such responses, especially if the circumstances are at all unusual, must depend upon the building up, prior to the need for action, of a 'reservoir of understanding' which enables a snap response to be – sometimes, though not necessarily always – both new and adaptive.

3. POSSIBLE EMPIRICAL SUPPORT; AND FURTHER CLARIFICATION

Shortly after the basic argument of Chapter III had first been formulated, a friend and former student, after reading the draft, drew my attention to a paper by Russell and Russell (1957) in which mention was made of factors of intelligence some of which appeared, superficially at least, to be similar to the three that I had derived from the evolutionary argument. The Russells, it turned out, were following up in different modalities the neuro-logical-behavioural work of Halstead (1947), and were using his terminology regarding a four-factor intelligence. Halstead had published an account of his theory of 'biological intelligence' in a book *Brain and Intelligence*, the main findings of which were apparently regarded as worthwhile from the point of view of Freudian analysis by Claire Russell, and from the point of view of ethology-zoology by her husband, W. M. S. Russell. (See also Russell and Russell 1961, which became available subsequently.) This in itself lent added plausibility to Halstead's theory, in my eyes, and made it necessary for me to come to grips with the difficult presentation of Halstead's book.

Halstead arrived at the characterization of his factors through an investigational process very different from the abstract evolutionary argument of this present book. He was attempt-ing to assess, and to develop better methods of quantifying, the behavioural impairment of brain-injured human subjects, especially frontal lobectomies whose behavioural capacity as indicated by then current I.Q. tests was supposedly in some cases unaltered, though their everyday behaviour had generally suffered a substantial decrease in adaptiveness Halstead based his four-factor 'biological intelligence' upon analysis of the results of a large battery of tests of diverse kinds used upon groups of brain-injured and behaviour-impaired subjects and a matched control group of normal subjects. Some of the details of this work will be discussed later in this chapter; the possibilities to be revealed for exploration at this point are that the differences in methodology and technique between Halstead's work and my own might on the one hand render any apparent similarity in our views of intelligence a matter of mere verbal similarity, a matter of quite superficial convergence. On the other hand, should any similarities prove to be more than superficial, the diversity of evidence and of derivation upon which they would then be based would confer upon them an ontological status and significance which they could scarcely attain from more restricted grounds of support.

The question then is, to what extent can the three factors inferred from the evolutionary argument be equated with the three of Halstead's 'biological intelligence' factors to which they bear at least a superficial resemblance? Assuming that a plausible case can be made for regarding them as being substantially the same, one final and most searching question would be left, apparently in refractory isolation: What about Halstead's *fourth* factor?

If there is any substance in the claim for four factors being discernible in present-day human intelligence it would appear somewhat odd, to say the least, if an argument from evolutionary considerations could produce only three! Would one be forced to claim that the fourth factor had appeared *sui generis*? Or what other explanation could be given?

I shall argue in a later section of this chapter that the seeming anomaly can be given a perfectly straightforward and simple explanation; before that, however, it is necessary to establish the similarities between the several factors apparently common to Halstead's account and my own argument.

Since I have already given an amplified account of my central memory store, it will be convenient to begin by comparing this with the apparently corresponding factor from Halstead's theory.

Halstead calls his C-factor, the one which appears to equate with the 'memory store' adduced in the course of my own argument, the 'central integrative field factor'. 'This factor represents the organized experience of the individual. It is the ground function of the "familiar" in terms of which the psychologically "new" is tested and incorporated' (p. 147). His account of the ontogenetic development of the central integrative field will help to illuminate some of its features and functions:

> Even before the child has reached an age for formal schooling, myriad experiences have already registered and become integrated as part of the reality-testing and reality-extending processes of a developing ego. His world of the senses is already well differentiated perceptually. Significant 'signs' have become effective in influencing his behaviour in much the same way as in his distant and less-distant cousins, the rat and monkey. But a world of symbols (of relations-among-relations) has only just begun to take form. As problem situations arise for him with their *ad hoc* rules and *a priori* standards to which he must conform, new realities emerge which are foreign to his world of senses. But he must garner from this new world of symbols a basis or framework of orientation or security. He must do this at the same time that he is extending his world of senses. The resultant of this dynamic contribution is a matrix or ground process of the 'old' against which the figured 'new' may be tested without undue degrees of anxiety. Chaos is avoided because the dominant term in the prevailing psycho-physical relation is stable. Stimuli in the outer world which once would have been explosions in his world of senses can now be tolerated because a new threshold of awareness has been established. The child has become less stimulus-bound in his motilities. A central integrative field has been differentiated from the products of a quasi-amorphous past (p. 43).

The natural complexity and subtlety of the subject-matter, plus the different methods of approach – the zoological, the diverse psychological, the neurological – each with its own sub-language, characteristic jargon, and, even more important, its own manner of statement, make it difficult to attain absolute certainty of communication. I suggest that the passages from Halstead, plus much of his explanation of details which I have not so far quoted, come down to something very similar indeed to my own outline of what is logically required in a biologically-efficient 'memory store'. The Russells (1961), discussing Halstead's C-factor, say that it 'hinges on a complete availability of all data, old and new, for comparison with each other and rearrangement in ever more realistic groupings' (p. 35). This supports my assumption that Halstead's C-factor, or central integrative field factor, can be regarded as interchangeable with what I have proposed as an efficient 'central memory store'.

Discussion of a seeming anomaly may further illuminate the nature of the C-factor and its similarities to the 'memory store'. In examining the types of behaviour involved in tests which were found to have a high factor-loading for C, Halstead remarks:

> An item analysis of the errors made on this test by brain-injured and by normal individuals reveals an interesting fact. Whereas the normal individual threads his way through the various items, familiar and unfamiliar alike, grouping them readily into appropriate categories, patients with certain types of brain injury tend to adopt *a priori* sets or attitudes towards the items and especially toward items with unfamiliar content. These sets persist in the face of mounting contradictory evidence in the form of errors.

One might conclude from cursory reading of this that Halstead is mistaken in attributing the poor performance of brain-injured persons to deficiencies in the C-factor. Talk of 'grouping . . . into categories' would seem to indicate that it is the 'abstracting and generalizing' factor rather than the memory store which is involved. Then again, other statements seem to suggest implication of the postponement-factor (power to delay response, to 'pause and consider'). 'Whereas one error is usually sufficient to point the normal individual toward the correct hypothesis, an initial hypothesis adopted by a brain-injured individual may persist as an *idée fixe* through twenty or thirty errors or until help is given . . .' (p. 44–5). This might be taken as implying deficiency of either or both of A- and P-factors. It must be emphasized that deficiencies in both Abstracting and Postponement factors would undoubtedly contribute (as Halstead's analysis demonstrates) to the ineffectiveness of such performance. Over and above this, however, argument along the lines indicated below convinces me that Halstead is correct in implicating his C-factor. If we characterize the behavioural deficiency in question as 'inability to group to a criterion', it can be seen that in ascribing the deficiency to the Abstracting and Postponement factors we are concentrating attention upon the *activity* of 'grouping', we are putting all emphasis upon the world 'group' as a verb. But emphasis could also, or alternatively, be laid upon 'criterion'. The relationship involved is essentially a triadic one, between the logical entities comprising the 'things grouped', the 'grouping', and the 'criterion', which, in the situation in question, could only be part of the 'knowledge' or 'organized past experience of the individual', i.e. the C-factor or Central Memory Store. 'Rigidity of grouping' may result simply from having *few criteria*, which in turn may be interpreted as 'few (relevant) items in the memory store'. Putting the same consideration round the opposite way, we might interpret 'flexibility of rearrangement of memory items' as involving a quantitative increase in the *number* of items. ABCD and ADBC can be taken as 'four items arranged in two different ways' – but they can also be taken as 'two items' (i.e. the two 'groupings'), and also as '*six* items' (if both individual letters *and* their groupings are taken as 'items'). It is in this last sense that the *flexible* memory store can also be interpreted simply as the *larger* memory store. Thus Halstead's ascription of deficiencies in 'grouping' performance to deficiencies in C-factor (no doubt along with deficiencies in the other factors) is justifiable, and its justification involves giving an account of the C-factor in terms virtually identical with those used to describe the Central Memory Store.

Further argument could be given, along similar lines but relating to other aspects of Halstead's experimental findings, which would strengthen the case for regarding the C-factor and my Central Memory Store as substantially identical. I feel that additional argument would be superfluous at this time, however, and that any divergencies can be left to be argued out as they become apparent in particular contexts. The general similarity must, I suggest, at this point be accepted as proven, and con-vergence between my own account and Halstead's accepted as a real and not merely a verbal one. The other factors have yet to be investigated; and further inquiry into the 'central integrative field', Central Memory Store or C-factor (I shall henceforward use these terms interchangeably) will be in the context of specific issues of function or explanation.

4. ABSTRACTING AND GENERALIZING

As argued in Chapter III, a great accumulation of past experience or 'knowledge' in the 'memory store' cannot increase the adap-tiveness of present behaviour unless there is some way of determining the relevance of particular items of memory-information to the situation in which present action is to be performed. There must be an ability for seeing similarities and differences, if some memory items rather than others are to be selected to act as modifiers of present behaviour. Such perception of similarities and differences depends, logically, upon a power of abstraction and generalization which is found in some form or other, as pointed out in Chapter III, in most animals. It is most notably developed, of course, in 'higher' forms, and indeed generally constitutes part of the criteria for their being 'higher'.

There seems to be an exact correspondence, again, between this Abstracting-and-Generalizing Factor as inferred from evolutionary considerations, and Halstead's A-factor. To quote first from Halstead's summary (p. 147): 'A factor of abstraction, A. This factor concerns a basic capacity to group to a criterion, as in the elaboration of categories, and involves the comprehension of essential similarities and differences.'

'Grouping to a criterion' is peculiarly a human activity, we assume, insofar as it is conscious, but what is logically the same function must be presupposed as being performed by most infra-human animals. As Halstead remarks (p. 96): 'Biologically speaking, this ability or factor is an old one indeed,' i.e. it appears to be phylogenetically old.

It may be useful to give an outline of one of the tests used by Halstead on human subjects, which shows a substantial dependence on the A-factor. The test

. . . is a performance (non-verbal) test for general intelligence which has been standardized on hospital populations. It consists of twenty combinations of three or four irregular-shaped wooden blocks, each set of which, when properly assembled, yields a $4\frac{5}{8}$-inch-square pattern. Each block has straight and bevelled edges which must be taken into account in assembling the combination. The score on this test is influenced both by the total time required for successful solution of each combination and by the total number of trial moves made toward assembling the blocks.

For each combination the subject is faced with the problem of comprehending the relatedness of marked differences in shapes and sizes, their critical positions in space, and their appropriate alignment of edges to yield the constant square pattern. From the marked variations in content presented, he must abstract unique groupings to a general criterion (square). To cast the statement in language which we employ elsewhere in describing abstraction, he must grasp essential similarities in the presence of apparent differences, and vice versa. He must accomplish this without very much to go on in the way of past experience. Not only is the general problem of the test an original one, but the solution for each combination is unique. Thus, even as the test progresses, he must still rely upon abstraction rather than upon recent learning for continued success (p. 97).

The schematization of perceptual elements involves abstraction, as pointed out earlier, and this may occur at a relatively peripheral level. The question of differentiating between peripheral factors which might be subsumed as mechanisms of perception, from a central factor (or factors) which might best be regarded as a factor of intellect, seems at the present time to have little point. Halstead argues: 'It seems probable . . . that neural integration involves progressive stages of abstraction, from the receptor through an intermediate centre to the primary sensory field. But the phenomenal properties of the outer world conveyed to the various sensory fields of the cortex must in turn give rise to phenomenal neural events which likewise require abstraction for integration of the whole. It is this terminal stage in the neutral train of events which corresponds with the A-factor as isolated here' (p. 98). In other words, the A-factor operates between and upon the primary sensory areas or centres of the CNS; and from the neurological point of view it may be desirable to localize the functioning of the A-factor in this way.

I feel it may be best to emphasize an element of provisionality in any attempt at localization, however, until the requirements put forward by Gregory (1961), regarding the clarification of *functional* relationships, have been met. From the functional (i.e. behavioural) point of view it matters little where, in the chain of progressive abstraction or 'coding' between sense organ and highest brain centre, the boundary between 'peripheral' and 'central' mechanisms is set up. The only proviso is that the boundary should not be too near either end: 'somewhere in the middle' is a perfectly adequate demarcation at this stage, since this investigation is aimed towards satisfying Gregory's demand for identification and clarification of relationships at the functional not the neurological level. (One hopes that the functional relationships adduced will eventually be found to have neurological 'validity' – nevertheless it is necessary to stress that only confusion can result, if it is thought that the present arguments are intended to relate as they stand directly to present neurological findings.)

A general identity in terms of function can therefore be established between Halstead's and my own Abstracting-and-Generalizing Factors. A great many distinctions can be made within the overall functional category in question; and it is assumed, certainly by myself and apparently also by Halstead, that the variety of functions subsumed under 'abstracting and generalizing' will be performed as a result of the operation of a considerable diversity of detailed neurological mechanisms. These are not in question for the moment, however, so we may proceed to examine Halstead's next factor, the P- or Power-factor, for detailed similarity to my own hypothesized 'ability to withhold or delay response', the Postponement or Withholding Factor.

5. POSTPONEMENT OR WITHHOLDING OF RESPONSE: THE P-FACTOR

In Chapter III I argued for a power to delay or withhold the instinctive responses as an essential precondition for the emergence of adaptive variability from within the rigidity of instinct-systems. Halstead does not give his account of the P-factor within this evolutionary frame of reference. He builds it up on the basis of behavioural comparisons between normal and defective human subjects. The contextural difference makes it especially difficult, in this case, to be sure of exact correspondence. Nevertheless I suggest that the correspondence can be established.

Again using Halstead's 'Summary' statement on the P-factor as the lead-in towards detailed discussion, we find that he describes it thus: 'A power factor P. This factor represents the undistorted power factor of the brain. It operates to counterbalance or regulate the affective forces and thus frees the growth principle of the ego for further ego differentiation' (p. 147).

I propose to neglect for the moment the clause about ego differentiation, and to concentrate on elucidating what is involved in the 'regulation of the affective forces'.

McDougall (1926) appears to equate 'affective forces' with 'instincts' (or perhaps with 'instinctive drives' or 'instinctual motivational complexes' – the terminological jungle in this area is extremely thick). 'Our primary emotions (affective forces) are rooted in our instinctive dispositions,' he remarks (p. 271), 'and . . . instinctive striving and emotional expression are but two inseparable aspects of one activity.' That emotions are 'rooted' in instinctive dispositions might be merely a metaphorical way of saying that they are derived from them, or, as a weaker possibility still, that they are merely related to them in some way or another. The metaphor does suggest a pretty strong relationship, a pretty close derivation, however, and to say that instinct and emotion 'are but two inseparable aspects of one activity' goes very much further. It comes very close to asserting their actual identity. As Thorpe (1963:34) remarks: 'Psychologists have always been struck by the appearance of intense emotion which animals and humans display when suffering from thwarting of "instinctive" drives. It has indeed been said that emotion is the boiling over of a heated instinct. This . . . fits in beautifully with the concept of the damming back of specific action potential – if for 'emotion' we substitute some term less saturated with qualities of human mental life.' This last remark is actuated no doubt by the fact that Thorpe's emphasis is on the infra-human animals: he does not exclude human behaviour from consideration by the terms of reference of his book *Learning and Instinct in Animals* put 'animal' rather than human behaviour at the centre of the stage – as indeed the title itself tends to suggest. His intention to minimize misleading connotations of commonly used terms is admirable, but of course my present purpose is simply to establish, if possible, a substantial identity of reference between the terms 'instinct', 'emotion', 'affective forces', and so on.

The picture which seems to emerge is that 'affective forces' and 'instinctive drives' refer to the same inferential or theoretical entities (or processes or events), and that these are expressed *either* in the form of overt observable activities *or* in the form of emotions which may but need not be observable; *or* in both (see Lorenz 1950; Fletcher 1957: 281 ff.). Either way, it is the motivational background rather than its 'expression' with which we are concerned; and it is this motivational background, whether called 'affective forces' or 'instinctive drives', which for Halstead is controlled and regulated, it appears, or by the Power- or P-factor. (Fletcher 1957 gives an excellent discussion of historical-methodological issues relating to the instinct-emotion problem, indeed his book provides an exceptionally valuable analytic overview of the interdisciplinary confrontations and affinities between 'learning' psychology, behaviourism, instinct theory, psychoanalysis, and sociology.)

Thus my Postponement Factor, 'power to withhold or delay the instinctive response' and Halstead's P-factor do indeed appear to be the same in substance. The convergence between us is not at the merely verbal level. There are many issues relating to this P-factor, and indeed to the C- and A-factors already discussed, which remain to be elucidated. Attempts will be made, later in this and in succeeding chapters, to raise and discuss some of the more important issues. But there remains Halstead's fourth, the D-factor, to be reconciled with my account, before it will be appropriate to move on to further specific investigations.

6. APPARENT ABSENCE OF HALSTEAD'S D-FACTOR

Quoting again from Halstead's Summary (p. 147), we find the following as a statement of the nature of the D factor: 'A directional factor D. This vector constitutes the medium through which the process factors, noted here, are exteriorized at any given moment. On the motor side it specifies the "final common pathway", while on the sensory side it specifies the avenue or modality of experience.'

This as it stands appears somewhat obscure. A statement taken from a summary may be excused a certain amount of obscurity, however, due to the need for compression, and it may be that a clearer statement can be found in the main body of the work.

In the chapter (Chapter X) devoted to the D-factor the following statement is found (p. 84): 'At any given moment, intelligence is expressed or applied via a particular medium, be it writing or reading, listening or speaking, reasoning or thinking, music or painting. These are but examples from the gamut of human abilities which may be utilized at will. On the motor side each constitutes an organized motor outlet, a final common pathway. On the sensory side they comprise the mediums of experience and thus are basically perceptual.'

Russell and Russell (1961) interpret the D-factor as follows (p. 35): 'The third factor, *specific expression*, needs less comment. Abstraction and integration are general concepts, but intelligence in practice always entails the use of some specific skill or executive ability. It may be expressed in painting a picture or composing music, in working with wood or metal, in the green thumb that makes plants grow or the greasy thumb that makes machines go, in preparing a sauce or cutting human hair. Specific expression thus implies a whole galaxy of different skills, at least one of which is needed in any behavioural context.'

The term 'specific expression' appears to cut out all relevance of the factor concerned on the input or sensory side. That Halstead regards input as of commensurate importance with output is attested by the quotation already given, also by his accounts of the experiments which are the basis for the detection and characterization of the factor in question. In discussing the two tests showing appreciable loadings on this factor, for example, Halstead remarks (p. 85): '. . . in both instances there is present a factor of strangeness of modality. The tactual form board must be solved in three different trials while blindfolded. There is little doubt that deprivation of vision under these circumstances, forcing the subject to rely upon his tactual discriminations, results in considerable feelings of tension and awkwardness.' Later, he states that: 'The agnosias afford valuable evidence for the existence of the directional factor D; for in such individual we find a loss of ability to recognize objects or symbols through hearing, or touch. The loss is here produced by focal brain damage which has not impaired memory functions, general intelligence, elementary sensations, or ability to execute simple to complex voluntary acts' (p. 88). Further quotation at considerable length could be given – but it seems clear enough already that Halstead's D-factor relates to both motor and sensory efficiency and might be regarded as a resultant or product of their several efficiencies. It is clearly to be understood as a central rather than a peripheral factor; though again, a sharp boundary need not be assumed to separate the one from the other. I suggest that the factor might well be termed the 'Sensori-motor Efficiency' factor – but I shall use this and 'D-factor' interchangeably henceforward.

Having elucidated the general nature of the D-factor, is it possible to explain its absence from the group of factors inferred in Chapter III as necessarily involved in the evolution of intelligent from instinctive behaviour?

As was indicated earlier, this problem, which at first glance might appear likely to undermine the very structure of the evolutionary theory presented in this book, is transformed into positive support for the theory once the nature of the 'missing' factor has been explicitly stated. For the truth of the matter is that the D-factor was not missing at all: it failed to appear in the argument on the transformation of instinctive into intelligent behaviour, simply because it had been present right from the start! The D-factor is present in 'pure instinct' animals. It is a factor indicating the general level of efficiency of the individual's sensory and motor intercourse with the environment; and it is obvious enough, even at the 'commonsense' level of biology, that the general biological efficiency of any animal, whether its behaviour be purely instinctive or largely intelligent, must be increased in proportion as its sensory range, acuity, etc. is improved and as its motor activities are made more precise, more powerful, more rapid, etc. . . . As Harlow (1958) has said, probably the initial impetus in the evolution of greater brain size came from the need to provide better sensory equipment and the 'sorting' centres to handle the increased range and quantity of information acquired as a result. (In fact, the very evolution of 'intelligence' as Harlow sees it (cf. Chapter III, p. 57) seems to involve little more than the evolution of the D-factor. C- and A-factors would probably be involved to some extent, in his picture; but the P-factor, which is central to my own theory, is completely missing from his.)

Thus the evolutionary development of increasingly complex and efficient instinctive behaviour must necessitate the elaboration of the D-factor. This would have been well advanced before the first stirrings of true intelligence became apparent, and in fact must have been simply 'taken over' by the developing intelligence system when it appeared. This is not to say that it would have been lost to the remaining instinctive mechanisms. Undoubtedly the various neural activities which make up the several factors or factor-groups in biological intelligence can contribute to behaviour patterns of both 'intelligent' and 'instinctive' type – just as motor organs like hands or legs can be used either 'instinctively' or 'intelligently'!

There are good grounds for believing that an increase in selection pressures favouring the D-factor must have occurred in the early phases of human evolution. Its overall importance probably increased; and undoubtedly there were specific changes which made a vital contribution to what are now peculiarly human cultural characteristics. Some of these changes, as they relate to the early evolution of *Homo sapiens*, will be mentioned and discussed in Chapter VII.

7. EVOLUTIONARY INTELLIGENCE: PRELIMINARY OVERVIEW OF SUPPORT FOR THE FOUR-FACTOR THEORY (ESPECIALLY THE POSTPONEMENT OR P-FACTOR)

In the foregoing sections of this chapter it has been demonstrated fairly conclusively, I think, that the conceptual structure of intelligence argued in Chapter III can be extended to accommodate the fourth (D) factor distinguished by Halstead (1947); and that the four factors then to be regarded as making the basis of intelligence in my own theory equate severally with those of Halstead. The empirical evidence which supports Halstead's picture of human intelligence thus supports mine also. It is true that Halstead's techniques were criticized by Hebb (1949), but then Hebb's own earlier work had been criticized by Halstead and others; and the criticism of Halstead, even if completely valid, would result merely in demoting the status of his evidence, from being *necessary and sufficient* for the theoretical inferences drawn from it, to being only *strongly supportive* of them. (Hebb's main criticism of Halstead, like Vernon's of the 'creativity researchers' which will be discussed in Chapter V, appears to rest on a methodological 'double standard'. Perhaps Halstead and the creativity researchers did argue beyond their evidence, but unless people are prepared to do this, there is little opportunity for progress in science.) I feel that I can fairly claim, therefore, that the mass of Halstead's evidence can be counted as positive support for the four-factor 'evolutionary intelligence' here presented: it supports the four-factor theory, even if it does not necessitate it.

It may also reasonably be claimed, I suggest, that the A-, C-, and D-factors are so widely supported, in one form or another, that they could be said to be generally accepted. A mass of provisos about nomenclature, distinctions to be made within factors, and so on, could be formulated and discussed at this stage. There seems little point in doing so, however, having regard to the level of generality at which we are working. I am only too ready to accept, indeed I hope and expect, that further research and deeper understanding will result in finer distinctions being made within each of the factors of intelligence here proposed. Nevertheless it seems desirable to emphasize that major features of the landscape need to be surveyed, their relationships determined and entered upon the map, before it becomes economical

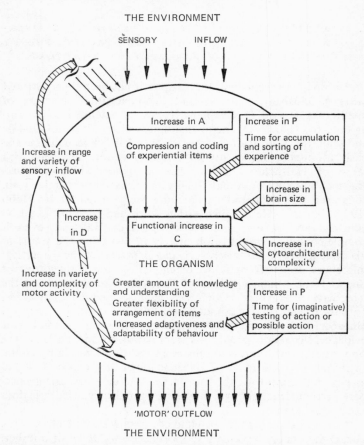

Figure 8. The functioning of four-factor biological intelligence in evolution (*from Stenhouse 1965*, Nature *208: 815*).

to get down to detail. At the moment I am concerned only with three of the major features, namely:

the Central Memory Store, or C-factor;
the Abstracting-and-Generalizing- or A-factor; and
the Sensorimotor Efficiency or D-factor.

For convenience of argument, I shall assume that the existence of these factors has been established at least in a general way;

and this will allow me to concentrate upon the new factor, the Postponement or P-factor.

It is interesting that in a paper on the ontogeny of behaviour, mainly on the interaction of instinct and learning in birds, Thorpe (1965) should argue thus: 'But positive response is not the whole of learning, and in such a task as string-pulling it is necessary also for the bird to learn to *refrain* from responding in situations in which reinforcement is absent or withdrawn. This ability to refrain is, as Vince shows, dependent on something similar to what Pavlov called internal inhibition. It is weak in very young birds, develops as a result of age, and, as a result of experience during the juvenile stage, is still being strengthened at an age when positive responsiveness is well past its peak and later weakens again slightly. Consequently, tasks that depend less on the level of activity and more on the ability to control behaviour by precisely timed imhibition are more likely to be mastered quickly and efficiently by older birds than by younger ones' (p. 495 – original emphasis).

Several inferences can be drawn from this:

(a) The *effective* potentiality for refraining from or withholding a particular response varies during the lifetime of the individual.

(b) There appears to be a general pattern in this variation which indicates the likelihood of an endogenous factor in its appearance and control.

(c) In addition, individual experience has a substantial effect on the actualization of the 'refraining tendency'.

(d) Behind all this, at the strictly genotypic level each individual clearly has the same potentiality throughout life from conception to death.

It is clear, therefore, that the 'potentiality' mentioned in (a) and (b) is itself phenotypic, i.e. it is dependent upon maturational processes and interaction between organism and environment (see Lorenz 1965). It is also clear that Thorpe recognizes that the cessation of a response, in situations where the response in question was previously reinforced, is a phenomenon which demands explanation in terms of some sort of positive causality. In the absence of definitely negative reinforcement, there is no *prima facie* reason why a habitual response should not continue indefinitely (provided the individual is healthy, adequately fed, etc.). In the lower animals, it appears that responses do often continue for an inordinate period (Thorpe 1963). The fact that non-reinforcement quickly leads to non-response, in the higher animals, therefore suggests an endogenous refraining or with-holding mechanism – Thorpe cites Pavlov's 'internal inhibition' – and this, it is suggested, is homologous with and is the precursor (at least in part) of the P-factor of intelligence. Selection pressures favouring a power to refrain from maladaptive perseveration in either instinctive or learned activities would therefore also be selection pressures for the P-factor of intelligence.

In fact Thorpe goes beyond the mere refraining from persevera-tion, and cites the work of Kavanau (1963, full reference in Thorpe 1965) which may be interpreted as demonstrating a definite negating tendency, a contrariness, in some animals. Thorpe is rightly cautious about conclusions from laboratory-conditions experimentation. The Deermice (*Peromyscus crinitus*) in Kavanau's experiments, if given the power to react against a *change* in the behavioural regimes imposed upon them, often did so, and brought back the 'non-imposed' regime, i.e. the one they had been experiencing prior to the change. This contrariness might, of course, have been a result of captivity, the conditions of the experiment; but at least the presence of the *capacity to be contrary* has been demonstrated.

Reverting to the question of perseveration, Thorpe (1965: 499) cites Lawicka and Konorski (1961) as having shown that a maladaptive increase in perseveration is associated with prefrontal lesions in cats and dogs. Ellen and Wilson (1963) have given similar findings from rats subjected to hippocampal lesions. This accords with the type of evidence offered in relation to frontal lobe damage by Halstead (1947) for humans. I have no intention of suggesting that the neurological site or 'centre' of negating, refraining, or withholding functions is locatable by ablation of any one particular part of the brain – tracking down the site and the details of functioning is the job of the neurologist and the neurophysiologist (see Lorenz 1965 on methodology). I am con-cerned, however, to emphasize that this sort of *function* seems quite definitely to have been established. It seems in general to be associated with the frontal lobes of the brain and/or their connecting commissures; but the location is irrelevant to the present argument. The existence of a refraining or withholding function within the CNS is supported. Chapter III and Fig. 8 provide a theoretical context, approximating to the 'block diagrams' demanded by Gregory (1961), within which such a function becomes more intelligible as it leads through from instinctive to intelligent behaviour and especially human creativity. Chapter V deals with various issues regarding human creativity, and argues for the significance of the P-factor within it. (On the 'negating' function in creativity, argued without explicit reference to P-factor, see Stenhouse 1971.) A theoretical context is also provided, I suggest, for the interpretation of studies on the functioning and phylogeny of the brain (e.g. Jerison 1963, 1971 and references therein; Nyberg 1971); and this theoretical context results, it is worth pointing out, from following Tinbergen's (1969) precept to give adequate attention not only to the causation but also to the functional/adaptive *outcomes* of behaviour. The evolution and phylogeny of intelligence is discussed in relation to various types of evidence in Chapters VI and VII. Before attempting to sketch in the beginning of the story of human intelligence, however, I propose to make a preliminary examination of the relationship between biological intelligence and a phenomenon which occupies a huge share of learned and popular attention at present, in both psychology and education: human creativity. If it is accepted that there is substantial agreement between the model of biological intelligence derived from evolutionary argument in Chapter III, and the model of 4-factor intelligence derived by Halstead from his investigations of behavioural impairment in brain-injured humans, the next step is to ascertain with what plausibility the 4-factor theory can be used to explain some of the refractory vortices among present notions of creativity and of intelligence at the operational level. Having considered the general theory in relation to two distinct and separate areas of evidence, we shall formulate a few deductions from it which will relate to this third and quite different area, and see whether they match up with any of the evidence available in this area. Even if a 'correspondence' test of the truth of the deductions cannot be made, it should be possible to make an assessment in terms of 'coherence' with the already existing body of theory relating to creativity. To put the matter bluntly, we can try to ascertain whether the 4-factor theory of biological intelligence enables us to make a sensible, even if not a decisive, contribution to the understanding of creativity, and in particular, to the resolution of present controversies between 'intelligence testers' and 'creativity researchers'.

V

Intelligence and Creativity: extending the perspective

INTRODUCTION

Early notions of intelligence were all founded on what was thought to be the distinctive and unique nature of intelligent behaviour in humans, as contrasted to the non-intelligent behaviour of 'animals', i.e. the infra-human animals. Thus historically a recognition of human intelligence led to Linnaeus' (1758) bestowal of the adjectival designation *sapiens* to denote our species of the genus *Homo*. It was not until the construction and use of intelligence tests, however, and the development of the concept of the intelligence quotient (I.Q.) by Binet and Simon (1916), Stern (1914) and their successors (a century and a half later) that intelligent behaviour was 'objectified' and quantified for scientific investigation. Intelligence testing became thereafter a major industry. Its benefits have been very widely appreciated. Extravagant claims have been made for intelligence tests as for all successful innovations, but, even apart from the easily-exposed excesses, intrinsic limitations in the best-designed testing procedures have been attracting attention for many years. Most recently intelligence testing and I.Q. ratings have come under fire from the growing army of devotees of the concept of 'creativity'. It is my intention in this chapter to attempt a sketch of the salient features of some creativity researches, and then to embark upon an analysis of some of the apparently antagonistic presuppositions of what have become two major 'schools': the 'intelligence testers' and the 'creativity researchers'. Upon this foundation I shall present arguments designed to show that the conflicts between 'creativity' and 'intelligence' schools can substantially be resolved by an enlargement of the theoretical framework to include evolutionary considerations hingeing upon the concept of adaptiveness or general adaptation (see Pittendrigh 1958: 391–2).

In short, it will be argued that the concept of 'biological intelligence' derived, as shown in Chapter Three, in terms of evolutionary mechanisms, can accommodate both 'high I.Q. intelligence' and 'creativity'.

Investigations of creativity have been mainly of two types, those done on adults of proven creativity (with or without comparison with non-creative 'controls'), and others done on children. It will be convenient to consider the latter first.

GETZELS AND JACKSON: CREATIVITY IN SCHOOLCHILDREN

Within a group of gifted American high-school children studied by Getzels and Jackson (1962), a significant number of those obtaining top ratings at their academic work were labelled 'over-performers' by their teachers. These children were achieving academic results considerably beyond what would have been predicted on the basis of their I.Q. ratings. The use of the term 'over-performer' is noteworthy as presupposing an acceptance by the teachers of the I.Q. rating as being the 'real' measure of individual ability, and were regarding the academic 'over' performance as the anomaly – an interesting reversal of the original role of the I.Q. test, which was developed by Binet in an attempt to find causal factors to account for 'under'-performance.

A supposition mentioned by Getzels and Jackson is that the 'over'-performance might be due to greater diligence. Their findings were in flat contradiction of this, however, a majority of the 'over-performers' achieving their academic results on less rather than more effort than that of their higher I.Q. classmates. From this it would appear that the phenomenon requiring explanation is not so much the high level of academic performance of these individuals as the low level of their I.Q. ratings. (See p. 102.)

All the students in the classes being investigated were given tests to determine their level of 'creativity', and it was found that the 'over-performer' group (i.e. those whose academic achievement was above I.Q.-based expectation) was included within the 'high creativity' group. Getzels and Jackson separated out two extreme sub-groups for detailed study:

(a) 28 individuals who were in the top 29 per cent of the study population for I.Q., the sub-group average being about 150, but whose creativity scores were relatively low. These were denoted the 'high I.Q.' group.

(b) 26 individuals not outstanding in I.Q. (sub-group average about 127) but whose 'creativity' ratings put them in the top 20 per cent of the study population on this parameter. These were denoted the 'high creativity' group.

The two groups formed the bulk of the top academic achievement group; and they were roughly equal, it should be noted, in such achievement. It must also be noted, however, that the whole study population was attending schools designed and run specifically to cater for children of high ability, with curricula, teaching methods and staff specially selected to provide interest and stimulation for such children. These points must be kept in mind when considering some of the criticisms of the Getzels and Jackson research, e.g. that by Vernon (1964), see p. 96ff.

It would appear, then, that in the Getzels and Jackson study the I.Q. tests used (which were probably not the most efficient available at the time (Vernon 1964)) were able to predict the academic achievement of only about half the top achievers. If academic achievement of the sort in question is to be regarded as a function largely of intelligence, then intelligence tests which even on a group basis perform their predictive function at a level of only about 50 per cent success must be considered substantially deficient. It becomes a pertinent question, whether over and above any possible inefficiency of administration and scoring of the tests there may not also be a deficiency in principle, i.e. they may be measuring the wrong sorts of ability. Burt (1962) indeed tacitly assumes that there is such a deficiency when he suggests that improved 'general intelligence' I.Q. tests should be constructed, incorporating a number of test items of a 'creativity measuring' type.

It is worth noting that if the creativity tests used by Getzels and Jackson were to be regarded as predictors of academic achievement, they would have to be condemned equally with the particular I.Q. tests which were used. The two types are about equally inefficient. But criticism along these lines would be to some extent mistaken, since although I.Q. tests are intended to allow (among other functions) prediction of academic performance, the creativity tests are quite specifically designed to serve a different purpose. Whether or not they do in fact achieve the intention of accurate prediction of 'real life' adult creativeness is irrelevant in this particular context. The only point is that they were *not* intended to predict academic performance in school.

Although their work has been a valuable stimulus to controversy and further research, the immediate results of Getzels' and Jackson's research have in a sense been negative. The deficiencies of I.Q. tests for the prediction even of school performance have been dramatically proclaimed; but creativity tests have not proved any better for this sort of prediction, and their claims as predictors of adult real-life creativity are as yet unsubstantiated. The interrelated questions, of prediction and/or explanation of intelligent/creative behaviour, seem thus far to be in no immediate danger of being settled.

HUDSON: CONVERGENT AND DIVERGENT SCHOOLBOYS

Hudson (1966, 1968 and further references therein) avoids for the most part the somewhat tendentious use of the term 'creativity' in connection with tests used on schoolchildren, and instead adopts Guilford's (1950) dichotomy of 'convergent thinking' and 'divergent thinking'. The convergent thinker or 'converger' is a person who prefers and is efficient at solving problems to which there is a single answer, the 'correct' answer, upon which his lines of information and inference can converge. The 'diverger', on the other hand, tends to prefer and to be most efficient in the solution of problems of a different sort, where the 'problem' is to give structure to an amorphous complexity and where there can be no question of obtaining a single or 'right' answer. (Experience has shown that it is necessary to emphasize that the term 'diverger' as used here has no connotations of social or political divergence, of rebellion or nonconformity. The diverger is simply a person who is better at open-ended tests than closed ones.) Not being over-anxious to prescribe immediate changes in educational policies and practices, hence to categorize and pigeonhole, Hudson frees himself to do justice to the facts of individual variation on the one hand, and on the other to speculate about the nature and causation of creativity. Some of these latter speculations and arguments are discussed below, (see p. 118f); at this stage we shall concentrate on a few points regarding the tests used and their immediate results.

Hudson adopts a similar initial basis of grouping to that used by Getzels and Jackson but uses a different and less question-begging terminology. 'The "High I.Q." I shall call a *converger*,' he says initially, (p. 38), 'the "High Creative", a *diverger*, and the two styles of reasoning, convergent and divergent, respectively. The new tests themselves I shall call open-ended.' He uses a greater variety of such tests than many American workers; including, besides the ubiquitous 'Uses of Objects' and 'Meanings of Words', also a 'Drawing' test in which the testees are asked to draw whatever they wish to a given caption, and a 'Controversial Statements' test to which any comments at all may be given. But his final converger/diverger grouping is based only on the results of the 'Uses of Objects' and 'Meanings of Words' tests in comparison with the score on an I.Q.-type test. 'The converger is the boy who is substantially better at the intelligence test than he is at open-ended tests; the diverger is the reverse.' 'As a matter of convenience, I define 30 per cent of my usual schoolboy sample as convergers, 30 per cent as divergers, and leave the remaining 40 per cent in the middle as all-rounders' (p. 41).

As he himself points out, Hudson's converger/diverger distinction depends upon the bias and not the level of an individual's ability (p. 41). Thus a converger might have a higher score on open-ended tests than a diverger, if his own I.Q. score were very high or the diverger's very low. It might be expected, then, that the converger/diverger distinction would cut across the lines of creative/non-creative prediction: if open-ended tests do measure creative potential, then anyone who scores well on them, converger equally with diverger, should be capable of subsequent adult creative work. That the adult creative cannot simply be equated with divergence, the non-creative with convergence, is attested by substantial evidence. Roe (1952) found strong converger-type traits in her sample of highly creative scientists; MacKinnon's (1962) findings complicate the picture, as his 'more creative' groups tended to have more diverger and fewer converger characteristics than corresponding 'less creative' groups. Nevertheless, as Hudson suggests (p. 110), the apparent conflict may perhaps be resolved by assuming that each profession or field of activity has its own characteristic 'band of wavelengths' within the converger/diverger spectrum, the physical scientists, say, tending from the nature of their subject to be in general more convergent than architects or biologists. Thus a creative physicist might well exhibit fewer diverger characteristics than a non-creative architect.

Several issues may be mentioned at this point. It would be interesting to obtain data on the actual performance on open-ended tests ('creativity' tests) of our creative physicist and non-creative architect. Could 'absolute' measures of creativity as against the essentially relativistic 'converger'/'diverger' ratings, give any clue to real and proven creativity irrespective of the field in which it occurs? Could there be a threshold level of 'open-ended' performance similar to the apparent threshold of I.Q. rating at roughly the 120 level?

Other issues, to be mentioned here but discussed later (Chapters IX and X), are: that the 'band of wavelengths' for a particular field of activity can be regarded as fixed only over a short time-span; that the 'band' for a particular field changes with time; that its rate of change varies; and that these changes may be affected by factors extrinsic to the nature of the field itself as an intellectual activity of a particular sort. It is implicit in the opening paragraphs of this book that the 'band' of personal characteristics of those engaged in the investigation of behaviour has changed markedly in the last thirty, even the last ten, years; and I have suggested (Stenhouse 1965a, 1970) that the personal and intellectual characteristics of biologists as a group may be altered over a period of time by differential recruitment due among other things to differences in the style and manner of approach of biology teaching. Similar mechanisms would no doubt be operative in other fields.

Reverting to Hudson's investigation of intelligent schoolboys the most striking empirical finding is of the very great diversity of personality and intellectual features exhibited by the boys. In particular, there is a sometimes marked discrepancy between the various predictors of performance and the actual level of real-life performance (albeit restricted to academic achievement at school and university) itself. Hudson's objective was the prognosis of arts/science specialization, to which questions of ability-levels in different subjects are clearly relevant. Prediction of type of specialization were found to be reasonably reliable in a majority of cases. The minority in which it was not reliable, however, turned out to be exceptionally significant. Hudson remarks: 'Contrary to expectation, one can measure the bias of an individual's ability, but not the level' (p. 28). Yet it would be ridiculous to counsel a boy to take up a mathematical career because he was 'strongest on numerical intelligence' if his mathematics, though less abysmal than his other 'abilities', was in fact very weak on any absolute scale. Hence a measurement of bias alone is not enough: an 'absolute' indication of ability, i.e. the ability of the individual relative to the range of ability-level of the population, is also necessary. And this, as Hudson demonstrates, has no validity whatever in some of the most important cases. To illustrate, let me quote Hudson on two of his most gifted subjects:

'Take Bailey, for example. When I tested him as a Fifth Former, he was judged one of the school's best young scientists, and in his general demeanour one of the liveliest boys of his year: inquisitive, enthusiastic, musically gifted. He fulfilled his teachers' hopes, won an open scholarship in natural sciences at Cambridge, and took a first in Part II of the Tripos. Yet his scores as a schoolboy were uniformly unimpressive. His vocabulary and general knowledge scores were presentable; but I.Q. and accuracy scores were lower than some 80 per cent of his contemporaries. Certainly there was no sign in his scores of the abilities he unquestionably possessed. Another example – one of Bailey's contemporaries, Gardner. His scores were quite good by the standards of his school, but his I.Q. was lower than those of a third of the Fifth Formers who later became Cambridge under-graduates. Yet Gardner won an open scholarship in mathematics at Cambridge, and took Part III of the mathematics Tripos, acquitting himself with distinction and carrying off the top prize. In other words, he is one of the ablest young mathematicians of his age in England (indeed, in the world) yet from his scores one would never guess it.

Again, the prediction and/or explanation of these high-level activities seems not as yet to have been achieved.

VERNON'S CRITIQUE OF CREATIVITY RESEARCH

A critique of American work on creativity as compared with 'orthodox' work in terms of I.Q. tests was published by Vernon in 1964. Following a brief survey of some of the more important researches on creativity (Taylor and Holland 1963; Terman and Oden 1947; MacKinnon 1962; Guilford et al. 1951–6; Getzels and Jackson 1962), Vernon embarks on a general discussion which is well worth dissecting for the presuppositions behind it as much as for its own sake. Since the tendency of the bulk of my argument will appear to be against Vernon I must emphasize at this point that, in my own view, many of the criticisms were much needed. His paper is a valuable corrective to 'creativity faddists'; and that it is based on a particular set of presuppositions which are them-

selves susceptible to criticism detracts not at all from its value –
indeed one of its chief merits of which I shall make use, is that it
gives a perspective of the underlying 'presuppositional geology'
of both the 'creativity' and the 'I.Q.' positions.

On the low correlation between creativity and I.Q. tests revealed
by Getzels and Jackson's work (correlations around 0.3), Vernon
remarks that 'obviously the correlation would have been very
much higher with (a) a more representative range of ability, (b)
a more up-to-date and reliable battery of g and v tests' (p. 167).
But a more representative range of ability would merely have
caused the differences between 'high I.Q.' and 'high creativity'
groups to be swamped by a mass of 'neither creative nor intelligent' individuals, and while this would undoubtedly serve to give
a picture of the population more 'true to life' in a purely descriptive sense, it is very much open to question whether or not this
picture would reveal significances – and what 'significance' would
mean in this context is in any case relative to the purposes of the
investigation – a point which will be discussed further (see pp. 110f).
That 'a more reliable battery of g and v tests' would improve
correlations is problematical, too, since it would appear that
'reliable' merely indicates a confounding of the 'creativity' and
'I.Q.' parameters: Vernon remarks (p. 168) that 'a well-balanced
set of intelligence tests . . . includes divergent as well as convergent
thinking items' and cites Burt's suggestion that 'creativity tests
might be quite useful as component items in a new general
intelligence scale'.

The foregoing helps to illuminate another point. 'I would
suspect then', says Vernon 'that one might get at least as good, if
not better, predictions of performance on adult creativity tests
from a reliable all-round measure of g and v at [age] 13 or 11+
than one could get from Getzel's tests' (p. 168). Clearly, if the
'reliable all-round measure' includes creativity-measuring items
within a variety of others, it is likely to be a better indicator of
total individual capacity than is any more specific test of whatever
type; and since a composite indicator must be presumed to be more
stable over reasonably long periods of time (especially when these
cover the years of adolescence, as Vernon apparently envisages)
than any single specific indicator, it follows that the correlation
between a composite indicator at time A and a particular indicator
at time B five to ten years later may well be higher than the
correlation between the particular indicator at time A and the
same particular indicator at time B. The only conclusion to be

D

drawn is the far-from-novel one that the reliability of prediction tends to decrease with its specificity.

It seems true, nevertheless, that Vernon's criticism of the long-range reliability of creativity tests is well taken. The importance of this criticism recedes, however, under the advance of more fundamental considerations relating to the purposes of the research on creativity and to the ultimate biological significance of the phenomena involved.

REAL-LIFE CREATIVITY

The most important issue relates of course, to 'real', 'proper', adult creativity. Let me denote this 'real-life creativity'. It has been the subject of study by Anne Roe (1952) and MacKinnon and associates (1962); the long-term longitudinal study by Terman is also relevant (see Terman and Oden 1947). That Vernon accepts this as the basic issue is attested by his criticism of the American creativity test work on these grounds: 'The most serious flaw is . . . that Getzels, Guilford and others provide no real evidence for the assumption that their tests are valid tests of creativity in general. They give us . . . no follow-up data . . .' Note that I am not suggesting that this criticism should hold – I point out merely that it reveals the criteria which Vernon holds to be relevant. And in terms of these criteria his criticisms, as I shall show, lose much or all of their force.

Real-life creativity is to be determined by the actual production of significant novelty. What is to count as 'significant novelty' must vary, of course with the field of activity: the criteria in poetry, for example, must needs be very different from the criteria in engineering or ecology (cf. Bruner's generalized categorization of a creative activity as one that produces 'effective surprise', Chapter 1 in Gruber et al. 1962). We can designate as creative, then, a person who has been judged, if necessary by those of appropriate specialist knowledge, to have produced 'significant novelty' (which is to be recognized, on the subjective side, by the elicitation of 'effective surprise'). Important limitations inherent in this criterion are immediately apparent. It is notorious, for example, that in the history of any branch of learning, in any type of human activity, the greatest and the most important creatives have often been recognized and accepted only after a long period of rejection, e.g. Rembrandt, Mendel. The rewards of fundamental creativity are often posthumous. Lack of recognition is usually not due to any failure to perceive novelty in the work in question, but to a judgement, explicit in the case of Whistler ('a pot of paint thrown in the face of the public') implicit in most (Mendel being a good example), that the novelty is not of any worth or significance. (We can also see, all too often at the present time, the opposite tendency to regard *any* novelty as significant and worthy *per se*!) That novelty, that a particular innovation, is *not* regarded as being significant, cannot therefore be taken as conclusive evidence that it will not be so regarded in the future – 'it is *not yet* significant' seems the sound formulation, emphasizing as it does the provisional element in all such judgements. On the other side of the coin, too, it must be recognized that many contributions once regarded very highly, as having great significance, have been demoted – not necessarily to the category of 'totally lacking in significance', but certainly to the category of 'being of very little significance', e.g. the evolutionary views of Cuvier (see Toulmin and Goodfield 1965), the philosophy of Dugald Stewart (Passmore 1958), and the poetry of 'Ossian' (MacPherson, 1762).

It appears, then, that while the complete absence of innovation of any kind must debar in individual from being regarded as creative in the sense of being a *proven* creative, it does not necessarily tell against the assessment of the individual as being *potentially* creative. And it is potential creativity only which is the object of diagnosis by creativity tests. The crux of the creativity problem as seen by societally-orientated writers is that a reservoir of potential creativity, a social and national asset, might be under-utilized. It is beside the point that potential for creativity might not be actualized: everyone recognized that it very often does fail to be actualized. The 'boom' in creativity investigations has occurred precisely because it was felt that the extant social mechanisms, notably in education, might be of sub-optimal efficiency in fostering creative talent towards actualization in activities thought to be desirable. And in order to determine whether or not this was happening it was necessary to find ways of testing for creative potential. To devise tests which seemed likely to give an indication of this sort of potential was the first phase of the operation; this is the phase in which the work of Getzels and Jackson in America, for example, and of Hudson (1966) in Britain, has been carried out. A second, and in terms of logic a quite different, phase is that concerned with the actual relationship between the 'potential creativity indicators' and 'real life creativity'. To call this 'validation' of the 'potential creativity indicators', the creativity tests, could be rather misleading. Talk of 'validation' presupposes that the criteria for validation are known, understood, and generally accepted – and that such criteria actually exist. A critical scrutiny of possible criteria for the 'validation' of creativity tests reveals that this is almost certainly not the case. 'Follow-up data' as demanded, apparently, by Vernon, would appear to involve a large-scale programme of creativity testing with subsequent case-history and retest data being obtained over a long period as in the famous longitudinal study by Terman (see Terman and Oden 1947). Substantial validation could hardly be expected in less than, say, thirty years, bearing in mind the spread of the 'productive period of life' over different fields of activity (Beveridge 1951), and since investiga-tions of potential creativity in young people were commenced in force only about the mid-1950s the suggestion that follow-up data should have been given to validate the creativity tests used for example by Getzels and Jackson can be seen to be unrealistic. It is true that the need for validation must be kept in mind – but to emphasize this is to do something different from criticizing research for an alleged failure to provide such validation – and, as we have seen, Vernon talks of the absence of follow-up data as 'the most serious flaw' in the creativity-testing programme.

Individuals who score high on the creativity tests 'show some of the personality characteristics and attitudes which are consonant with the characteristics of the truly creative individuals studied by MacKinnon and Roe', and Vernon remarks (p. 168) that this is 'interesting and suggestive'. But to say that it constitutes 'no real evidence' is to make an implicit appeal to a 'something more' that just is not there. 'Real' evidence in the form of follow-up data, something along the lines of Terman's long-term study, cannot be expected for many years yet; and may anyway have to be accepted, when it comes, in a form rather different from what seems to be envisaged.

THE 'BOGUS PRECISION' PRINCIPLE

The reason for this last somewhat cryptic statement I shall attempt merely to indicate in general terms for the moment. The Heisen-berg Principle of Uncertainty is familiar from sub-atomic physics: its tenor is, roughly, that you can determine the position *or* the velocity of a sub-atomic particle at a given time, but not *both* position *and* velocity. The reason for this is that position and velocity can be discovered only through the use of some form of 'radiation', and that if the radiation has impinged on the particle with sufficient 'density' (i.e. if there has been *enough* radiation) to give an accurate fix of position, the velocity with which the particle *was* travelling will have been affected by the radiation and is likely to have been altered; while if a 'small amount' of radiation has been used, though the velocity is probably unaffected, the position of the particle can be fixed only approximately. In a closely analogous way, the behaviour of the higher animals, especially the higher mammals, is subject to an uncertainty principle which in terms of experimentation and observation we might call the Bogus Precision Principle. It depends upon the fact that the very observing and/or measuring of the behaviour of the higher animals is itself liable to cause distortions of the behaviour. In a limited form the Bogus Precision Principle is well enough known. It is usually formulated in terms of the presence of the observer affecting the behaviour under observation – and elaborate precautions are taken, in psychological laboratories, to hide the observer behind one-way mirrors or other forms of screening. But the mere invisibility and inaudibility, etc. of the observer may be of very minor importance. The major factor may be the total context in which the behaviour is to be observed. In some instances it is possible to demonstrate that the context of observation has been such that the behaviour observed has been substantially atypical. Zuckerman's (1932) data on reproductive behaviour in primates, for example, were precise and quantified, but recent work in the field, though comprised of much less precise and detailed observations but of behaviour *under natural conditions*, indicates more and more strongly that the reproductive activities investigated by Zuckerman were in fact to a considerable extent aberrant (see *Symp. Zool. Soc. Lond*, 1963; also Morris and Morris, 1966). Thus precise and detailed 'objective' observations can be made of behaviour which may be aberrant, perhaps very much so; or more generalized and perhaps 'impressionistic' or even anecdotal (but it depends who tells the anecdote!) observations can be made under natural conditions and often over a very long term, of behaviour which is more likely to be the 'real' or typical behaviour of the organisms in question. Precision *or* typicality, but not both – this is the dilemma behind what I propose to designate as the Bogus Precision Principle.

PRINCIPLE OF 'TEST-SITUATION TRIVIALITY'

Getzels and Jackson note the pejorative nature of the term 'over-performer' as applied to those whose academic performance outruns I.Q.-based expectations. They suggest that the discrepancy in performance, so far as 'over-performing' creatives are concerned, is to be ascribed to the greater challenge, the greater complexity, of the real-life as against the test situation. One cannot quarrel with this formulation, yet I feel that it can be given more point, its implications can be pushed closer to the surface and its heuristic value increased, if it is put in inverse form. Thus from saying that 'Real-life situations (as against test situations) tend to enhance performance' I suggest we should move to the statement: 'Test situations (as against real-life situations) tend to depress performance.' This formulation is not intended to suggest that there is any positive objective factor in the test situation which actively tends to depress performance – as a description of the causal situation *external* to the testees, Getzels' original statement may be more accurate than my suggested emendation. In terms of interpretation, however, attention is now focused on the functional aspects of behaviour, i.e. its results, and from this point of view the depressant implications of the test situation are worth considerable emphasis.

Not that all individuals, or even many, are likely to be affected by test-situation inhibition; but those that are are likely to comprise a minority that may on a number of grounds be highly significant. This is discussed on p. 113f, (Empirical research into the incidence of test-situation inhibition would appear to hold interesting prospects.)

In its functional aspects the behaviour of any organism is purposive in that behaviour patterns have been selected in the course of evolution for their contribution to the survival of the individual and the species (Pittendrigh 1958; Lorenz 1965: 8–9). As Lack (1954) has demonstrated, almost any causative mechanism which gives rise to the 'required' results may be selected: the implication being that out of a number of equally efficacious mechanisms the one actually selected will have been singled-out by chance factors in themselves quite irrelevant to the perpetuation of the species concerned (see Lederberg 1958). From this point of view it is the *results* of behaviour that count. The means whereby the results are attained are, other things being equal, of negligible importance. Following from this, as the transition from mere purposiveness to sometimes-conscious purposefulness took place in the course of human evolution, the results of behaviour continued to exert and still exert a determinative effect. This is now mediated, of course, in human behaviour by the mechanisms of expectation and intention, by the results the agent *hopes to achieve* – and the results achieved are not always the same as the results intended! Nevertheless it is vital to our understanding of human behaviour not to be distracted by the expediencies of experimentation into a neglect or underestimation of subjective 'intentions' and motivations orientated towards results. It is difficult to take account of these intentions in an experiment – but they are neglected at our peril. In particular, it appears that the results of test-situation behaviour are not regarded as important by some individuals, therefore their performance in tests is depressed by the lack of result-orientated motivation. Getzel's 'over-performing' high creatives were stimulated in their academic work by its unusually high interest-level (it must be remembered that these children were in schools especially tailored for gifted pupils), whereas the I.Q.-test situation must have appeared relatively devoid of possibilities of significant result. The earlier parts of most I.Q. tests are deliberately made easy, so as to allow low-ability individuals to 'get on the scale'; and this very feature makes them less attractive, less stimulating, to persons who seek and need a substantial challenge to bring out their best (see Chapter Eleven and Epilogue for further discussion of motivational mechanisms). This defect of triviality is likely also to be found, of course, in creativity tests – and here we return to the crucial question of their usefulness in actually predicting real-life adult creativity.

Interest and exploration are activated by novelty. What distinguishes the real-life creative from the butterfly-minded individual who superficially may appear rather similar (a wide diversity of interests carried on in alternation is characteristic of many individuals of both types), is the fact that while *mere* novelty is sufficient for the butterfly, the true creative requires *significance* in his – and 'significance' in this context implies something familiar to be discovered within the novelty, unity within diversity, inter-connections lurking below the surface of a variegated dispersion. This is what lies behind the point made earlier about tests: they are trivial precisely because they do not link up with the things that matter, things that are important. Thus while a true creative may respond well to the first test-situation he encounters (the exact nature of the test probably does not matter), once he has categorized tests as holding little of long-term interest or significance for himself he is likely to treat them in cavalier fashion and hence to perform in them less well than he might.

'EVOLUTIONARY INTELLIGENCE' FACTORS IN CREATIVITY

Familiarity within novelty, 'unity within diversity', 'inter-connections' – this is very much the perception of similarities and differences noted as the diagnostic feature of the Abstracting or A-factor in Chapter Three. Is it possible, we may now ask, to resolve some of the discrepancies between the performances of 'high I.Q.' individuals, 'high creatives' distinguished by performance in tests of creativity (let us call this group the 'high-test creatives') and 'real-life creatives', in terms of different endowments or 'quantities' of the Abstracting Factor? In other words, are the differences between these three groups to be explained by their possessing different 'powers of abstraction'? Is the real-life creative to be distinguished from the 'high-test creative' by a greater ability to abstract and generalize?

Without attempting to argue the matter in detail, I feel that the answer to these questions must be 'No.' Undoubtedly there must be a considerable range of variation in abstracting ability within each of the three groups in question; undoubtedly also, all three groups would rate high for this factor in comparison with the rest of the population. But if we consider an hypothetical comparison between a real-life creative who is also relatively low in I.Q. (say around 120, a rough lower limit for real creativity according to MacKinnon 1962; Taylor 1956), and a high I.Q. non-creative, it is clear that neither of them can be considered poorly endowed with abstracting ability. From the nature of the case they would both have to be regarded as highly endowed – the difference between them obviously arises in the different situations and different ways in which abstracting ability is used, rather than in a quantitative difference in the ability between the two individuals. The high-I.Q. person sees similarities and differences, inter-connections and implications, in the material presented in the I.Q. test situation; while the high creative does not see them in this situation, but does in real-life situations to which the high-I.Q. individual does not react.

It is implausible to suggest that the differences in question could be explained by differences in Sensorimotor Factor (Halstead's D-factor), since equivalent ranges of variation in this factor would appear to be found in all three groups being considered. Similarly for the 'Memory Store' (Central Integrative Field or C-factor of Halstead) – 'good memories' are possessed by high-I.Q., high-test and real-life creatives. Again, the differences would seem to demand explanation not so much in terms of presence or absence, greater or lesser endowment of these factors, as in terms of the *ways they are used* and, specifically, in the *types of situation* which elicit or stimulate their use.

There remains the Postponement-and-Pertinacity Factor (Halstead's Power or P-factor): can we differentiate our three groups on the basis of differences in this? This factor, to recapitulate, is the one postulated as conferring the ability to delay or postpone a previously normal response and thus to allow opportunity for a different and possibly improved response to be substituted. If the analysis given in Chapter III is accepted, the Postponement Factor must be seen as an especial prerequisite for the original evolution of intelligent behaviour – the other factors could have been, and undoubtedly were, elaborated first within a context of purely instinctive behaviour. Thus the Postponement Factor would appear to be peculiar to intelligent behaviour in the past early stages of its evolution – is it also operative, and is it of any importance, in the present-day behaviour of humans? Specifically, can we differentiate between real-life creatives and other apparently intelligent individuals on the basis of differences in power of Postponement, Power of Abstention? To achieve a fundamental understanding of the answer to this question it is necessary to digress into areas which may superficially appear to be unrelated.

THE FALLACY OF 'THE UNKNOWN', AND THE POSTPONEMENT FACTOR

If we recognize that curiosity, a drive to explore, tends to push human individuals into asking questions about the multifarious phenomena among which they live we must recognize also that to the vast majority of questions that can be put (not necessarily explicitly or even consciously), there are answers which can be and are given. These 'answers' are sometimes 'bare facts', answers given ostensively, involving direct sensory experience and little else. In most cases, however, a cultural element is involved, even if it is only a knowledge of the appropriate 'language game' (Wittgenstein 1953) of the culture. The culture has an answer to most questions, which is not surprising in view of the fact that the same questions tend to occur to each new generation. Answers are handed down along with the questions as part of the cultural accumulation of the social group.

It is important to realize that the notion of discovery and innovation as 'pushing into the unknown', 'extending the frontiers of knowledge' is substantially misleading in several ways. It is in most instances *not* pushing into the 'unknown' – it is pushing into the '*known*.' The 'known' has to be pushed aside and an 'unknown' revealed; once this has been done the 'unknown' can be touched and felt over and its contours mapped. Kuhn (1962) nearly but not quite captures the essential point: 'Because the old must be revalued and reordered when assimilating the new', he says, 'discovery and invention in the sciences are usually intrinsically revolutionary.' But 'almost none of the research undertaken by even the greatest scientist is designed to be revolutionary.' The crucial words in these quotations are 'assimilated' and 'designed'. It is true that *conscious* revaluation and reordering occurs only when the new is being assimilated, worked into the fabric of accepted theory – nevertheless the first and logically the most important revaluation occurs, implicitly and unconsciously, when the bare possibility of the intrusion of a 'new' is first dimly apprehended. Similarly, it seems likely to be descriptively true that revolution is not the *conscious* intention or design of the innovator in science – yet provided that we exclude connotations of 'extensive scope' which 'revolution' has acquired from its hyperbolic misuse is popular writings about science, it is clear that every significant theoretical innovation is necessarily revolutionary. In fact, many a creative innovator has testified to feeling pushed, almost against his own will, into discovery and creation. 'Extending the frontiers of knowledge' has an essentially retrospective reference, one can use this sort of language only *post hoc*. The real achievement is in blasting out of the retarding atmosphere of the 'known', in seeing that the subjectively accepted 'known' of the social group has been wrongly or, more commonly, unnecessarily accepted. Rejection of the current orthodoxies is the first, the most difficult, and the most essential step towards innovation, creation and discovery.

The logical structure of the situation may be illustrated by reference to the popular (though quite erroneous) version of the discovery of America by Columbus. According to this story, Columbus was a brave pioneer who sailed westwards across the Atlantic 'into the unknown'. The point which must be kept in mind is that, so far as Columbus' contemporaries were concerned, he was *not* sailing 'into the unknown'. He was sailing into the known. They knew very well what would happen to him: he would sail on until he reached the edge of the world, and then he and his ships would fall over the edge and be destroyed!

In short, Columbus had to *create an unknown* before he could sail into it on a voyage of discovery. The extensive opposition against which he had to struggle in order to embark upon his voyage of discovery was, in fact, simply the social-psychological emanation or symptom of the more basic logical opposition of 'the known' towards its own negation. The innovator and discoverer has always had to overturn previously-accepted beliefs.

We tend to see this most clearly, of course, with regard to past and established discoveries. We appreciate the percipience and courage of Galileo in his fight with the authorities of his time; and since we now take for granted the views which were so revolutionary when he formulated them, we imagine that there is no orthodoxy to fight against nowadays – or, that we have a new 'orthodoxy' which, in contrast to the old, is *favourable* to change. This last assumption is, or rather has been, partly true. The advances in physics, for example, over the last century or more have been made, in general, within a climate of opinion receptive, indeed expectant, of change. But while socially and psychologically the times have been favourable to advance, it is necessary to recognize two qualifications which modify this general picture in important ways even with regard to physics. Firstly, while it is true that innovations regarded as advances have been welcomed, the word 'advances' embodies presuppositions of direction and evaluation – and it may turn out, at some time in the future, that our directions and evaluations have been in some cases mistaken.

The second proviso with regard to creative innovation is this, that even granted a general receptivity to change, there is still the logical problem of 'deciding' which part of currently accepted 'knowledge' is to be set aside, which knowledge is to be thrown out. And this pleasantly aseptic and abstract logical formulation of the problem turns into the worrisome business, in terms of real life, in terms of the actual creation of novelty within a social context, in terms of what actually happens, of *disbelieving what one has been taught*. A young real-life creative has to say, in effect, to the eminent Professor in whose classes he has learned, within whose laboratories he has worked and upon whose research money he has fed: 'You have been mistaken, Sir, all these years, in your views as to the nature of your subject. Come, I will show you the true story. . . .' Clearly, to do this demands self-confidence almost appalling to contemplate; and clearly, many creatives must shrink from the act, usually well before the new theory has been properly formulated even in their own minds.

(Hudson (1966:146) suggests that the innovator 'attacks the older generation and their theories much as he would like (in psychoanalytic caricature . . .) to have attacked his father.')

While the psychological trauma of promulgating a discovery has been dramatized often enough, the point to which I am leading the argument is this, that an even more radical psychological *tour de force* is required to reject or suspend belief in the previously-accepted knowledge in the first place, before any new discovery is or can be made. We could state the position in exactly the same terms as were used to describe the break away from instinct towards intelligence: *the individual must have the ability to withhold the previously-normal response*, if he is to have any chance at all of elaborating a new response. 'Response' here would have a largely or purely intellectual denotation – new discoveries and inventions seldom now involve an immediate change in overt motor activity. And 'the previously-normal response' would not now be an instinctive but an intellectual or cognitive one.

Thus the Postponement Factor or something very like it appears to be a necessary prerequisite for real-life creativity. The immediate problem from which we started in this chapter, however, was that of distinguishing between real-life creatives on the one hand, and high-test creatives and high-I.Q. individuals on the other. Can we establish a distinction in terms of the Postponement Factor? It appears that we can. Real-life creatives can, as we have argued, be assumed to have a high level of P-factor – I hope now to show that we can assume a relatively low level of this P-factor in both of the other groups to be considered.

We have seen that for the real-life creative the operation of the Postponement Factor results in the provisional rejection of the accepted answer to a question (or, which is logically the same thing, putting in question an accepted 'theory' – the 'theory' not usually being recognized as such, but rather taken as factual). To attain a high score on an I.Q. test, on the other hand, no rejection of existing knowledge is required. On the contrary, many items within I.Q. tests involve definite acceptance of specific culturally-determined views or 'answers' – this is true irrespective of whether the item asks explicitly for the information or leaves the information-requirement implicit – and even when antecedently acquired 'factual knowledge' (i.e. accepted 'theory') is not involved there is still the necessity for procedures and concepts to be known. (This is what makes I.Q. tests culture- and language-specific, in practice.) The testee has to know and accept the standard meanings of the instructions, and must, of necessity, use standard procedures if his responses are to be meaningful, and especially if they are to score well.

It might be thought that the open-ended tests used in identifying 'high-test creatives' and 'divergent thinkers' (Guilford 1950; Hudson 1966) would allow scope for, or even be dependent upon, the exercise of the Postponement Factor. To a limited extent it would appear that they do. To score highly on a test like 'thinking up as many uses as you can for a blanket' (Hudson 1966: 165), the testee must be able at least to reject the obvious and standard uses as the *only* uses: he must see other uses besides the usual ones. But the usual and standard uses are not in fact *rejected* by most high-test creatives. They still appear in the lists, even though their relative importance and conspicuousness may be much diminished (see Hudson 1966: 75). What happens is that other uses are added to the standard ones; and this must be done if the testee is to score well, since the main scoring systems give points for the gross number of uses, or the number of different types or categories of use, and under either system the omission of the standard uses must reduce the test score. It would be interesting to know whether any of the people so far tested have actually omitted standard uses from their lists; and if any have, whether there are indications of real-life creativeness to be detected in the individuals concerned. But probably the present manner of administration of the tests pushes the testees towards inclusion of standard uses. To avoid this, it should be possible to alter the wording of instructions so as to leave the requirements more open. Between the two extremes of 'Think of as *many* uses as you can . . .' and 'Think of *new or unusual uses* . . .' some subtly neutral or ambiguous formula could probably be devised which would allow but not openly suggest the omission of standard uses. Even if this were done, and even if a number of individuals did omit standard uses, however, the emended test would still reveal merely one of the (apparently) necessary but not sufficient conditions for real-life creativity. An I.Q. over 120 seems likely, as we have seen, to be another necessary but not sufficient condition – it does seem plausible, however, to assume that an ability, or at the behavioural level an actual tendency, to reject the 'standard answer', should lie rather closer to the heart of real-life creativity than does a mere 120+ I.Q. rating.

SOCIAL AND BIOLOGICAL SIGNIFICANCE OF TESTING FOR CREATIVITY

At this point it will be useful to revert to some considerations arising from Vernon's critique of creativity testing. There is no justification, says Vernon (p. 168), 'in trying to diagnose and select the potentially most creative . . . on personality grounds, since there are lots of rebels, delinquents, beatniks, introverts, CND's, idealists and so forth with very much the same personality pattern who never create anything worthwhile.' Several criticisms seem to be called for here. The tendentious nature of the passage is obvious enough; what is not obvious at first glance is that if things were to be contrary to what Vernon complains about, the actualization of creative potential would have to occur completely independently of environmental (i.e. social) influences. If creative potential were not to be found sometimes in beatniks who did not give evidence of real-life creativeness (and can we be sure that all beatniks are non-creative anyway?), we would have a situation in which potentialities were not only always actualized, but always actualized in socially acceptable ways! This, of course, is just not so. What Vernon is really complaining about is the fact that tensions exist, and perhaps must always exist, between individual and social groups.

I might be accused of 'quoting scripture to my purpose', of wrenching a passing and perhaps facetious remark out of context, when I criticize Vernon's statement about the rebels and beatniks. This may be so. But the statement in question does tie in with other presuppositions revealed in his paper. 'The correlations between I.Q. and creativity tests were around 0.3,' he says (p. 167). 'Or to put it in a more tendentious way as Getzels does – had the pupils been selected by I.Q. 70 per cent of those with high creativity would have been missed. Obviously the correlation would have been very much higher with . . . a more representative range of ability. . . .' Now 'a more representative range of ability' can only mean a sample with more individuals from the lower parts of the I.Q. range; in fact an *absolutely* representative sample can easily be defined, in terms of the validation procedures for I.Q. tests, as one giving a mean I.Q. rating of 100! It is perfectly clear, since Vernon himself apparently subscribes to MacKinnon's (1962) view that a cut-off of creativity occurs around the I.Q. 120 level, that in a sample representative of the full range of I.Q.'s there *would* be a fairly high correlation between high I.Q. and high-test creativity. But what would this prove? The *significance* of correlations, etc. is a function of the hypothesis that an investigation is designed to establish. 'Significant for what?' we can always ask. Significance depends on the purposes of an investigation, on the evaluative decisions implicit within it, on the basic attitudes of the individual who thought of and planned it. Similarly with criticism and evaluation of results. Fundamental and usually unconscious attitudes, 'pro' and 'con' (see Nowell-Smith 1954), do not usually achieve explicit statement in a discussion – yet they determine the statements that are made (Toulmin 1961). Vernon's suggestions about more representative samples, for example, reveal what I take to be an egalitarian outlook, very natural and praiseworthy in a member of one of the several generations that were responsible for correcting many of the gross inequalities of social and educational opportunity existing in earlier times. But in any social organization there is, besides the question of fairness and equity, also the question of long-term efficiency. The question of efficiency is ultimately a biological one. A society that loses its efficiency will sooner or later cease to exist (but see Chapter IX for some of the complexities involved). The American research on creativity was greatly expanded (though it was not initiated) as a result of concern with national efficiency. The specific objectives of this research may perhaps turn out to have been mistaken in detail, but the general aim, of investigating creativeness as a national asset susceptible to more efficient utilization is one which makes sense from the biological point of view. Again, provisos need to be inserted: one feels that emphasis on *national* requirements can easily be overdone; and how creativity is to be 'utilized' is an obstreperous problem . . . But it seems clear that the attitudes and aims of the creativity researchers have been substantially different from those of Vernon and perhaps most of those who have been part of the intelligence-testing movement. The aim of the creativity-testers has been to identify the minority of individuals potentially capable of making a social contribution in terms of *new ideas*. In this they have adopted, perhaps unconsciously and perhaps mistakenly in detail, an orientation towards advance, innovation, progress – all of these words have acquired unfortunate emotive connotations – which must be based on an essentially evolutionary foundation. The rationale of creativity research must include argument along the lines that, just as ordinary biological (neo-Darwinian) evolution proceeds by the action of Natural Selection upon the 'raw material' of individual variation, so psychosocial evolution (Huxley 1964) proceeds by cultural diversification and natural and social selection within and between societies – and cultural diversification must depend upon innovations made by real-life creatives. The problems are to identify these people, and to make use of their creative abilities *in the right way* – this last being the most tricky and the most important part of the business. (It is discussed in Chapters IX and X.)

The long-term social aim of creativity research could be represented, then, as the improved utilization of creativity. This need not entail the segregation of a 'creative elite' in special schools, there to be 'forced' (the metaphor is polyambiguous) like hens in a battery, so that they produce a copious emanation of ideas for 'consumption by society'. The role of education in relation to biological intelligence, creativity and human evolution is discussed in Chapters IX and X. At this point I shall remark only that 'utilizing creativity' should involve not early selection and special treatment for high-test creatives, but a reshaping of educational procedures for *all* young people to avoid stultifying whatever creativity may be present, and a sharpening of general social awareness (this itself must be largely the result of educational policies) so that, if and when new ideas *are* produced, in science, technology, politics, the arts or commerce, they may be tested and utilized in the most effective way. The aim of creativity research need not, I repeat, be visualized as the creation of an elite separated from the common herd of humanity – but it must involve the discovery and elaboration of methods for identifying creative individuals, not so that they can given special opportunities, but so that the needs of creative individuals can be studied, in their persons, for the eventual benefit of everyone. The concluding sentences of the follow-up study of gifted children by Terman and Oden (1947) are worth quoting in this context: 'The fruits of potential genius are indeed beyond price. The task . . . is not simply that of finding how gifted children turn out; it is the problem rather of utilizing the rare opportunities afforded by this group to increase our knowledge of the dynamics of human behaviour, with special reference to the factors that determine degree and direction of creative achievement' (p. 381).

Much of the subsequent work on creativity might be seen, in the light of this passage, as a continuation, in a slightly different direction, using different techniques, and based on rather different theoretical assumptions, of the work started early this century by Lewis M. Terman.

Vernon's criticism of the sampling techniques of Getzels and Jackson and other creativity researchers, on the grounds that their samples are not representative of the total population, can thus be seen to be beside the point. Evolutionary advance does not work through averages, though it does result in a change in the average. It does this by spreading the characteristics of a minority through the whole population. If Getzels and Jackson want to identify exceptional individuals for intensive examination, they have the backing of evolutionary theory for concentrating on an atypical minority rather than an 'average' sample. This is not to say that the persons actually chosen for investigation are necessarily the 'right' ones from the evolutionary point of view. I have argued in this chapter, indeed, that in some respects the methods used so far to detect creativity are likely to be inconclusive and even, perhaps, misleading. High-test creatives appear to possess some but not all of the qualities required for real-life creativity; they may be lacking in others, notably the power to reject or substantially to suspend belief in currently-accepted theories. This 'power' may be equated with the Postponement Factor revealed as necessary for the emergence of intelligence from instinct, and with the Power Factor discerned by Halstead. But the questions of how to decide on the qualities to be sought in the minority, and how to detect the qualities decided upon – these being the motivating questions of the creativity research movement – are subsidiary to the question of whether the general objective should be the investigation of qualities restricted to a minority or of qualities to be found throughout a sample representative of the whole population. On this issue the verdict in terms of evolutionary theory is clear: characteristics significant for future evolution, for survival in a changing environment, are more likely to be detected in a minority than they are in the full population, The intentions of Getzels and Jackson and others, in searching for the identifying characteristics of a minority group are not, then, less justifiable on biological grounds than are those of intelligence testers like Vernon who seek to describe characteristics found throughout the population. It is not being asserted that work on minorities is *more* justifiable – I am merely exposing and attempting to contravert Vernon's implicit assumption that it is *less* – a balanced overall view does of course demand information both on the characteristics common to all sub-groups within a population and on the characteristics specific to the various sub-groups. Both orientations are necessary, though their relative practical emphases must change from time to time. The part of the present general theory which seems most likely to be significant in this respect is that concerned with the Postponement or P-factor; and the specific improvement in creativity investigations which it suggests is that attention should be focussed upon the ability of the individual to reject or hold in abeyance the current theoretical 'knowledge', the 'ideals of natural order' (Toulmin 1964), which are inculcated by the society in which he grows up. The power of disbelief, of resistance to education in its narrower and more immediate form, is the heart component in real-life creativity.

P-FACTOR IN IMPROVED CREATIVITY TESTS

Suggestions for the incorporation of tests of sensorimotor efficiency and other physiological abilities within I.Q. test procedures have been made from time to time (Guilford, 1967). As indicated earlier, there seems little reason to suppose that positive sensorimotor efficiency has any close and significant relationship with general real-life creativity among normal individuals. Deficiencies, it is well recognized, are another matter, and research on and remedial work with the handicapped are extensive and most important. Sensorimotor efficiency of one sort or another must obviously be of relevance, too, where specific types of creativity, e.g. in the graphic or performing arts, are in question. But for general creativity the Sensorimotor factor, like the Abstracting and Memory factors already being measured by I.Q. tests, appears a necessary but not a uniquely important pre-requisite. The Postponement factor, however, while not to be regarded as in the strict sense a sufficient condition for actual real-life creativity (if it were, no influence would be allowed to the effects of environment, individual experience and personality factors) may still in another sense be regarded as completing the set of conditions necessary for *potential* creativity. It is also peculiar to creativity, as against 'high I.Q. intelligence', for example. In this sense the Postponement factor could be said to be 'sufficient'. Irrespective of this terminology, however, it is clear that the Postponement factor is uniquely important to creativity, and it is desirable that it should be measured. To what extent is an objective assessment possible? And can it be obtained in the convenient form of some sort of 'test'?

At this point we may revert to the extensive and diversified testing programme upon which Halstead (1947) based his four-factor concept of 'biological intelligence'. Out of the twenty-six tests used to obtain a spectrum of performance over a range of brain-injured, psychiatric, and normal subjects of various ability levels, three were found to show substantial loadings for the Power factor, P, which is assumed here to be the same as the Postponement factor of the present study. (That the factors of intelligence can be detected and measured only as loadings on the different tests, and not as 'pure' measurements, is in accord with the methodological assumption that they are facets or components of the functioning of an essentially unitary organism: they are distinguishable but not functionally separable.) The three P-loaded tests were:

(a) The Halstead flicker-fusion test (for details of this and the following tests see Halstead 1947, p. 171–2);
(b) Halstead's Dynamic Visual Field test (Central Form); and
(c) Halstead's Dynamic Visual Field test (Central Colour).

These tests will be referred to, for convenience, as 'flicker-fusion', 'D.V.F. form', and 'D.V.F. colour' respectively.

The D.V.F. form and colour tests appear, so far as the P-factor is concerned, to measure the subject's ability to withstand the temptation to concentrate on only one part of his visual field. In other words, to score high on these tests, the subject must maintain for an appreciable period a breadth of attention sufficient to perceive and note down correctly diversities of form and colour presented simultaneously at the centre and at the periphery of his field of vision. He must postpone 'closure' until all relevant information has been obtained. Thus a reasonably direct and obvious measure of the P-factor can be gained by these tests. The flicker-fusion test, on the other hand, though not giving so high a 'loading' for P (0.54 as against 0.64 and 0.61 for D.V.F. form and colour respectively, p. 41) does, along with various other findings discussed by Halstead (see Chapter IX), give independent evidence for a central factor operating largely in conjunction with cortical regions of the frontal lobes. It is not clear how different levels of flicker-fusion frequency are related to different levels of the Postponement factor as formulated here in terms of abstract evolutionary requirement. For Halstead, of course, 'P' stands for 'Power' not 'Postponement' (see Chapter IV): 'This factor reflects the undistorted power factor of the brain. It operates . . . to regulate the affective forces . . .' (p. 147). Differences between Halstead and myself on this point are a matter of emphasis only. It is implicit in my own account that 'power' is required to effect the all-important postponement or withholding of response; while it seems apparent that Halstead visualizes the power being used, substantially at least, for the functions to which I have given more prominence than he. My intentions, in directing emphasis away from the notion of 'power' and towards the functional and adaptive aspects, have been multiple. The word 'power' has a vast array of connotations, augmented even in the twenty years since Halstead wrote, and many of them aversive. Rather than buck the tide of associations streaming from an undifferentiated use of 'power', especially in titular denotation, I have chosen to keep my feet on the safer shores of limited and specified usages, as in 'power of postponement'. Apart from this Halstead, like the 'intelligence testers', was interested in practically measurable differences between individuals over extensive ranges of diverse performance – and 'power' is useful to summate overall measurable differences which one may not wish or be able to specify in detail; while on the other hand my own interest is all in questions of what *sorts* of specific differences there are, irrespective of whether they are measurable or not in practice. Thus while Halstead's measurements of flicker-fusion might appear to have little *functional* relationship to Postponement of response, if they give an indication of the level or 'amount' of 'power' available to the individual for effecting postponement they must be regarded as effective measurements of the Postponement factor itself.

No data are available, so far as I know, as to possible correlations between real-life creativity and a high flicker-fusion frequency. This could readily be explored empirically, given the requisite apparatus and a group of proven real-life creatives. Halstead has demonstrated, at the other end of the scale of human behaviour from creativity, that 'frontal lobectomies' and subjects exhibiting severe organic behavioural disturbance give flicker-fusion frequencies significantly lower than normal controls. Not only are the frequencies lower, but the average deviation in fusion rate adjustment (over five trials) is significantly lower in these two groups. (All significancies at .001 level; for actual figures see Halstead Table 7 p. 76; Table 17, p. 177.) Halstead suggests that these findings may be explained in terms of differences in 'power', on an explicit 'machine' analogy. Two machines differing only in power will have similar performance under light load, but will 'become markedly differentiated . . . as they are put under critical load. The one with inadequate power reserve will fail acutely within a narrow range of variance, whereas the other with good power reserve, fails only gradually over a somewhat wider range of variance due to the operation of factors other than power' (p. 78). I would expect, at the upper limits of human behaviour where the 'load' is one of creativity, that while a lot of 'power' may be available, the nature of the activity is often such as to cause a sudden build-up of 'load' or 'resistance' with the concomitant likelihood of sudden failure – but of course this 'real life' possibility would not be expected to show up in a laboratory test of flicker-fusion frequency. (Hudson (1966:97) describes an instance of breakdown of a converger's 'defence system' – which may be an instance of, or analogous with, a breakdown of P-factor 'power'.)

Although flicker-fusion appears superficially to have little functional relationship to creativity, Halstead remarks on the fact that considerable diurnal and longer-term variations in flicker-fusion 'seem to be related to "work efficiency" ' (p. 79). Probably flicker-fusion frequency is best regarded as an emanation of activities of the CNS which are also, but collaterally, causally involved in creativity. If this were the case it might be expected that measures of flicker-fusion frequency would give an indication of potentiality for real-life creativity not in a simple and direct way but in terms of some relatively complex and subtle interaction. The real-life creative might be detected, for example, by a particular pattern or 'profile' in the variation of his flicker-fusion frequency over a period of several weeks, rather than by an 'absolute' value determined at one time. If this were shown to be the case as a result of comparing proven real-life creatives with a non-creative control group, the practical consequence for creativity-testing programmes on children would not need to be an impractically-laborious repetitive flicker-fusion testing of all the children in every sample. It should be possible to use present creativity tests, supplemented perhaps, as screening rather than as determinative measuring devices. And if this sort of procedure were to prove practicable it might ultimately be possible to sort very large samples or even a whole school-age population, into 'creative' and 'non-creative' categories. The ethical, sociological and political implications of this are obviously important and problematical. Some of them are discussed briefly in Chapter XII.

MECHANISMS OF CREATIVITY: FURTHER DISCUSSIONS

Recent speculations about the causal mechanisms of creativity have emphasized a probable dependence on personality rather than intellectual factors. Hudson (1966) remarks on a number of personality traits which seem to be common to real-life creatives: persistence, self-confidence, predatoriness, crisis-seeking, rebellion and sexuality. Before attempting to discuss in detail the relevance of such personality factors to creativity, a general comment may indicate my own position and help to clarify for the reader the ensuing discussion.

It seems to me that a shift of attention away from the purely intellectual factors in creativity, while understandable historically as a reaction against the over-intellectualization characteristic of early approaches to the question, must not be allowed to obscure the necessary dependence of actual creation upon factors of the 'intellect' or the mind. Personality traits may directly, or indirectly by generating situational or contextual pressures, bring an individual towards a need to create, a desire, conscious or more usually unconscious, to assert himself by means of some *uniquely* personal – and therefore, of necessity, novel – enterprise or construction. But while personality traits and 'situations' may provide a 'motivation' (to be understood in a very broad sense) for creativity, they do not of themselves result in it. Over-simplifying, we might say that the individual *wants* to create, but unless he has the right sort of ability he cannot actually *do* so. 'Ability' here must be intellective, again in a broad sense: a function of intelligence, of the activity of the brain, or of the conscious or unconscious mind. To make the same point in a different way, we might say that an individual can exemplify all the characteristics mentioned by Hudson *without* being creative. It is surely a matter of common observation that a person may be persistent, self-confident, predatory, crisis-seeking, rebellious and highly sexed, yet not be creative.

Yet undoubtedly it is true that creative individuals do in one form or another exhibit the traits listed above. What then can make the difference? How can the apparent paradox, of identical traits in the social outcast and the creative scientist, for example, be resolved? (Hudson says (p. 150): 'In searching for crises . . . the individual plays a solitary game in which, in a sense, the integrity of his personality is at stake.' This could apply to anyone who in any way sets himself up over against his social group, e.g. the criminal, the outstanding competitive sportsman, as well as the creative intellectual.) We must seek for a differentium; and it is to be found, I suggest, as a predisposing though not a determinative factor, in the very same attribute that is necessary for transforming creative motivation into real-life creative performance: the Postponement factor.

The Postponement factor contributes to the genesis of creative motivation, I suggest, along the following lines.

A child with high P-factor endowment is likely to display not only 'childish curiosity' about all things that are unfamiliar to him, but also a 'critical' attitude to the information given him by adults. Besides asking 'What are the red and green and yellow lights for?' when he sees traffic-control lights for the first time, he may well ask, *after* it has all been explained to him, 'But why have red for "Stop"? Why not have green for "Stop" and red for "Go"?' Logically, he is then asking, over and above the question: 'What is the convention in this matter?', the much more general and more searching question: 'What is a convention?' He is asking whether there is some objective feature which necessitates or makes desirable the use of a red light to signify 'Stop!'; or whether the actual convention is based on a purely arbitrary decision. Two difficulties arise when children ask questions of this sort. In the first place the child usually lacks the linguistic and logical sophistication which would enable him to say clearly what he means; and in the second place, even if the child did manage to formulate his question properly, he would find few adults capable of giving him a reasoned answer. Questions as to the natures of social and intellectual conventions are notoriously tricky (see Nowell-Smith 1954; also Toulmin 1961. His *Ideals of Natural Order* equate to some extent with what I have here generically labelled 'conventions'); and most people are apt to become emotionally disturbed if their 'hammocks of convention' are given a twitch. This being so, the unfortunate high-P child tends to be regarded as cheeky, insubordinate, 'a trouble-maker'. Getzels and Jackson (1962) remark that their high-test creatives were unpopular with teachers; and MacKinnon (1962) notes that the real-life creative individuals in his sample reported frequent unhappiness at school due to friction with teachers. The tendency to query established views thus sets up a conflict situation for the high-P child. Expression of P-factor tendencies comes into conflict with the desire for social acceptance. The child is faced with the alternatives of either foregoing a considerable measure of social acceptance, or of suppressing an intrinsic part of his behavioural repertoire. The dilemma of *either* social acceptance *or* personal integrity – *but not both* – is a fundamental one, and painful even for an adult. Its sharpness has achieved recognition and discussion many times in literature and drama. The existence of this problem for the adolescent has long been acknowledged, but it seems likely that the embryonic horns of the same dilemma make life uncomfortable for some children at a much earlier stage than has hitherto been suspected.

If we grant that it is an integral part of the intellective functioning of some individuals to query the information given them, to be sceptical of current orthodoxies – that is to say, to have a high endowment of P-factor – and if the need to achieve a measure of acceptance or, more truly, *recognition*, is a universal of human motivation, it follows that there is only a limited number of possible strategies open to such individuals. At one extreme, social acceptance and/or recognition could be renounced; at the other the 'need to doubt' (i.e. to exercise the P-factor) could be suppressed or repressed. Between the extremes of social secession and complete conformity are several logical possibilities (and of course a multiplicity of psychological realities). A combination of the opposites may be attempted, a measure of conformity being intended to make acceptable a limited calling in question of the standard presuppositions; or the opposites may be embraced in temporal alternation, after the style of Jekyll and Hyde; or, finally, conformity may be used as a facade behind which the freedom to innovate and criticize may be attained. Each of these strategies has its virtues, and each its dangers. Outright secession from human society is still likely to make the individual's life nasty, brutish, and short. Suppression on the other hand is likely, if sustained, to slide into repression in the psychoanalytic sense: to consistently disown part of oneself can lead to neurosis.

The neurotic symptoms of just this sort of P-factor repression are not uncommon at the present time among scientists and other *professional intellectuals* (see below).

And the 'intermediate' strategies are all liable either to become futile and nihilistic or to drift to one or other of the extremes. As Packard (1962: p. 279) says: 'It has become fashionable to suggest that everybody is opposed to conformity in thinking but that conformity in behaviour is simply good manners . . . But how long in terms of years can a man act a role and mind his manners without his total personality becoming involved? How long can independence of mind withstand such muffling? It does not seem reasonable to assume that any profound spirit of independence can persist in such situations over decades.'

Few of us possess the hardihood to 'buck' society as did D. H. Lawrence, Gaugin, Beethoven and Semmelweiss, to name but a few examples; so perhaps the best strategy for the retention and exercise of a well-developed iconoclastic P-factor is to conceal it beneath a paint-and-canvas camouflage of orthodoxy. Parkinson (1958: 104) depicts the career of a capable individual within a moribund institution: 'An individual of merit penetrates the outer defences and begins to make his way towards the top. He wanders on, babbling about golf and giggling feebly, losing documents and forgetting names, and looking just like everyone else. Only when he has reached high rank does he suddenly throw off the mask and appear like the demon king among a crowd of pantomime fairies. With shrill screams of dismay the high executives find ability right there among them.'

With the proviso already made as to the danger of sustained mask-wearing, it is important to note that the very awareness of the need for a mask is apt to generate all sorts of motivations towards the exercise of creative ability. Chagrin (at mask-wearing) ferments into vindictiveness, and in this there is a logically essential component of agressiveness. McClelland (1962) remarks on the significance of aggressiveness and of the need for controlling it, in relation to creativity. Hudson (1966) develops this into a speculative theory about a 'killer instinct' in real-life creatives.

'Few people can express aggression fluently,' he says, 'even on a harmless substitute . . . The minority who have the "killer instinct" are the few who enjoy an aggressive fluency that the rest lack.' He goes on to suggest that: 'The good intellectual explorer is emotionally open. This may now mean one of two things. He may be open in the sense that he copes well with a mixture of strong and conflicting feelings – aggression and guilt, for instance. Or his openness may depend on a lack of guilt. In this second case, he would be able to entertain the feeling of straightforward aggression which most of us find shocking. The second interpretation seems to me the more likely . . .' (p. 145).

If Hudson's first alternative is transposed into ethological terms, it would appear that creativity could be regarded – or should we say the *act of creating something* could be regarded – as a displacement activity. The analogy appears reasonable: two sets of motivation, equal and opposite and at a high level of intensity, resulting in an activity which is quite different from both of the normal outcomes of the motivations (see Chapter Two). But the predictions which might be made on the basis of this analogy do not seem to hold. Creative activity should occur, on this line of argument, only in conflict situations, when aggression and guilt feelings (or some other set of incompatible drives or emotions) are both strong. This does indeed seem to occur sometimes. Some of Van Gogh's pictures, for example, seem to have been painted under the influence of strong aggressiveness held in check by strong religious devotion to non-violence. Chopin's 'Polish nationalism' music, again, was allegedly composed under the influence of strong emotion, but this creative activity could with rather more plausibility be described as *redirection* rather than displacement activity. Chopin's aggressiveness could not be discharged upon the Russian overlords of his native Poland since they were well out of range of his Paris domicile, hence its expression in the form of musical composition. Suggestive as this simple analogy (or homology?) might be, however, especially with regard to some of the more spectacular instances of creative action, it certainly fails to equate with the accounts given of a majority of creative occasions. One of the most famous definitions of poetry – 'emotion recollected in tranquillity' – stands against it. The story of Archimedes leaping from his bath with the glad cry 'Eureka!', and numerous incidents mentioned by Beveridge (1950), indicate that scientific discovery or creativity occurs generally without concomitant strong or immediate affective motivation. True creativity in science or any other field does not seem usually to depend on an obvious conflict of powerful motivations, or on an externally-imposed thwarting of motivation; and this appears to be true irrespective of whether the motivation be aggressive, sexual, or any other.

Turning now to the second of Hudson's alternatives, that the 'openness' of the real-life creative may be due to a 'lack of guilt', i.e. lack of motivation conflicting with aggressiveness, it hardly seems likely that creativity should result from the mere *presence* of aggressive (or sexual) motivation. Hudson seems almost to suggest that 'mere presence' of this motivation actuates creativity when he says that the presumptive real-life creative is 'able to entertain the feeling of straightforward aggression'. But surely if this were the whole story the aggressiveness would be expressed in some recognizably aggressive way? The tendency to reductionism, to say that creativity is 'nothing but' aggression, must be resisted. As Hudson himself points out, 'The core of the present idea is that some basic emotions get translated into work, and that these inhibit discovery in the majority and facilitate it in a few.' But again, what makes the difference?

I suggest that the differentium is again to be found in the identical mechanism to that proposed (see Chapter III) for the transition from instinctive to intelligent behaviour: the operation of an inhibitory or postponement mechanism which holds in check the affective motivation and allows scope to the operation of the other factors of intelligence, notably the Abstracting and Memory factors. This mechanism is identifiable within – it is not, of course, in any sense separable from – intelligent and/or creative activity, as the Postponement factor. In this context, where we are considering the 'upper' levels of human behaviour involving creativity, it is appropriate to use Halstead's name for the P-factor: the *Power* factor (see Chapter IV). Thus we can talk of Hudson's 'the few' as those who have the power to direct and utilize their own strong 'basic emotions'.

It would appear, then, that real-life creativity is a product of a high level of affectional motivation held in check, controlled and channelled by a high endowment of P-factor. Probably some degree of conflict within the motivational complex is necessary, since 'simple' or 'pure' motivation would probably tend to be expressed in autochthonous activity in relationship to its motivation. Leaving this particular question aside, however, the position of the general theory at this point might be summarized by saying that *both*

(a) a high level of motivation; and
(b) a high endowment in P-factor

are necessary prerequisites for real-life creativity, a deficiency in (a) leading to passivity, a deficiency in (b) to a confused turmoil possibly involving innovation but this remaining at the level of 'mere' novelty. (Perhaps the beatniks and hippies to whom Vernon objects are activated by strong and conflicting emotions rampaging uncontrolled by a sufficient development of P-factor: perhaps their behaviour is to be regarded as displacement activity in the strict sense!)

The above account of a possible mechanism for real-life creativity is, as is desirable, abstract and simple. The realities of the creative person and his creating must, however, be excessively complex in most instances. Darwin, for example, would appear to have been actuated, in the course of the production and elaboration of the Theory of Evolution, by no very violent emotional motivation, certainly by no turmoil of conflicting forces. Yet as Hudson points out (p. 121): 'He lived in awe of his gargantuan, twenty-five-stone father; and . . . published nothing of a controversial nature until this domineering creature was dead. The nearest we can come to a psychological explanation of Darwin's creative energies is through consideration of his adolescent bloodthirstiness (hunting and shooting), a quality which gradually disappeared as his life's work began.' Here perhaps we do have a transition from one redirection activity – aggression against the father being taken out on wildfowl – to another, 'predation' on ideas. The first redirection could be explained, perhaps, in terms of a conflict between aggression and a fear-respect complex; for the second redirection the *absence* of the father could explain why a *redirection* was necessary, but this does not explain the *form* of the activity, i.e. that it was a manipulation of ideas, something superficially quite different from aggression. Accepting the assumption that the basic motivation *was* aggression, why did Darwin's expression of it not take a recognizably aggressive form? It is well known that far from participating in the strident polemics which erupted after the publication of the *Origin*, Darwin avoided them and left his side of the battle under the brilliant management of T. H. Huxley. Why was this? My answer is, of course, that Darwin's behaviour was intelligent and in particular that his aggressive motivation was checked and controlled by a vast P-factor endowment. This answer can be amplified and made clearer if we reformulate Hudson's notion of the predatory nature of creativity. (But while we are on the topic of Darwin's *Postponement* factor, note the twenty years' postponement of publication of the Theory of Evolution and the fact that the delay would have been longer but for Wallace, Lyell, and Hooker!)

Lorenz (1965) has shown that there is a fundamental difference between predation on the one hand, and the various forms of intra-specific fighting on the other. Predation is essentially an interspecific transaction, and from the predator's point of view is simply food-getting activity: bird eats insect, lion eats zebra, man eats sheep. Fighting on the other hand is intra-specific, and while it may sometimes relate to food-getting (see Gibb 1954), in most instances it is connected directly or indirectly with reproduction. Male animals fight with each other, ultimately for the right to mate, but immediately for the possession of a territory or a place on the social hierarchy; females may do likewise, to varying and usually lesser degrees. For the individual the immediate objective of fighting is dominance, whether locational in the case of a territory, or in relation to the other members of the group in 'social' animals. Teleonomically (Pittendrigh 1958), we can say that the purpose of fighting, of aggression, is to achieve dominance – and on the assumption of continuity between the human and other animal species we should expect to find the same sort of purposiveness, even if not necessarily of conscious purpose, in human aggression. Thus if we think of creativity not as predation (which would imply that the creative *feeds* on what he creates) but as a means towards a dominance relationship, some possible anomalies can be cleared up and the psychological background to creativity becomes somewhat less opaque.

Let us postulate that motivation for creativity must be aroused by the frustration of some 'dominance' activity. (Hinde (1966: 289) discusses and appears to support the view that 'frustration' may be regarded as a general stimulus for aggression – the implications for the present argument are obvious.) In individuals of low P-factor, a straightforward and immediate reaction follows. Its nature will be determined by the cultural background of the individual, especially the patterns of reaction which have been built up during childhood. The immediate reaction may thus be openly or covertly aggressive, or may be a 'redirection' or 'sublimation' activity: the rejected lover and the non-promoted subordinate, for example, console themselves with sport, gardening, drink, or 'good works'. In high-P individuals, on the other hand, the reaction will tend to be delayed, thus allowing time to 'think things out' and take long-term considerations into account. This increases the chance of the individual choosing, whether consciously or not, to attempt a 'constructive' response to the challenge of frustration. Now in a high-P individual one possible interpretation of 'constructive' will tend to be ruled out: the attainment of satisfaction, of a feeling of dominance, by means of assiduous conformity to established social and/or intellectual patterns will tend to be uncongenial, precisely because of the 'rejecting' tendency within the P-factor. Creativity in one form or another will, by contrast appear relatively more attractive, since along with its penalties there are possibilities of several different levels of reward. Successful innovation, i.e. an innovation that has been accepted, brings vastly greater material compensation than do most 'conformist' activities; fame and position are at the command of the accepted innovator; and beyond all these there is the prospect of a reward which is independent of the whims of public fancy, a reward which often enough is the *only* one vouch-safed to the creative innovator, the reward of *knowing his own achievement*. This knowing is itself a product of the individual's knowledge and judgment (Memory and Abstracting factors) operating under the self-doubting influence of the P-factor. The inchoate awareness of this dimension of satisfaction, fermenting secretly from the early trivialities of adolescent creation, provides a feed-back of motivation which can result, ultimately, in an *addiction* for the doubting confidence which is creativity. Hudson (1966: 145, footnote) reports: 'Einstein once said, ostensibly about Planck, but evidently about himself too, that his devotion to work sprang not from discipline but from an "immediately felt need". He likened it to the behaviour of a deeply religious man, or a man in love.' Thus arises a life-strategy, or a habit or style of individual behaviour, which is orientated towards creative activity. Its initial motivation may come from an accident of circumstances which frustrates the drive to dominance – but the likelihood of such an 'accident' is increased by the doubting 'daemon' of high P-factor. Given the motivation, P-factor comes in again in channelling the individual towards long- rather than short-term adjustment; and it operates in yet another way to increase the working efficiency of the other factors in intelligence in seeking appropriate goals and strategies. Thus it tends to direct the individual towards non-conformity (in a broad sense); and it operates directly within the 'creative process' itself, in a way already outlined (see pp. 105 ff.), to reject established answers and clear the way for the Abstracting factor to reveal within the furniture of the memory new and hitherto-unsuspected signifi-cancies. Finally, by directing the scrutiny of P-factor doubt upon the products of his own creativity, the real-life creative eliminates the 'merely novel' among them, to leave behind only that which is *both* new and significant – and this he may eventually decide to make public. But even if he does not he has already achieved the reward of knowing his own worth. The paradox of combined self-doubt and self-confidence, in the real-life creative, is really no paradox at all, for the confidence is the *result* of the doubt: creativity is a self-testing, self-correcting activity, and we have confidence in something in proportion to the tests it has survived.

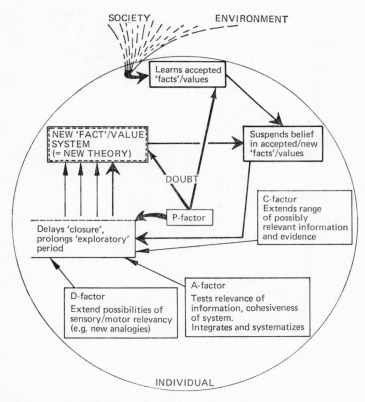

Figure 9. Creative activity of four-factor intelligence.

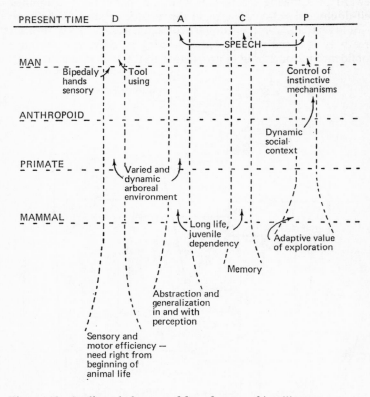

Figure 10. Outline phylogeny of four factors of intelligence.

VI

Towards a Phylogeny of Intelligence: intelligence factors and selection pressures in infra-human animals

1. TYPES OF EVIDENCE: METHODOLOGICAL DISCUSSION

If it can be accepted that the coherence test of the abstract general theory for the evolution of intelligent behaviour (Chapter III), with a variety of empirical support, notably the clinical findings of Halstead (1947) on the one hand (Chapter IV), and with various findings relating to human intelligence and creativity on the other (Chapter V), has yielded a positive result, it will now be appropriate to attempt to lay groundwork for a reconstruction of the phylogeny of intelligence. This must be done in terms of three categories of evidence:

(a) The accepted phylogenies of the various groups of animals, which have been derived principally from structural investigations using fossil in comparison with contemporary material.

(b) The corpus of evolutionary theory. Included here is a great diversity of knowledge, ranging from matters of direct observation to speculative syntheses, and in another modality ranging from genetical, embryological, and histological findings to the results of work in ethology and ecology. The whole may be given generalized summary description as: the mechanisms of individual variation and of natural selection.

(c) Our knowledge of the actual behaviour of living animals. Of greatest significance must be the behaviour of animals:

(i) that are phyletically close to the human lineage (or to any other species known to be intelligent);

(ii) whose natural way of life is well enough known so that the action of selection pressures may be estimated; and

E

(iii) whose behavioural capacities have been investigated sufficiently for indications of the several factors of intelligence to be obtained.

The great apes (family Pongidae) are of course man's closest relatives, and some of their behavioural capacities have been investigated fairly intensively under laboratory and other experimental conditions. One of the classical early studies was that of Köhler (1925) on chimpanzees. This has been followed by a variety of investigations, among which those of the Yerkes (1929) Zuckerman (1932), Kohts (1935), and Hayes (1952) may be mentioned. Thus some of the behavioural capacities of apes are known relatively well – but unfortunately, since little has been discovered until very recent times, within the past decade, in fact, about the natural behaviour of these animals in the wild, it has been difficult to assess the adaptive significance of the behavioural capacities revealed by laboratory study. Even now, with the benefit of excellent work on the behaviour of wild animals under natural conditions by Goodall (1963, 1965), Schaller (1963) and others, it is still impossible to do more than guess at the adaptive significance of much of the behaviour of the higher primates. As Bolwig (1960) and Kortlandt and Kooji (1963) have have pointed out, it appears that the great apes possess behavioural capacities far beyond what is required by their way of life. The corollary of this is that these capacities could not have been produced by selection pressures operating in their present complex of habitat-plus-way-of-life. It follows that *either* the 'excess capacity' has resulted from orthogenetic processes of some sort, *or* that the excess capacity was built up while the ancestral populations followed a different way of life, presumably in a different type of habitat. Since to invoke orthogenesis in the absence of any positive evidence of its occurrence is to give a 'blanket' or circular explanation – which is no explanation at all – we must assume that the ancestors of the present forms lived a life which differed in significant respects from the ways they live now. The generally accepted view is that the evolution of intelligence in man was due to the demands of ground-living in relatively large social groups in savanna-type country (Etkin 1954; Kortlandt and Koojj 1963). It would therefore appear reasonable to assume with these authors that the ancestors of chimpanzee, gorilla, and orang-utan lived for a time a life similar to that of our own acestors, and that they reverted to forest-living only subsequently to the development of a degree of intellectual sophistication far in excess of that demanded by the latter way of life. (This is known as the 'dehumanization hypothesis'.)

Leaving for the moment, then, the question of the possible adaptive significance of the various components of great ape behaviour, we may proceed to the direct question: Are there indications, in what is already known of the actual behaviour of the several pongid species whether in field or laboratory, of the four factors of intelligence inferentially isolated in the argument of this book?

The proviso must be reiterated before we attempt to answer this question that since the observations and experiments to which we must refer were undertaken within a theoretical context quite different from that elaborated in this present work, it will be necessary in many cases to reinterpret findings in a way not envisaged by the persons who actually carried out the researches. This is unavoidable. It is necessary, however, that some reinterpretation should be attempted, albeit on a provisional and tentative basis, if only to point the way to further researches. In this connection a general proviso may be made. The delineation of the actual course of evolution of the several primate groups is outside the scope of this book. I do not pretend to bring a scrap of new empirical evidence relevant to its elucidation. But what is to count *as* evidence must depend upon one's theoretical background, one's methodological presuppositions. As J. Z. Young argues (1963: 207–8) it is a difficult but necessary task 'to provide the appropriate methods of measurement *and reasoning* to answer these difficult questions' of primate phylogeny (my emphasis added). Presumably 'reasoning' covers the essentially philosophical operation of elucidating criteria appropriate to the heuristic needs of the situation; and this in turn must depend upon exploration of the interrelationships between possible meanings of the terms used. It is at this level that the arguments of this book must be judged: not to bring new facts, but to provide a new basis for selecting the facts.

2. c-, a-, and d-factors in hominoidea

The higher primates have long been recognized as coming closest among all the animals to the human level of intelligence. The emphasis, in accounts of their behaviour based upon observations of wild or captive animals, and also in the experimental evidence which has been amassed, has been on performance in which the C- and A-factors of the present study have been predominant. This has no doubt been due to the preconceptions of 'learning theory', perhaps with a residuum of anthropomorphism. Since human behaviour is characterized by the possession and use of abstract knowledge, the features of ape and monkey behaviour which have received most attention have been their powers of abstraction and of 'knowledge' – the latter being understood largely in the sense of 'memory'.

From the classical studies of Köhler (1925), the ability of monkeys and apes to perceive the relevance to a contrived problem-situation of various pieces of apparatus (or potential apparatus) such as boxes and sticks, has received a great deal of attention. A bunch of bananas is hung beyond the reach of a chimpanzee. The animal solves the problem by moving tables or boxes, from the positions in which they had been set, and piling them on top of one another so that he can climb up and reach the fruit. Or else he fits pieces of bamboo together, to make a utensil long enough to sweep the fruit within grasping distance. Either way, the primary requirement is that the potential relevance of the equipment should be apprehended. This must largely involve the A-factor. The accounts of the experiments strongly suggest a connection between A- and D-factors rather closer than is found at the higher levels of human performance, since it seems that unless the various 'aids and implements' could be seen within the same visual field as the 'problem situation' itself, the animals were unlikely to appreciate their relevance to the problem, hence were likely to leave it unsolved.

It will be convenient to quote in full a short account of an experiment performed by Köhler in 1914, using the chimpanzee Sultan.

'A long thin string is tied to the handle of a little open basket containing fruit; an iron ring is hung in the wire-roof of the animals' playground through which the string is pulled till the basket hangs about 2 metres above the ground; the free end of the string, tied into a wide open loop, is laid over the stump of a tree-branch about 3 metres away from the basket, and about the same height from the ground; the string forms an acute angle – the bend being at the iron ring opposite. Sultan, who has not seen the preparations, but who knows the basket well from his feeding-times, is let into the playground while the observer takes his place outside the bars. The animal looks at the hanging basket, and soon shows signs of lively agitation (on account of his unwonted isolation), thunders, in true champanzee style, with his feet against a wooden wall, and tries to get into touch with the other animals at the windows of the ape-house and wherever there is an outlook, and also with the observer at the bars; but the animals are out of sight, and the observer remains indifferent. After a time Sultan suddenly makes for the tree, climbs quickly up to the loop, stops a moment, then, watching the basket, pulls the string till the basket bumps against the ring (at the roof), lets it go again, pulls a second time more vigorously so that the basket turns over, and a banana falls out. He comes down, takes the fruit, gets up again, and now pulls so violently that the string breaks, and the whole basket falls. He clambers down, takes the basket, and goes off to eat the fruit.

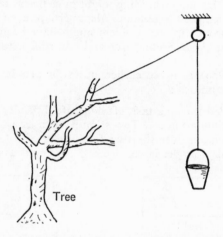

Tree

Three days later the same experiment is repeated, except that the loop is replaced by an iron ring at the end of the rope, and the ring, instead of being put over the branch, is hung on a nail driven into a scaffolding (used for animals' gymnastics). Sultan now shows himself free from all doubts, looks up at the basket an instant, goes straight up to the scaffolding, climbs it, pulls once at the cord, lets it slip back, pulls again with all his might so that the cord breaks, then he clambers down, and fetches his fruit (pp. 14–15).'

The operation of the A-factor is clear enough here, as is that of the C-factor as revealed by the highly organized performance after the three-day interval. The fact that the *modified* problem was tackled immediately and successfully is the clearest possible demonstration that the understanding gained in the earlier experiment had been retained (C-factor) and perhaps even clarified (A-factor), and that the similarities and differences between the original and the modified situations had been assessed and appreciated (A-factor again).

It is noteworthy that, as Köhler himself points out, the best solution was not found. Sultan should have lifted the ring off the nail, to release the basket, instead of pulling on the string so that it broke. Implicit in Köhler's remarks is the suggestion that Sultan might not have intended to break the string, thus allowing the basket to fall to the ground and its contents to become available, but might instead have been pulling the basket *towards himself*. Thus his understanding might have been rather less than we might be tempted to assume. Also, perhaps even more significant, his approach to the problem might have been more direct and immediate, less circumspect and controlled, than we might wish to think.

Harlow (1958), as noted earlier, poses the problem in terms of 'learning capabilities':

A well-studied learning task of moderate difficulty is the object or cue discrimination . . . Two stimuli, such as a triangle and a circle, are placed over the two food wells of a test tray. One stimulus, the triangle, is consistently rewarded as indicated by

Trial 1   Trial 2

Figure 11. Representative object-discrimination trials (*after Harlow*).

the enclosed +, although it is on the right side of the tray on trial 1 and on the left side on trial 2. The position of the triangle varies in an irregular but balanced manner during the learning trials.

Now it is obvious that on a single correct trial both the triangle and the position it occupies are simultaneously rewarded, and because a particular position as well as a particular object are rewarded, the nature of the reward is ambiguous rather than differential. During the many trials, however, the triangle is rewarded on every trial, and each of the two positions, right and left, is rewarded on only half the trials. The inconsistent reward of the ambiguous position cues apparently leads to their elimination, and learning the object discrimination problem may be described as the inhibition or suppression of the positional response tendencies.

From the point of view of the complexity of the ambiguity of cues, the object discrimination problem is relatively simple. Only one condition of ambiguity of reward exists, ambiguity between the object and the position rewarded. This comparatively easy problem can be solved by a wide range of animals, including fish, mice, rats, pigeons, cats, and dogs, as well as monkeys, apes, and men (pp. 279–80).

Here again we have tests orientated heavily towards the A- and C-factors of intelligence. However, when Harlow comes to talk about the basic processes underlying 'learning', he states: 'I believe that *inhibition* is *the single* process accounting for all learning' (p. 282, my emphasis). 'It is presumed that this unitary inhibitory process acts to suppress the inappropriate responses and response tendencies operating to produce error,' he continues, immediately rendering his argument implausible by insistence upon a 'unitary' inhibitory process. There seems little doubt that, far from being 'unitary', at the neurological level there is a considerable diversity of mechanisms all resulting in inhibition; there is also a great deal of evidence to suggest that much of 'learning' is not inhibitory at all (see Thorpe 1963, Chapter VIII, and references therein; also Eccles 1953). But at a classificatory rather than a causally explanatory level (i.e. the C-3 rather than the C-2 level of Nissen 1958), Harlow's views can be given a rather different significance. Interpreted in this way, he may have been led to his emphasis on 'inhibition' through a possibly subliminal awareness of the operation of P-factor 'withholding' in the performance of his monkeys. Much of the 'inhibition' in the test performance of lower animals would be due to other mechanisms besides those of the P-factor – Harlow's 'inhibition' is undoubtedly a heterogeneous category – but the 'inhibition' sensed in the performance of monkeys and apes is undoubtedly a reality (as will be shown by further evidence below), and is to be equated, I suggest, with the functioning of the P-factor of intelligence.

From this suggestion – which at this stage is only a suggestion – it will be appropriate to move to a brief consideration of some of the evidence for P-factor functioning to be found in the literature on the natural behaviour of primates in the wild.

3. MECHANISMS FOR SELECTION OF P-FACTOR

(a) The 'impassivity display' in social dominance

Of the four factors in intelligence, whose existence is inferred in this book on grounds of evolutionary theory, three have been recognized in one form or another in previous theories. The evolutionary production of these same three factors can likewise be given a reasonably straightforward explanation. The D- or Sensorimotor Factor is under positive selection in instinct-controlled animals, so no radical change was required in building this factor into the developing intelligence mechanisms. The A-factor (abstracting) and the C-factor (central integrative field or dynamic memory-store) have obvious survival value in allowing learned variations to be incorporated within and modifying the instinctive patterns, with an increase in plasticity of species though not necessarily always of individual behaviour. The factor not adumbrated in previous theories touching upon the evolution of intelligence is the P-factor: the Postponement or Procrastinatory factor, the factor allowing the withholding of the 'normal' instinctive (or learned) response in a particular situation. This factor is peculiar to the present theory, and its great importance has been argued in preceding chapters. One problem which has not so far been posed and dealt with is that of accounting for the build-up *of* this P-factor. Once it has been elaborated, along with the complementary A-, C-, and D-factors, its adaptive value becomes the adaptive value of the total interacting complex, of intelligence itself – but it must be assumed that in sub-optimal development it would merely interfere with the previously-existing mechanisms and would therefore tend to be selected *against*. Assuming that the first appearance of the 'embryonic' P-factor was due to random changes in the genotype, is it possible to suggest a mechanism which would account for the preservation and elaboration of the P-factor from its first appearance up to the point where it could make a positive contribution to total adaptiveness?

One mechanism which can be suggested, which is in fact a particularization of the general heading 'social factors' in natural selection, is associated with social dominance within groups of individuals. As Russell and Russell (1961: 144) have pointed out, the mark of the socially dominant male in a group of monkeys is his demeanor of relaxation and impassivity (see Van Hooff 1962 on facial expressions in anthropoidea. The homologues in human behaviour are discussed in Chapters VII, IX and X). The completely dominant male owes his relaxation and impassivity to the fact that he has no need of threat displays: the use of displays is actually an indication that dominance is incomplete. But impassivity of demeanor, involving the absence of threat displays facial or general, can presumably come to function as *itself* a dominance display at second remove, a second-order 'display'. This would simply be a form of ritualization (see Blest 1961). It seems usually to be assumed that this sort of 'display' of dominance is entirely dependent on learning, on the previous experience of the dominant individual in encounters with the subordinate individuals of the social group. The dominant animal *has learned* he has nothing to fear from his subordinates, therefore he does not threaten them – so, implicitly, the argument appears to run. But it seems rather to beg the question, so far as the causal mechanisms of interpersonal transactions are concerned. Why should the first-order threat displays be suppressed? If it is the mere presence of another individual which elicits them, then we should expect to find either the displays themselves or some phenomena of habituation, extinction, etc. If their suppression previously resulted from the perception of appeasement displays of the subordinates, why is complete dominance, the completely stable dominance hierarchy, not accompanied by profuse appeasement displays on the part of all the subordinates? Why do not the various individuals threat and appease, just to be 'on the safe side'? (see discussion of Chance and Jolly 1970 below, Appendix). What actually happens is that there is for considerable periods a virtual absence of both threat and appeasement displays. This appears to be especially marked in the gorilla. Schaller (1963: 240–4) notes the 'relative infrequency of direct dominance interactions' and states that in situations where overt dominance displays might be expected 'the dominant animal frequently does not assert its rank'. Yet the full first-order displays can be resurrected in an instant, if a subordinate tries even a hint of challenge to the overlord. To assume that the normal absence of display within a stable dominance system is due to the subordinates having learned that the overlord *as an individual* 'will not attack unless provoked' is to set up a model which is both too stable and too chancy for the individual subordinate. It would neither perform the function required of the system nor would it correspond with our observations of real systems. The system would be too stable, in that once it had been learned that 'Overlord A is harmless, with provisos x, y . . .' then it should follow that behaviour would be completely free, outside the provisos, until some new proviso was erected by means of a threat display (or attack) by the overlord. But this line of argument reveals the other inadequacy of such a system, its 'chanciness' for the individual. The subordinates, in a system dependent entirely on learning, will never be free of the chance of surprise attack by the overlord, provoked by some inadvertent and novel action. The remedy for the 'extremism' inherent in a system purely dependent on learning is to retain the essential features of the instinctive social-regulation systems using signals of one sort or another. There is of course no need to assume that learning is totally to be excluded from such a system. It is not excluded from perhaps the majority of instinct-systems especially in the 'higher' animals. Functionally it makes little difference whether subordinates recognize the overlord by his impassivity and relaxed attitude as a result of having learned the meaning of 'impassivity' as a signal, or of having its recognition 'built in' in the way that 'recognition' of releasers is 'built in' in, say, stickleback courtship sequences. It would appear likely, indeed, that learning does play a significant part in such recognition, in primates anyway – but the learning must be based upon instinctive tendencies (Lorenz, 1965).

Van Hooff (1962) in discussing facial expressions of the higher primates distinguishes what he calls 'The Relaxed Face' in the following terms: 'Common to all monkeys, this is not an expression in the proper sense, as it does not indicate a particular tendency or a conflict of tendencies. *Nevertheless this negative condition as such is expressive.* . . . The expression is seen when the monkeys are not engaged in any particular activity, but are just sitting or walking around.' (pp. 104–5. My emphasis added.)

The assertion that the Relaxed Face is 'expressive' is not amplified, but it appears possible that Van Hooff may have had more in mind than merely to describe in negative terms the absence of any of the *other* expressions. Possibly he sensed something approximating to what I have proposed as the 'impassivity display', but lacking a theoretical context within which significance could be ascribed, felt disinclined to speculate as to its function and even its existence *as* a display. It must be remembered, also, that the bulk of Van Hooff's work was concerned with cercopithecoid monkeys (Macaca-type monkey), rather than the great apes which are likely to exhibit the impassivity display to a greater degree.

If there is a case, then, for regarding impassivity, the suppression of overt first-order facial and postural dominance displays, as being itself the result of an instinctive proclivity, there is a case for assuming a tendency towards the emancipation of the second-order 'impassivity' display. In other words, any factor which tends to cause the 'impassivity' display in appropriate circumstances will be under positive selection pressure. For example, if an overlord when being mildly provoked by a subordinate can suppress his tendency to respond with first-order overt threat – which as Russell and Russell (1961) point out, always has a fear component in its motivational background – his second-order 'impassivity' display will come to function as a high-intensity display. It will come to have significance to the other individuals of the social group as an indicator of the complete absence of fear. (That it might indicate no more than the absence of all motivation is ruled out, I suggest, by the fact that at least the possibility of attack will always be contextually implied: it is implicit in the nature of the total situation that a dominant animal will attack if sufficiently provoked.) The complete absence of fear is itself an implicit or latent threat; and beyond this, the apparent absence of fear must have a similar effect. It follows, then, that an individual who is able to suppress all overt signs of first-order threat display, *irrespective of his real state of internal motivation*, can achieve the high-intensity 'impassivity' display even when, 'inside', he is afraid. The individual who has the ability to appear impassive and relaxed, even when fear and attack drives are actually strong, must tend therefore to be at an advantage, other things being equal, in contests for social dominance. He must tend to rise in the hierarchy, and once he attains the position of overlord he will tend to retain it with a minimum of friction within the group. Two consequences follow from this situation.

In the first place, since the more dominant male or males within a primate group tend to have a greater chance of siring offspring than do the less dominant (Morris and Morris 1966; Chance and Jolly 1970), the qualities that have enabled the dominant male to reach his position are likely to be passed on (with all the usual provisos regarding 'nature' and 'nurture') to a majority of the next generation. Whatever genetic factors contribute to the mechanisms for suppression of first-order dominance displays will therefore tend to be preserved and spread through the social group. Ultimately, the capacity for suppressing first-order displays and substituting for them the second-order 'impassivity' display could be expected to be enhanced. (Not that first-order displays would always be suppressed; on the contrary, they would continue to be used when appropriate, but their relative frequency could be expected to decline. And the point being made relates only to the capacity or potentiality for suppressing first-order displays, which does not have to be invariably exercised in order to have adaptive value.)

The second development consequent upon the substitution of the 'impassivity' display for some of the usages of first-order displays is in the likelihood of enhanced efficiency of the social group as a whole. If greater security in the overlord does in fact result from his suppression of first-order dominance displays, if his seeming to be secure tends to increase the chance that he is secure, the resulting relative stability within the social group could be expected to allow greater efficiency in a variety of other activities. Over and above the likelihood of increased efficiency in individual activities such as foraging and feeding, caring for young, grooming, nest- and shelter-making, and so on – activities of direct and immediate importance to group and individual welfare, hence providing channels for the operation of selection pressures – over and above these is the 'multiplier' benefit that a stable group in which amicable relationships are the rule allows for the development and elaboration of whatever propensities may be present towards *co-operative* enterprises.

It seems likely that in the evolutionary history of man and his precursors a substantial amount of competition and selection has been between social groups rather than individuals.[1] Certainly this has been so at the cultural level of competition as depicted in the records of history: the 'superior' society supplants and extinguishes less powerful competitors. Not that every individual of the vanquished group is invariably killed. The more attractive females, and tractable individuals of both sexes, may be incorporated into the victorious group. Genetically, the vanquished

[1] This is an implication of the promiscuous reproductive behaviour observed by Van Lawick-Goodall 1971 in chimpanzees.

need not be utterly destroyed – nevertheless even in terms of gene frequencies the subsequent generations must usually bear the predominant mark of the victorious group (cf. Darlington 1969). We know little in detail about the mechanisms of natural selection even among our nearer ancestors, the Neanderthalers and Austrolopithecinae for example, yet the weight of informed opinion seems to regard inter-group selection as being of at least equal importance with inter-personal (inter-individual) selection. Insofar as inter-group selection is important, social co-operation and anything conducive to it must be important. Social co-operation once initiated may be expected, other things being equal, to generate selection-pressures in its own favour. Tasks can be performed on a co-operative basis which would be quite impossible for individuals in a Hobbesian 'state of nature', and a cultural tradition of co-operative activity provides a context for further selection *of individuals*. If social co-operation becomes a cultural norm, the unco-operative individual is likely, unless he has strong redeeming features, to incur violent disfavour. This may take any form from mild ostracism to killing. Whatever happens the genetic basis of non-co-operation must tend to be filtered out from the group gene-pool. In terms of phenotypic behavioural characteristics, the individual who is unable to suppress his agonistic instinctive displays is likely to have a short life and few descendents.

That most intra-group transactions should be on an amicable rather than an agonistic basis allows, then, for the development of co-operation. This in turn provides a context for individual selection in favour of co-operativeness, or at least of the suppression of agonistic displays (and especially of actual fighting). But the foundation of all this is the reduction in dominance contests, i.e. the formation of a *stable* dominance hierarchy; and this is likely to be much facilitated by the suppression of the first-order agonistic displays associated with dominance contests. In particular, the ability to suppress first-order displays and to use only the second-order 'impassivity' display is advantageous to the overlord of the social group. Whatever genotypic features help him to do this must be under very powerful positive selection-pressure. Finally, it would appear necessary to postulate that, since first-order displays are organized and activated by 'higher' brain-centres, the ability to suppress first-order displays must operate through appropriately developed centres (here I must reiterate Tinbergen's (1951) emphasis that 'centre' must be understood functionally not structurally – a 'centre' need not be located at one single locus in the brain) would be detected through changes in the behaviour of the individual, the causative agency of these changes being termed a 'factor' causally affecting the behaviour.

If the preceding argument is recast in the terms just derived, we find that we have outlined a mechanism for the selection, through the adaptive value of stable social dominance systems leading on to co-operative activity by a number of members of a group acting in concert, of a neutrologically 'central' factor which operates to suppress or withhold particular first-order instinctive responses, namely, agonistic social-dominance displays.

In short, we have outlined a mechanism for the evolutionary development of the P-factor.

Several provisos must be added to the above bald statement. The P-factor as originally proposed allowed for the withholding or postponement of *any* instinctive response, whereas the preceding argument in this chapter refers only to first-order social dominance displays. A suppressor mechanism for the latter need not affect any other of an animal's instinctive activities. The 'dominance-display suppressor' is not, then, necessarily identical with the P-factor; and the demonstration of a possible channel for the selection of the one need have little bearing on the other. Several considerations tell against this, however, if it is advanced as a substantial weakness in the overall picture. There seems no reason to suppose that a factor developed to withhold dominance-display must be exclusively confined to this function: there are many examples of phenotypic features being elaborated for one function and then changing to another (or enlarging their range of functions to include new ones). Over and above this, there are likely to be selection pressures acting towards the withholding of other instinctive activities besides social-dominance displays; and if this is so, then on the principle of economy it could be expected that there would be secondary pressures to utilize a single withholding or suppressing centre, a single P-factor, to override all the various instinctive centres and activities. It is known that a hungry rat, for example, if presented with food within terrain unfamiliar to it, will withhold the feeding response until the area has been explored to a certain extent (Barnett 1958). Clearly there is a mechanism for 'withholding of feeding response' here, clearly it is well established, adaptively and phyletically, in the rat; and equally clearly, the causal factor on which the withholding depends is so similar to that involved in suppressing dominance displays that the two might well be one, and might indeed finally become part of a single 'withholding mechanism' quite general in its application: namely, the P-factor.

Thus in terms of overall evolution, it may be suggested that the P-factor is the last of the major factors (or factor-groups) to be developed, and that its appearance and elaboration are dependent upon the prior functioning of the other three factors. To say the same thing in other words, the P-factor has and can have adaptive value only in an animal in which the D-, A-, and C-factors have already surpassed a threshold of efficiency, the threshold being unspecifiable but necessarily to be postulated. In anthropomorphic and teleonomic terms, it is *worth* postponing one's instinctive response only if one has already built up a sufficient fund of knowledge and understanding, and sufficient aptitude at using it, to make it likely that one's newly-formulated response will in fact be useful. (Also, one must already possess appropriate sensory and effector equipment and its analysis and control systems – the D-factor background. An outline of the phylogenetic sequence of the four major factors in intelligence can thus be suggested (see Fig. 10).

Direct evidence of the several factors in the hominoid phylogeny is, of course, abundant, especially as regards the D-, C-, and A-factors. Most extant observations upon and discussions of primate 'mental abilities' have dealt with these factors. The Postponement Factor has so far escaped notice. Evidence of its operation is available, surprisingly enough, in the accounts of those who have become familiar with the normal behaviour of the 'higher' primates under natural conditions; though it is perhaps not so surprising, after all, since theory-based expectaions or preconceptions are likely to be overborne, in all but the most bigotted, by sustained experience of the realities. Schaller (1964) after describing an association between two groups of mountain gorillas lasting over several days, during the course of which a certain amount of inter-group rivalry and conflict might have been expected, remarks: 'I wondered why they had not mingled, when they had seemed so calm about the whole meeting. But gorillas, I knew, are introverts, who keep their emotions suppressed. They retain their outward dignity when in fact they may inwardly be seething with excitement and turmoil' (1967: 203–4).

This is a remarkably forthright indication, from a man who

has intimate experience of the day-to-day life of these animals, of the operation of P-factor control among our nearest relatives. It is also notable, from the accounts he gives of the details of intra- and inter-group interactions, that his 'keeping the emotions suppressed' is most strongly developed in the dominant males, the 'overlords' – which is, of course, precisely where we should expect to find such 'outward dignity retention', i.e. 'impassivity', on the basis of the theoretical argument already given. Consider the following, which relates to the early phases of an encounter between two groups headed by 'Mr Dillon' and 'Mr Crest' respectively: 'At 10.30, when only one hundred feet separated them, Mr Dillon rose, took a few abrupt steps downhill, and squatted on a hummock, only to beat his chest and retreat uphill a few minutes later. Then without warning, he walked rapidly at an angle towards Mr Crest. The two huge males then sat thirty-five feet apart, backs towards each other, the most flagrant case of studied indifference I had ever seen' (p. 203).

The first half of this passage shows fairly clearly that agonistic motivation was appreciable, the second that its expression was rigidly suppressed. Again the overt behaviour was 'impassivity', 'studied indifference'.

This characteristic of gorilla behaviour was noted as early as 1927 by Yerkes, who remarked, of the behaviour of a captive female mountain gorilla: 'Whether lacking in affectivity or merely inexpressive of her emotions, she appeared to be strangely calm, placid, even-tempered, and self-dependent. It seemed at times as though she were repressing or inhibiting acts. Especially when confronted with trying situations, such as insoluble problems, disappointment, or disagreeable stimuli, she exhibited often a degree of self-control which was suggestive of stolidity. Her aloofness and air of independence suggested also superiority. She seldom acted impulsively, and a fit of temper such as young chimpanzees and orang-utans frequently exhibit, was never observed' (quoted by Schaller, 1963: 81).

Remarkably similar mechanisms of social dominance have been reported in Woolly Monkeys (*Lagothrix lagotricha*) by Williams (1965, 1967).

In free-ranging captivity, the relationships of the two senior males of the colony, Jess and Pepi, are illuminating.

'Jess was the least aggressive of all the monkeys. Although a powerful animal, he was not as strong as Pepi, and he was five pounds lighter in weight [i.e. over 20 per cent difference]. He took no special interest in Liz or Samba [females], both of whom were available and never shunned him. He took no special interest in any one. Why was he almost *encouraged* by the others to be the leader?' (p. 61)

'Pepi had protruding canines that overlapped and interlocked. Jess's canines were small by comparison' (p. 61). Yet 'even Pepi, a stronger, larger, sexually mature animal . . . accepted Jess as ruler' (p. 60).

This does indeed call for explanation.

Williams does report that 'Jess was very good at in-fighting' (p. 62). But it is clear that this is not accepted by Williams as an adequate explanation, for he goes on to report: 'Pepi's fight-sounds of "aarrk" were double the volume of Jess's, but his displays of defiance and intimidation, calculated to alarm Jess, had no effect at all. When Pepi, drooling with saliva and flashing his canines, made an awe-inspiring leap, Jess would sit there, his arms stretched out and ready, waiting for the clash that never came. . . .'

It is possible that this reaction could have resulted from an earlier agonistic victory by Jess over Pepi. Williams gives no indication of such a background, however; and even in the passages quoted it is clear that he himself does not entertain this as a possible explanation. He had had an intimate and continuing knowledge of all the animals right from the formation of the colony, after all, and would certainly have known of a fight had one occurred. My suggestion is that Jess, for one reason or another, exhibited the 'impassivity display', and it was this meta-threat that gave him (unsought) dominance. His case is particularly interesting in that there are indications that his 'impassivity' was not solely due to a P-factor analogue: having been kept exclusively with humans from infancy his reproductive behaviour had failed to achieve normal development (p. 60), and it is possible that his other instinctive patterns were also deficient. But this seems unlikely to be the whole story; and in any case the reaction of the other animals, notably Pepi, is not explainable except in terms of some sort of impassivity display as a standard part of the species' repertoire.

It might be argued that, even admitting the evidence of P-factor impassivity in, say, the gorilla, this same P-factor withholding tendency generates its own negative selection pressures. Schaller (1963: 283–4) describes incidents in which the withholding tendencies of dominant males appear to result in their missing out on opportunities to impregnate females. If this were of frequent occurrence, the greater an individual's P-factor, the fewer the offspring he would sire. Thus there would be selection against, rather than in favour of, the P-factor. Two points may be made against this line of argument. In the first place, the number of copulations observed by Schaller was very small; and the dominant males may in fact effect the greatest number of impregnations. Not all copulations are equally likely to lead to fertilization. Washburn and DeVore (1962) describe a pattern of mating behaviour in female baboons in a species where the harem system apparently does not operate: The female 'is receptive for approximately one week out of every month, when she is in oestrus. When first receptive, she leaves her infant and her friendship group and goes to the males, mating first with the subordinate males and older juveniles. Later in the period of receptivity she goes to the dominant males . . . and the female and a male form a consort pair. They may stay together for as little as an hour or as long as several days.'

These copulations in the later stages of oestrus are of course the most likely to result in fertilization; and since they involve the dominant males, the traits contributing to dominance must be under strong positive selection. It may be that in gorillas the dominant males enforce copulation rights at times most likely to lead to fertilization, thus providing for the perpetuation of dominance traits (or, more strictly, the perpetuation of the genetic basis of the dominance traits). But an alternative to this – and the second major point against the *general* argument of P-factor dysgenicity – is that the behaviour-system of the gorilla may in fact be dysgenic. This may actually be the case. We may be observing, in the gorilla, the closing stages of a slump towards extinction. There are grounds for believing that human interference has accelerated rather than initiated a decline in numbers and a contraction of range (Bolwig 1960). If this is so, the causes of the decline are open to speculation. It could be that sheer body-size has pushed the gorilla into dependence on a habitat where very large quantities of vegetable food are easily obtainable; and that this maintenance of large body-size has been the result of a behaviour system (say for mating preference) which could not be 'put into reverse'. Large size plus vegetarian feeding habits may have pushed the gorilla into a way of life in which previously developed intelligence could not be utilized. There may have been selection for P-factor-induced impassivity to be exaggerated into lazy amiability ending up in sheer sloppiness. Thus the P-factor in the absence of various other features may indeed by dysgenic for the gorilla. But if these 'various other features' were not absent in the human line of descent – as we must assume there were not, since intelligent adaption did unquestionably result – then there is no need to accept the suggestion that the P-factor *as such* is harmful or maladaptive. The effect of P-factor in the gorilla, even assuming that its effect *is* to some extent dysgenic, would then be no more significant than that of any factor which becomes involved in and contributes to the distortion of a species' adaptive system. In the other great apes the P-factor may even now be of substantial positive adaptive value, along the lines suggested.

That a positive response to the configurational complex of impassivity-relaxation-'confidence' is endogenous rather than 'learned' is suggested by consideration of the functional requirements of the juvenile dependency period which is such a notable feature of the life-cycle of the hominoidea.

(b) *'Impassivity' in the parent-offspring and juvenile dependency relationships*

An incident related by Schaller (1963) will serve to introduce this topic: 'Mrs V. . . . attracted my attention because she seemed to have two infants. She carried one tiny gorilla, about four months old, to her chest, and another, about a year and a half, toddled at her heels, often holding on to her rump hairs with one or both hands. It was not unusual for youngsters to sit by or even be fleetingly held by a strange female, but such youngsters usually returned to their own mothers within an hour. At first I thought that perhaps Mrs V. had borne young only one year apart, but after watching for several days I noticed that when the group travelled rapidly or when evening was approaching the large infant returned to a certain female, most certainly its mother. Why this infant preferred the company of Mrs V. I could not fathom. Mrs V. was a rangy animal with a bulging belly that looked ludicrous, and her face had a gaunt ascetic mien. She was an efficient mother, with a certain unhesitating sureness in the handling of her own infant that bespoke of experience in baby care. She usually ignored the large infant, although she groomed it on occasion, yet the association persisted at least from November through May. The real mother of the infant never showed the slightest overt interest in the periodic absence of her young and this is perhaps a clue to the tenuous relationship between the two' (pp. 202–3).

Unusual 'looseness' in the mother-offspring bond explains only one facet of the situation. Why the youngster should attach to Mrs V. is also in need of explanation. I feel that the 'unhesitating sureness' of which Schaller speaks is probably the key to the surrogate relationship; and I suggest that it points to the existence of a behavioural mechanism involving a form of dominance, in which the subordinate or dependent individual responds to dominance or 'authority' signals of the 'confidence syndrome'. Here again, if there is any substance at all in this suggestion of intra-group relationships being facilitated by confidence signals dependent upon P-factor control, we have a means whereby natural selection could act to increase and further develop the capabilities intrinsic to such control.

Schaller seems implicitly to suggest that the 'confidence' display is due to prior experience, that it is a learned pattern, when he mentions 'experience in baby care'. This would bring in the C- and A-factors: the 'memory store' and the ability to see relevancies between memory items and the particularities of present experience. This need not be discounted. Undoubtedly 'confidence', and hence effectiveness in individual relationships and transactions, is affected by prior experience *and what has been made of it*, i.e. the extent to which it has been ordered, systematized, and incorporated in the central integrative field. It is clear, on the one hand, that *bare* experience, as it were at the merely sensory level, has no effect on later behaviour: there must be central changes, and 'storage' of information in some form or other, before subsequent behaviour can be affected, i.e. before there can be any learning. But it seems equally clear, on the other hand, that merely *having* certain items of information, and merely *seeing* their relevance to a present situation, still cannot exert a decisive effect on the action to be taken, unless there is also some agency for preventing a sort of 'random' embarkation on activity, for preventing the mere repetition of what was done last time. Unless there is a 'holding off' mechanism it would appear difficult to account for 'what was done last time' *not* being repeated. A deficiency in 'holding off' ability would appear to underlie the crass fact that certain human individuals, not 'unintelligent' in terms of I.Q. scores (which draw mainly on C- and A-factors, much less on D, very little indeed on P), are apparently quite unable to learn by experience in ordinary 'real life'. Perseveration can be maladaptive, even pathological; and problems of recidivism might well repay re-investigation in the light of P-factor implications suggested here. But to return to the possibility of selection for P having occurred through its enhancement of maternal efficiency in child-care.

Schaller's observations suggest that maternal effectiveness does increase with practice, among gorilla as among human mothers. Nevertheless this cannot be allowed to rule out the possibility or the importance of the role of the P-factor, for two reasons. First, there are great differences between individuals in the extent to which they do learn from experience. This cannot be shown as yet among gorilla mothers, since too few of them have reared offspring under observation; but it is notorious among human mothers, that some, even intelligent women under ideal circumstances, are never able to inspire calmness and confidence in their babies and young children, no matter how much practice they have had. And secondly, there is the undoubted fact that some women (men too, for that matter) do have the knack of establishing 'bonds' with children, even without prior experience of any kind. It is clear from Schaller's observations that the same sort of thing occurs among gorillas. Presumably it also occurs among the other great apes; and presumably the behavioural mechanisms are similar throughout. Personal observation and experience among humans suggests that young children respond best to 'undemonstrative goodwill'. They shy away from 'gush' and even from demonstrations of friendly affection which stop well short of the 'gush' level. Uncertainty, uneasiness, and most pretensions are detected very quickly, and are either avoided or attacked. Real indifference, on the other hand, often 'takes them in', just as it does in love affairs in later life: how often do children, and lovers, persist in adoration of some coldly uncaring god- or goddess-like figure within whose subjective universe they the worshippers do not even exist! Thus there can be a simulation of impassivity by sheer indifference – and the observable fact of the attractive power even of sheer indifference is itself evidence of a powerful mechanism usually actuated by impassivity.

The mechanisms by which a 'P-factor-imposed' relaxation-and-impassivity among hominoid mothers could be of biological advantage are worth brief theoretical exploration. The 'impassivity display' could be given, or rather, something superficially very similar could be given, by a mother who was *really* quite indifferent to her offspring, i.e. by a mother virtually devoid of the normal instinctive concomitants of maternity. Even though the offspring might respond positively to this display, however, there should be no danger of selection for the mere absence of instincts, since the material needs of the offspring would be most unlikely to be satisfied under such circumstances. Similarly for a mother who was, in anthropomorphic though not necessarily inappropriate terms, simply too stupid to rise above a level of vacuous indifference: the offspring might well become 'attached' to such an individual, but would be likely to suffer fatal neglect in the long run. The implications of having strong maternal (and other) instincts totally uncontrolled by P-factor are also worth examination. It seems likely that this would result in so much fuss being made of an infant that its chances of survival would be impaired. If it were always being fed, caressed, groomed, etc., it would probably get insufficient rest, would become fractious and upset, this would stimulate even more exaggerated parental reactions . . . a vicious downward spiral seems likely to result, with the death of the offspring as the end point. Uncontrolled alarm reactions, or exploratory drive, or uncontrolled reactions to the social interplay of the other individuals of the group – any of these in the mother would tend to interfere with the care of the young, hence would tend to be deleterious. It has been shown (Thompson 1957; Thompson *et al.* 1962) that mothers of high 'emotionality' tend to generate exaggerated emotionality in the young associated with them, irrespective of whether the young are their own natural offspring or not; and the extremes of emotionality are certainly dysgenic in a variety of ways (see Hinde 1966: 386). It could be expected, then, that the most desirable complex of traits would include *both* strong instinctive (maternal) drives *and* the P-factor to control them. The lengthening of the period of juvenile dependence, characteristic of the high primates and reaching its extreme development in the human lineage, would furnish increased opportunity for the operation of this complex, and hence increased opportunity for its selection and development. The conflicting requirements for encouraging independence yet maintaining control (discipline) in the juvenile, during the long period of dependence, would appear to enhance the premium on P-factor withholding – 'forbearance', it could most appropriately be called in this context.

4. EXPLORATORY BEHAVIOUR AND P-FACTOR SELECTION

Possibly the original selection in favour of a chance-produced P-factor came via the adaptive value of exploration. It has long been known that an animal placed in unfamiliar surrounding will usually postpone other instinctive activities, such as feeding, mating, etc., in favour of exploration (Barnett 1963). The only activity which seems to have priority over exploration, for an animal in unfamiliar surroundings, is 'escape from danger'. This may take a variety of forms (see Hogan 1965 on differentiation between 'fear' and 'withdrawal'; also Hinde 1966: 242) according to the species and the particular 'danger' involved; but apart from the various responses to 'danger', most other activities are held in abeyance when a disparity between the perceived situation and the 'neuronal model' of 'what is familiar' (Sokolov 1960) elicits exploration.

Since the proximate 'stimulus' for exploration is an internal and generalized one, and since the activities to be withheld comprise most of the instinctive repertoire of the animal, it would appear that the 'withholding mechanism' prerequisite to exploration must have most of the features inferred to be necessary in the P-factor of intelligence. In the higher reaches of human creativity there seems to be a degree of spontaneity in the operation of 'P-factor doubt': the questioning or holding in abeyance of accepted beliefs seems, sometimes, not to be dependent on the perception of a disparity or discrepancy. Indeed the perception of a discrepancy seems often to be the result rather than the cause of creative activity. This might appear, superficially, to constitute a difference between the 'withhold for exploration' mechanism and the human P-factor. Difference there certainly is – and it may turn out to be greater than suspected – but the relationship is anything but simple. Starting from the view that a withholding mechanism precursive to the P-factor of intelligence is to be found as a necessary prerequisite for exploration, and that selection pressures for the elaboration of P-factor withholding mechanisms would operate through the adaptive value in 'knowing the lie of the land' as a result of exploration, it may be illuminating to explore some of the complexities of detail of this relationship.

There must, on *a priori* grounds of evolutionary theory, be considerable individual variation in the threshold of the 'discrepancy mechanism' for exploration. In anthropomorphic terms, some individuals must feel the need to explore more frequently than others. This 'threshold' variation could come about in a variety of ways. It could be the result of a rapid waning of the 'neuronal model' of a previously explored topography, so that a discrepancy was produced between what was seen (smelled, heard, etc.) and what was remembered. Or it could result from a stronger than usual withholding mechanism, so that a relatively small discrepancy could trigger off exploration. Or it could be actuated by unusual acuity or sensitivity on the sensory side, so that even minute variations in the previously explored environment would be seized upon as indicative of discrepancy. Each of these possible mechanisms would produce an unusually strong tendency towards exploration, in the sense that exploration would tend to be of more frequent occurrence. The opposite mechanism in each case, would of course weaken the tendency to explore. And over and above these possibilities considered separately is the likelihood that there would be multiple interactions between them.

An enhanced appreciation of these mechanisms can be obtained if the possible interactions are formulated in terms of the four factors of evolutionary intelligence.

(a) Rapid waning of the 'neuronal model' would at the simplest level involve merely a 'decay' of the individual items in the memory store. An atomistic view of memory is, however, untenable as a general account. While it is convenient to hypostatise a 'memory store' (the C-factor), it is essential to keep in mind that the reality must equate with a process or activity rather than a 'thing' or object, and that the C-factor, while distinguishable logically, is not separable functionally from the operations of the other factors. Thus 'waning of the "neuronal model" ' must involve not only the C-factor but the A-factor as well, since the question of relevance must always be answered, not only with regard to the relationship between particular memory items and the perceived real situation but also within the dynamic memory store itself. In the continuing autonomous re-sorting of items the questions must recur, and must be settled, whether item A is similar to or different from item B, or whether within a group of items A and B are more similar to each other than either is to C, and so on. It follows from this that great *activity* of the A-factor in resorting items in the memory store might be expected to 'loosen' the bonds of association and thus allow a discrepancy to be generated between what was immediately experienced and what was remembered; while on the other hand great *acuity* of the A-factor might be expected both to confirm the accuracy of the memory items and thus reduce the tendency towards discrepancy, in the event that the previously-acquired information had been accurately recorded; and on the other hand to facilitate the detection of any discrepancy which did in fact exist.

(b) Turning now to the second general possibility, that an unusually strong withholding mechanism (P-factor) could, as it were, open the way for even a small discrepancy to trigger off exploration, it is clear that here again we are dealing with relativities. 'Strong' in relation to what? In relation, within the individual, to the other instinctive drives that would complete with exploration. But these drives are mediated in their coming towards expression by various central control centres (including those of the P-factor); and it seems reasonable to assume that the 'strength' of their competition depends, proximately, upon the representation of each drive as it appears within the central control areas rather than as it is 'in itself'. Thus the absolute strength of the P-factor at any one time would be modified by the relative strengths of the competing instinctive drives, and the representations of these would be affected by their 'ease of access' to the central 'decision centres'. This in turn would be a function, in part at least, of the D-factor of intelligence; so that the representation of a hunger state, for example, would be enhanced by increased sensitivity of the D-factor and depressed by lessened sensitivity, independently and over and above the actual physiological state of hunger, level of blood sugar, etc. It seems possible that there may be a connection between the effective level of D-factor and the activity of the reticular system of the brain. Hinde (1966: 161–4; see also references therein) suggests that 'behavioural arousal, cortical EEG arousal, and reticular activity usually go together, and much of the evidecne indicates that the cortical arousal, mediated by activity in the reticular formation, is a necessary accompaniment to a stimulus for a behavioural response to occur. Further evidence indicates that reticular activity can increase the efficiency of responsiveness'. Since 'the reticular system receives its input not only from the main sensory pathways, but also from the cortex, the limbic system, the basal ganglia and the cerebellum', this system – and hence the D-factor, on the assumption that they are but different nominate aspects of the same reality – can cause variation in responsiveness not only to external stimuli but also to changes in the internal state of the animal.

(c) The hypothetical relationships just discussed lead naturally towards investigation of the third general possibility, mentioned earlier, that unusual sensitivity or acuity on the sensory side might allow even small variations in a previously familiar terrain to function as 'discrepancies' and thus elicit exploration. Sensitivity or acuity in the receptor itself would of course be a prerequisite; but there may be greater variation between individuals, and in one individual between one time and another, due to variations in the central mechanisms collectively denoted in the present study as the D-factor. It must be emphasized that high D-factor endowment might be expected to facilitate the operation of a 'discrepancy' mechanism for exploration even in the absence of actual change in the previously-familiar environment. It is not uncommon in one's own experience to *feel* that there is something vaguely unfamiliar in a known environment, a room of a house for example – which proves, upon closer examination, to be objectively unchanged from one's previous occasion of experience. In such a case, if indeed no objective change had occurred, it would appear likely that information about some detail had come through to the higher centres of attention where previously it had been restricted to, held out at, the periphery, But while the 'letting through of more information' may be ascribed to D-factor acuity, it must be remembered that the establishment of a discrepancy must involve the comparison of the newly-transmitted item of information with the stored complexes of items of the C-factor, the comparing being itself a function of the A-factor. Thus the proviso must be reiterated, that while it is convenient to separate the possible interactions of the various factors and to consider them in abstraction, the factors can in reality never operate except in intricate conjunction.

Thus exploratory behaviour, which seems to be an important feature in the lives of most 'generalist', versatile, or 'euryethic' animals from the cockroach (Darchen 1952) through the rat (see Barnett 1963, 1967 and references therein) to man, would appear to furnish mechanisms of positive selection not only for the P-factor of intelligence but for the other factors as well. The enhancement in the general level of adaptiveness of an animal's

behaviour as a result of exploration would seem to be very substantial indeed, although the very generality or diffuseness of
effect must inevitably make quantitative assessment both difficult
technically and problematical methodologically. For example, in
the quantifying of overt exploration in the 'highest' animals, e.g.
hominoids, is 'more effective exploration' to be equated with 'a
greater number of movements from place to place within the
area in question'? This is the type of measurement which has
been used and found effective in lower mammals, e.g. the rat
(Mead 1960). It has the merit of allowing for quantification on a
straightforward basis. But in 'higher' forms it would lead to the
paradox that young and inexperienced animals 'explore' an
unfamiliar area 'more' than do older and sophisticated individuals – though the subsequent behaviour of the latter is
found to be not less, but if anything more, adaptive than
that of the former. Clearly, the older animals are to be
regarded as more sophisticated precisely because they can,
generally, assess their environment with less physical movement
than is necessary for young animals; and if this is so, if a greater
proportion of the processes of exploration are 'internal' rather
than overt, then obtaining separate and independent measures
both of exploration and of the adaptiveness of subsequent
behaviour is going to be problematical. Avoidance of circular
argument is likely to be difficult. It seems generally to be accepted,
however, that the adaptive value of exploration must be regarded
as high, even though this has not yet been extensively and conclusively demonstrated under experimental conditions.

An interesting parallel between the functioning of P-factor in
intelligent behaviour and in exploration may be pointed out. In
'adaptively variable (i.e. intelligent) behaviour', as was argued in
Chapter III, the operation of A- and C-factors (and the D- or
sensorimotor-factor, though this was not mentioned in the argument at that point) cannot lead to a gain in adaptiveness unless
the P-factor operates to 'hold off' the previously normal response
whether learned or instinctive. Similarly, exploration cannot occur
in competition with any other activity for which motivation and
releaser situation are present unless there is a 'holding off'
mechanism available, analogous or homologous with the P-factor.
Again, both the intelligent elaboration of a new response and,
on the discrepancy model discussed above, the initiation of
exploratory activity, are dependent upon the prior experience of
the individual (C-factor or memory store), organized and

mediated in terms of relevance to the present situation through the operation of the A-factor, all the interactions with the environment being modulated by the D-factor.

One highly significant point can now be made. Although the discrepancy mechanism for the actuation of exploratory activity (derived from Sokolov 1960 and outlined above in terms of the four factors of intelligence) seems to give a satisfactory explanation of one salient feature, namely the recurrence of exploration, periodically, within an environment which has been explored and which is changing but little, there is another feature for which a discrepancy mechanism is in principle inadequate. This is the ontogenetic first appearance of exploration. A newborn animal cannot be stimulated to explore through the operation of a discrepancy mechanism, it might be argued, since it has no 'neuronal model' of the environment against which perception could generate a discrepancy. But this, while it delineates very clearly the logical essentials of the situation, is much too simple to be satisfactory as a characterization of the problem of the actual ontogeny of behaviour. It can be argued, for example (see Lehrman 1953) that 'neuronal models' of the environment are never completely lacking, even in the neonate or the embryo. Counter-arguments to this are obvious enough. It might be pointed out, for example, against the contention that the response of the territorial male Stickleback to an intruding male is dependent upon a 'neuronal model' of the releaser situation, that only *part* of the neuronal model can be regarded as inbuilt (viz. the 'fish-shape, red underside, etc.' which has been determined through experimentation with models), that the rest of the 'situation' has been 'modelled' as a result of prior exploration, and that the inbuilt mechanisms so far demonstrated all seem on the one hand to 'ignore' discrepancies as far as possible (illustrated in anomalous instances of imprinting) and on the other to presuppose a modicum of previously-acquired information about the environment. Generalizing the problem, it might be said that all the mechanisms for ensuring adaptive response to environmental specifics *are* specific, their 'neuronal models' must be specific – while exploratory behaviour essentially involves going *beyond* the immediately-available specifics to encounter features which are novel, and for which, therefore, there can be no analogue within the neuronal model. It could of course be said that this 'going beyond' itself implies a 'discrepancy', a hiatus at the edge of the 'known' – but to say this is to make nonsense of language. A discrepancy is one thing, a hiatus something different. To extend the usage of one word to cover both situations leads only to confusion. It is necessary to recognize, then, that besides the discrepancy mechanism already discussed there must also be an endogenous drive towards the enlarging of experience, to *look for* novelty even when there may in fact be nothing novel to be found. There must, in short, be a drive towards exploration which is not actuated by the perception or apprehension of a discrepancy, which will operate even when the individual is in completely familiar surroundings, and which causes instead of being caused by the awareness of discrepancy. That this occurs in human creativity has already been remarked. The inference to be drawn at this point is that exploration must be an autochthonous activity, the drive towards it must be endogenous and spontaneous: it must be one of the major instincts in Tinbergen's (1951) sense (see Shillito 1963). And if this is admitted, along with the intimate involvement of the mechanisms of exploration with those of intelligence *per se*, we are led towards the superficially paradoxical conclusion that some aspects of intelligent behaviour itself must be regarded as emanations of the total complex of the instincts. If this is so, the situation of intelligence *vis à vis* instincts falls into one of the standard patterns of biological advance, where an advance is achieved by the economical device of turning *some* of the impeding forces against the others. Thus energy which might have hampered the advance, is made to facilitate it. (See Chapters XI and XII for further methodological discussion.)

5. EXPERIMENTAL EVIDENCE OF P-FACTOR

Experimental evidence of the importance of P-factor 'withholding' has been available for many years. In the absence of a satisfactory conceptual frame of reference, however, its significance has not fully been perceived up to the present time.

What Köhler (1925) characterized as 'roundabout methods' seem in general to be very clear exemplifications of P-factor operation. The experimental technique he used was as follows:

When any of those higher animals, which make use of vision, notice food (or any other objective) somewhere in their field of vision, they tend – so long as no complications arise – to go after it in a straight line. We may assume that this conduct is determined without any previous experience, providing only that their nerves and muscles are mature enough to carry it out.

... we may use the phrase 'direct way, and 'roundabout way' quite literally, and set a problem which, in place of the bio-logically-determined direct way, necessitates a complicated geometry of movement towards the objective. The direct way is blocked in such a manner that the obstacle is quite easily seen; the objective remains in an otherwise free field, but is attainable only by a roundabout route ... it is assumed that the objective, the obstruction, and also the total field of possible roundabout routes are in plain sight (Köhler 1925: 17).

Two experiments with a dog, a 'mature Canary Isle bitch' (p. 19), illustrate very clearly the operation, but also the limitations, of the P-factor in this non-primate species. The dog 'was standing at B near a wire fence (constructed as in Fig. 12) over which food was thrown to some distance'. The animal 'at once dashed

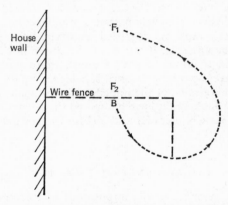

Figure 12. Experimental evidence of P-factor operation (*after Köhler 1925*).

out to it', coming *back* from position B, i.e. away from the food, and running 'in a smooth curve' round the end of the fence, as illustrated. This demonstrates the P-factor withholding of the obvious and direct approach geometrically straight toward the food. But Köhler goes on:

'It is worth noting that when, on repeating this experiment, the food was not thrown far out, but was dropped just outside the fence, so that it lay directly in front of her [at F_2], separated only by the wire, she stood seemingly helpless, as if the very nearness of the object and her concentration upon it (brought about by her sense of smell) blocked the "idea" of the wide circle round the fence; she pushed again and again with her nose at the wire fence, and did not budge from the spot' (p. 19).

Instead of talking about 'blocking the "idea" ', we can now ascribe the attempt at the direct approach simply to the fact that the 'food signals' were coming in at a strength greater than could be 'held off' by the P-factor, and that they therefore elicited the straightforward instinctive approach behaviour. (A closely parallel incident involving a hen is described below, p. 161.)

Experiments on 'delayed response' performance in mammals are subject to difficulties in interpretation, as Thorpe (1963: 127) points out. Nevertheless, there appear to be strong grounds for accepting impairment in delayed response tasks following pre-frontal lesions in dogs and cats as being due, not to loss of short-term memory, but to an increase in non-adaptive perseveration (Lawicka and Konorski 1961). Operated animals tend to repeat the last reinforced performance prior to the operation. These findings appear susceptible to reinterpretation in terms of P-factor functioning. Non-adaptive perseveration may be regarded as resulting from a sudden lowering of P-factor effectiveness consequent on the frontal lobe lesion (similar perseverative effects were found by Halstead 1947 in humans with frontal lobe injuries, see Chapter IV). The converse of this, in positive terms, is that abstention from non-adaptive (indeed under natural conditions definitely maladaptive) perseveration must be ascribed to the operation of the P-factor. A further means for its selection and enhancement is provided, if it does have this function. Other neurological findings tend to support this view. Ellen and Wilson (1963), for example, find perseveration to be increased in rats suffering from hippocampal lesions. Provided the temptation is resisted, to assume a localization of function corresponding to the site of some particular extirpation (see Gregory 1961), evidence of the type cited seems to support in a very direct way the supposition of a general behavioural capacity corresponding to the P-factor of intelligence. It must be emphasized that neither the P-factor nor the other factors necessarily confer 'intelligence' upon the animals in which they are to be detected. Neither is the behaviour from which their functioning is inferred, to be regarded as in every case itself 'intelligent'. On the contrary, one of the major strengths of the methodology followed in the present inquiry is that it allows for the separation of biologically plausible 'components' of intelligence, each of which can be sought, within the multiplicity of evidence, independently of the others.

Finally in this section, an account of an advanced 'roundabout' experiment with a Chimpanzee (from Köhler 1925):

The experiment is considerably more difficult when a part of the problem, if possible the greater part, is not visible from the starting-point, but is known only 'from experience'.

One room of the monkey-house has a very high window, with wooden shutters, that looks out on the playground. The playground is reached from the room by a door, which leads into the corridor, a short part of this corridor, and a door opening on the playground (see diagram). All parts mentioned are well known to the chimpanzees, but animals in that room can *see*

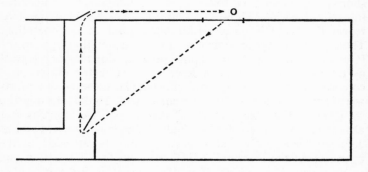

only the interior. I take Sultan with me from another room of the monkey-house, where he was playing with the others, lead him across the corridor into that room, lean the door to behind us, go with him to the window, open the wooden shutter a little, throw a banana out, so that Sultan can see it disappear through the window, but, on account of its height, does not see it fall, and then quickly close the shutter again (Sultan can only have seen a little of the wire-roof outside). When I turn round Sultan is already on the way, pushes the door open, vanishes down the corridor, and is then to be heard at the second door, and immediately after in front of the window. I find him outside, eagerly searching under the window; the banana has happened to fall into the dark crack between two boxes. Thus not to be able to see the place where the objective is, and the greater part of the possible indirect way to it, does not seem to hinder a solution; if the lay of the land be known beforehand, the indirect circuit through it can be apprehended with ease (p. 25).

6. INTELLIGENCE-FACTORS IN NON-PRIMATES

If the P-factor is, as has been suggested, the last of the factors of intelligence to have been evolved, the prediction could be made that behaviour dependent on this factor would be found at a highly developed level only in a few recently evolved groups. Conversely, the 'pause' or withholding of response which is the manifestation of P-factor endowment could be expected to be at a low level or absent in all other animals. A great many observations can be interpreted as fulfilling this latter prediction. The work of Révész (1924), for example (cited by Thorpe 1963: 366), showed that a hen would push aside a piece of paper under which food had been seen, immediately previously, to be hidden. The hen would not, however, push aside a piece of glass under the same circumstances, but instead would peck repeatedly at the food through the glass. It is suggested that this indicates among other things, a deficiency in the power to withhold the pecking response in the presence of the stimulus for it, viz. sight of the food. That there was no lack of 'understanding' of the basic spatial relationships is established by the fact that the visually opaque paper *was* moved aside. Thus sufficient endowment of C-, A-, and D-factors would appear to have been present to allow the animal to cope with the 'food under glass' situation – but these potentialities were vitiated by an inadequacy of P.

These observations are comparable with those of Köhler (1925), on the effect of food placed very close to a dog but separated from it by a wire fence (see above, p. 158).

Indications of the presence of the several factors of intelligence in fish are scattered throughout the literature. Aronson (1951) has shown that a species of *Bathygobus* living in tidal pools has such an accurate knowledge of the relevant spatial relationships that at low tide, when the pools are separated by dry areas of rock, the fish can leap from pool to pool without much risk of 'missing the target'. Clearly, relatively high levels of C-, A-, and D-factors must be involved here. Thorpe (1963: 301) in discussing

F

Aronson's findings, appears to accept the provisional conclusion that 'these gobies swim over the pools at high tide, and thereby acquire an effective memory of the general features and the topography of a limited area round the home pool, a memory which they are able to utilize when locked in their pools at low tide. . . . A glance at Fig. 13 will make clear the remarkable precision of knowledge which the fish must have in order to be able to proceed safely from one pool to the next'.

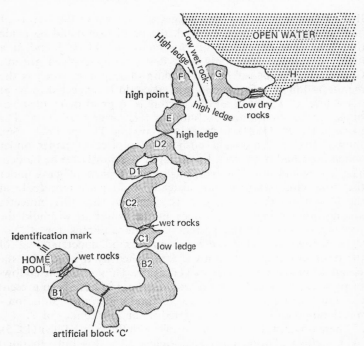

Figure 13. Tidal pools traversed by fish (*after Aronson 1951*).

Now although words like 'memory', knowledge', 'latent learning' used in this connection point directly to the C-factor of intelligence, the topographical knowledge or 'cognitive map' itself is not enough. The fish must also be able to locate itself in the 'map' with great precision – this involving A-factor activity – and this locating cannot be a static business analogous to a grid reference. For the fish successfully to leap from one pool to the next a complicated dynamic plot must be kept and used, involving momentarily changing patterns of movement, momentum, muscle tone and body position. Thus the D-factor must be prominent. P-factor operation is not superficially apparent; its prior importance can be argued, nevertheless, through the need for exploration of the topography. As in all exploration (see section 4 above), all the other instinctive activities except escape from predators have to be 'held off' while the animal undertakes the learning of the terrain – without this, the adaptiveness of later activities within a territory would be much less in general, and in particular the jumping between pools would be so hazardous as to be dysgenic.

It would appear unprofitable for a number of reasons, to attempt in this place a survey of the evidence for the four factors of intelligence among the invertebrates. No intelligence-dependent species has appeared outside the vertebrates; and the neural mechanisms in invertebrates are severally different from those in vertebrates (Thorpe 1963; Hinde 1966). Nevertheless it could be predicted on grounds of general principle that similar selection pressures on suitable invertebrates should produce behavioural propensities analogous to though not homologous with those found in vertebrates. This could be expected in species whose way of life involves coping with a high degree of variability in the environment. (All species have to 'cope with variability', but most species cope with it by ignoring it, i.e. by keying their reactions to specific regularities among the irregularities (cf. Lack 1954; von Uexküll 1934). What is meant here is that variability itself should be 'coped with'.)

Environmental variability is a problem which has to be faced *per se* not simply dodged, by the caterpillar-hunting wasp *Ammophila*, as Thorpe (1963: 259) in effect points out. After discussing the investigations carried out by Tinbergen and Kruyt (1938) and van Beusekom (1948) on the orientation and navigation back to the burrow by the digger-wasp *Philanthus*, Thorpe remarks:

> The problem encountered by some of the spider- and caterpillar-hunting wasps is even greater than that which *Philanthus* has to face. *Philanthus* can carry its prey in flight directly to its previously constructed hole, but some other insects have to drag the prey some distance . . . *Ammophila* hunts caterpillars which are too heavy to be brought back on the wing, and which may

thus have to be dragged for a hundred yards or more across
and through every imaginable natural obstacle . . . The original
learning of the territory has probably been in the main effected
by observation from the air, and yet the return has to be made
on foot. Although the insect may from time to time leave her
prey and take short survey flights, this is by no means invariable
. . . and quite often the wasp seems able to maintain orientation
while on the ground as a result of earlier aerial reconnaissance.
It is, of course, not easy to be certain that the terrain has never
been previously explored on foot as well as from the air, and
it is true that a great deal of apparently random exploratory
wandering is shown by members of this genus. But in so far
as the insect has appreciated the lay-out of the area as a result
of aerial observation and can then utilize this 'plan' while on
the ground, we seem to have an orientation ability of extra-
ordinarily high type (pp. 259–61).

If Thorpe's account is reinterpreted in terms of the four-factor
theory, it would appear that the 'plan' which the insect is to
utilize can be equated with the C-factor or 'memory store',
while the utilizing on the ground of an air-survey 'map' would
appear to necessitate a considerable A-factor power of perceiving relevant similarities. The use of kinaesthetic cues seems
to be largely ruled out; and even the use of, say, generalized
and configurational visual images as orientation and guiding cues
would seem to necessitate a considerable amount of 'interpretation' and sorting out, due to the difference in viewpoint between
flying over, and walking under and through, the substratum of
undergrowth. Thus if we assume a high level of 'D-factor' (the
central factor for sensorimotor efficiency), there are grounds for
asserting the presence of both 'C-' and 'A-factors'. (It seems
advisable to mark the probability of an analogous relationship
between the several factors, in vertebrates and invertebrates
respectively, by using quotes in the latter case.)

Experimental work with *Philanthus* showed that 'the wasp
orientates [when in the vicinity of the nest on a "return trip"
carrying prey] to a "nest hole + a particular configuration of
landmarks" ' (Thorpe: 258). The artificial landmarks round the
nest hole can be varied by human investigators to a certain extent
without disrupting the orientation of the wasp – but 'Tinbergen
has . . . found that when the training marks are moved so that
the *Philanthus* might be expected to show complete disorientation,

it will suddenly utilize new landmarks completely unrelated to the previous orientation marks on which it had apparently been trained' (Thorpe: 259). Clearly, there must be involved here a comparison of the experientally altered landmarks with a remembered 'plan' or image, the latter entailing a 'C-factor' and the 'act' of comparison an 'A-factor'. Further, the apparent suddenness and completeness of the switch argues for a rudimentary P-factor, the power to withhold the previously-established response. It appears either question-begging, or merely implausible, to suggest that the switch could be due purely to some threshold effect. The actuation of 'P-factor' mechanisms by 'discrepancy' has been discussed elsewhere (see pp. 151ff), and the complexity of causation would appear to demand an account in these rather than simple 'threshold' terms. Thorpe himself remarks that the 'switch of orientation-system' phenomenon 'has interesting and suggestive similarity to Krechevsky's work [on rats] on hypothesis in maze learning' (p. 259); and of this latter work, he says: 'In practice the animal [rat] often appears to select one type of cue, and if he succeeds on this basis will not use the others at all. If, however, he is unsuccessful, the less preferred responses are brought into use, and so we get the appearance of a systematic variability in behaviour' (p. 91). In the rat there are thus two 'withholdings':

(a) While the 'preferred' cue-responses are being tried, the 'less preferred' must be withheld; and
(b) The 'preferred' ones must be withheld while the 'less-preferred' are tried.

I interpret Krechevsky's results as being evidence for P-factor in the rat, albeit at a relatively low level for a mammal; and the findings on *Ammophila* and *Philanthus* make a plausible case, I suggest, for the existence of at least rudimentary 'P-factor' in these insects.

Although Harlow's (1958) suggestion that the 'evolution of learning' in the vertebrates was due to the need for efficient sorting mechanisms for the receptor systems, has been rejected, it appears likely that the initial impetus towards 'higher mental powers' in both vertebrates and invertebrates was due to the demands for efficient gathering and sorting of environmental information. Thorpe (1963: 230–1) reinforces the general point in his discussion of the behavioural capacities of insects:

The insects are of particular interest to the student of behaviour, primarily because by developing wings they have mastered another element, and this conquest has in its turn opened up immense possibilities for new ways of life and the colonization of new environments. It is chiefly this single step which has enabled them to become the dominant group they are. Another outcome of the development of flight has been the necessity for concomitant advances and improvements in the sense organs – since the speed and distance of travel can be so much greater in the air. Thus a premium is set on organs which give perception at a distance, upon organs which give a quick response, and also upon a variety of senses – so that if one sense fails to provide data for orientation to the nest or territory, the animal can fall back on another.

The demands made by the way of life upon the sense organs must be relayed back as demands made by the sensory apparatus upon the nervous system, especially when speed of response, and shifts from one sensory modality to another, are necessarily involved. And although, as Thorpe points out, 'There is hardly any precise knowledge about the relationship between brain structure and learning ability in the Arthropods' (p. 232), he also suggests that neural mechanisms may be infinitely more plastic and flexible than has generally been thought, and that 'some at least of the fixity or apparent fixity of arthropod behaviour is due to the rigidity of the exoskeletal structure rather than to the nervous system'. It may be, then, that appropriate comparative studies could reveal further illuminating parallels between the several vertebrate and invertebrate groups even with regard to the 'component' factors of intelligence.

VII

Towards a Phylogeny of Human Intelligence

FOUR-FACTOR INTELLIGENCE IN THE HUMAN PHYLOGENY

Since the argument of the present investigation abstracts its initial definition of intelligent behaviour – namely, that it is adaptively variable – from what is known of human behaviour, and since it is a basic presupposition that some at least of the total present range of human behaviour is intelligent, it will be appropriate at this stage to attempt a brief overview of what is known of the human phylogeny to see whether the four-factor theory can provide fresh illumination and an improved basis for the appreciation of significancies. Since the argument has not, up to this point, been developed in relationship with the particular factual information available on human evolution, the attempt to correlate theory with 'fact' can provide an additional test of the plausibility of the theory.

One of the most striking features in our knowledge of the human phylogeny at the present time is the enormous and so far inexplicable gap in direct fossil evidence between our branching-off from the general hominoid stock (somewhere between 12 and 18 million years ago) and the appearance of a fully hominid form, *Australopithecus* and its relatives, roughly one to two million years ago. Prior to the branching-off point, the hominid ancestor must, by definition, have been of generalized hominoid type; while in the earliest known hominids the characteristic features of bipedaly and an intelligent tool-using culture were already, it seems, fully developed (Dart and Craig 1959; Brace 1967). The transition from partly arboreal habits to terrestrial bipedalism must have taken place during the period for which we have no direct information. The precise nature of the various steps involved in the transition to the hominid type of organization must therefore remain, for the present, a matter of conjecture.

But we do know the final product; and we can build up an inferential picture of the ape-like starting point. The gross changes that must have been involved can therefore be indicated with some confidence, even though the detailed interrelationships between the various changes may have to be left problematical.

Evidence of the existence and the adaptive contribution of the several factors of intelligence in the great apes has already been indicated. It must be assumed that the same factors were present to substantially the same degree in the ancestral hominids, and that their contribution to overall adaptiveness of behaviour was at least broadly similar. If this postulate is admitted, we are enabled to set out the changes in structure, way of life, etc. between ape and man, and attempt to find the significance of those changes for the several factors of intelligence.

The development which has perhaps achieved most notice, and which relates in the most obvious way to one of the factors of intelligence, is the alteration in the function of the forelimb and hand from a locomotory to a manipulative function. This development could occur fully only subsequently to the achievement of bipedaly. It has been of the very greatest significance for human evolution, allowing as it does the development of a complex cultural dependence upon the use of weapons and implements. Undoubtedly the breakthrough from instinctive signal-systems to the abstraction and generality of true speech has been facilitated by the demands of an already-elaborate tool-dependent culture, indeed it is difficult to conceive that selection pressures in favour of speech could ever have become sufficiently powerful in the absence of such culture. The most fundamental changes in the transition from ape to man may therefore be represented as:

Bipedaly——→Complex tool-dependent——→Speech
Culture

The question of the use of fire seems to be a subsidiary issue. Undoubtedly fire has contributed very substantially to some cultures. Its use is significant in palaeontological research, too, since it leaves identifiable traces – but there seems little doubt that proto-hominid cultural traditions had become well established long before fire-using became a feature of them (Oakley 1962).

The relationship between the factors of intelligence and the development of true speech will be discussed below. At this stage the significance of the preliminary development of bipedaly must be assessed. Some fairly clear-cut inferences seem possible.

Firstly, the use of the fore-limbs for manipulation rather than locomotion must have demanded prior enhancement on the motor side of the D-factor, and must in turn have caused further development on this side. This has long been recognized (see Oakley 1961). But there must have been a similar reciprocal relationship on the sensory side of the D-factor, in association with two changes in the possibilities and patterns of sensory inflow:

(a) The use of the hand as a sense organ (as well as an effector). As well as direct tactile information from the receptors in the hand itself, the need to sort and interpret kinaesthetic information from fore-limb proprioceptors must also have had a substantial influence on D-factor build-up.

(b) Improved distance reception due to the elevation of the head, as a result of bipedal stance, must have had considerable effect on D-factor sensorimotor efficiency (see Fig. 8). Again, an amplified sensory inflow, bringing greater possibilities for the detection of predators, prey, and members of the individual's or competing social groups at a distance, must itself have had a considerable direct effect; but the indirect effect on the motor side, with an enhanced premium on prompt implementation of co-ordinated social activities, must also have been appreciable.

Secondly, from the considerations just mentioned, especially those relating to complex co-ordinated activities of the social group in combating predators, and in food-getting, it is apparent that the effectiveness of the moment will be enhanced if earlier experience can be remembered and appropriately applied. Thus both C- and A-factors are implicated. It must be remembered, also, in this connection, that 'earlier experience' can have a wider reference, even in the pre-speech stages of culture, than the direct *participant* experience of the individual. Mainardi and Pasquali (1968) have shown that the adaptiveness of certain performances even in mice can be enhanced as a result of *spectatorial* (vicarious) experience; and a great deal of evidence suggests that in the higher primates this form of 'cultural facilitation' can be of great significance.

Some of the points just made are adumbrated in general terms by Westoll (1962), who remarks that 'the attainment of an upstanding quadrupedal posture needs far greater and more precise control of balance and movement, and the new position of the head means that more and more of the surrounding territory is under observation. The recording and selection of such increased information, if appropriate action is to follow, necessarily implies greater demands on the brain. Memory and quick response to stimuli would become more important. I am myself inclined to believe that this change of posture may well have been correlated with the improvement in 'brain-power' of mammals. . . . Some other tetrapods that have 'raised their heads' have also shown great mastery of their environment – the dinosaurs, the birds, etc. – and at least in the birds increasing demands upon balance and neuro-muscular co-ordination have resulted in a liveliness and "mental agility" far above that of most reptiles' (pp. 57–8).

Later, in coming to a brief discussion of adaptive changes in anthropoids, Westoll is even more specific. 'I have already suggested,' he says (p. 59), 'that mammals had to acquire superior brains, capable of better neuro-muscular co-ordination and capable of receiving and storing more information, if their less sprawling posture were to be advantageous. I would now suggest that this is true to a still greater extent of man and similar anthropoids. The attainment of a bipedal gait, without a balancing tail, must have led to free use of the arms for balance; again, rapid postural adjustments need a more intricate set of neural controls, so that imbalance is translated into dexterity. The amount of information available from the environment is increased as the animal gets its head up in the air – except for olfactory information. Selection would therefore tend to favour the refinement and elaboration of the whole neural mechanisms for receiving, storing, and acting upon, information received from the world around.'

These indications of the roles of D-, C-, and A-factors are clear enough – but what of P?

The role of the P-factor in the transition from quadrupedaly/brachiation to bipedaly, and subsequently in the development of tool-using manipulative skills, is not obvious to superficial inspection; but can, as in other instances, be perceived through careful assessment of the prerequisites of what is actually observed to be the case. Consider the change from quadrupedal to bipedal movement on the ground. This change can take place, obviously enough, only in a suitable habitat, a place where bipedal stance and locomotion are possible. But not only must the change be physically possible, it must also be adaptively advantageous. The type of habitat which is generally accepted at the present time as providing the opportunities and selection-pressures required for the transition to bipedaly is the open sub-tropical savanna, with groups of trees to provide refuges and sleeping places separated by stretches of grass in which foraging and group social activities take place. Various means of selection for bipedaly are available – that is to say, it has a number of advantages. The head is lifted above the grass, for improved 'distant' reception. In locomotion, reversal to quadrupedaly means running 'blind'; whereas bipedal running gives a better chance of dodging a pursuing predator, and gives much more efficient hunting of prey especially if this hunting involves co-operative group action. The individual can see the movements of his fellow hunters as well as the prey, and can anticipate and organize. . . .

So bipedaly, once it had been tried, must have been found to confer great advantages. But the various selection pressures favouring bipedaly can operate only upon animals which attempt it – they cannot make the animals get up on their hind legs in the first place. The first occurrence of any phenotypic change cannot but be due to factors *other* than those which subsequently cause its retention and elaboration – the contrary view entails adherence to a Lamarckian and teleological position, that the 'need' for a function causes the wherewithal for the function to come into existence. So with bipedaly, its first appearance cannot have been caused by the advantages accruing from it. Assuming an anatomical structure such as to allow at least a primitive form of bipedaly, the problem remains as to why, at the behavioural level, an animal should forsake the habits of its upbringing and ancestors, and rise up on its hind legs. The acuteness of this theoretical problem must not be neglected on account of the degree of abstraction required for its perception. Random genetical changes generating a propensity towards bipedaly must no doubt be assumed to have occurred. It is easy to give this as a blanket answer, however, and to overlook the true complexity of the behavioural requirements of this phase of evolution. Bipedaly cannot have resulted simply and directly from genetic change, since such simple and direct causation would leave open no possibility of variation, no possibility of alteration from moment to moment between bipedal, quadrupedal, and, say, brachiating locomotion. The mechanisms of locomotion thus cannot be genetically fixed, neither can they be keyed in a fixed manner, to particular environmental or situational stimuli (see Lorenz 1965). We have here, in fact, in the problem of the change in stance and locomotion from quadrupedaly/brachiation to terrestrial bipedaly, a particular exemplification of the more general problem, posed in abstract terms in Chapter III of this book, of allowing for variability of behaviour to be adaptive to changing circumstances. And in terms of this particular problem, just as in the more general context, what seems to be required is a 'central' behavioural capacity to inhibit the previously-standard activity and thus to allow for the substitution of a new activity, namely bipedal locomotion. Thus the change to bipedaly would appear to be dependent, at least in part, upon the P-factor of intelligence. In active and particularistic terms, the individual must be able to resist the temptations to crawl and to climb trees, if he is to even begin to learn to stand and walk upon his hind legs.

On the basis of the preceding argument it is now possible to specify the periods and circumstances in the human phylogeny when selection pressures have operated most strongly to build up the component factors of intelligence. It may clearly be discerned that selection for P-factor must be especially powerful at any time of transition from one adaptive pattern of behaviour to another. One such transition is that from quadrupedaly to bipedaly. Another, as we shall argue shortly, is the transition from the instinctive signal-system to true speech. The A- and C-factors have probably been under steadier selection during the period since lengthened individual life-span, prolonged juvenile dependency, and an active and social way of life have been the rule, i.e. since the transition from a prosimian to an anthropoidean level of organization. Selection for D-factor has probably fluctuated according to the precise nature of the change in intercourse with the environment. The commonplaces of primate evolution – binocular stereoscopic vision, eye-hand co-ordination, judgement of distance, reduction in olfaction, and so on – have obvious implications for the D-factor. The main periods of D-factor selection must therefore have been those leading up to and immediately following the prosimian-anthropoidean transition, and that of the arboreal-terrestrial (i.e. quadrupedal/ brachiating to bipedal) transition.

In the absence of anatomical specializations for predation, it seems likely that the swing from herbivorous/frugivorous diet to a predatory and carnivorous way of life occurred *after* bipedaly was established. Reasons for this have been touched upon above. The hands had to be free to use weapons and implements; and the bipedal stance with head and distance-receptors elevated must have facilitated hunting, both directly in locating prey (and predators), and also indirectly in enabling social co-operation and co-ordination to be improved. Morris (1967), Dart (1963) and others have pointed out the importance, for the development of large complex social groups in which co-operative activity is carried on, of the change from vegetarian to carnivorous diet. Vegetarian animals need to remain relatively dispersed, need to keep on the move through relatively extensive tracts of country, in order to keep up a steady and large intake of relatively low-nutrient foodstuffs. They have little or no occasion for detailed and complex co-operation between reasonably large numbers of individuals, except in defence against predators; and as Chance and Jolly (1970) show, the appearance of co-ordinated action is in fact due to the interaction of a few relatively stereo-typed instinctive reactions. Hunters, on the other hand, since a big 'kill' can provide a relatively large quantity of high-grade protein at one fell swoop, have both the opportunity and the incentive to operate in large groups for the 'high dividend invest-ment' which becomes possible under these circumstances.

Accepting, then, that the achievement of bipedaly was followed by the enlargement of the 'unit' social group and the concomitant adoption of predation, what pressures can be postulated as operating upon the factors of intelligence in the new situation?

If social co-operation is to be superimposed upon the hier-archical organization characteristic of the vegetarian primates, one of the first desiderata must be the loosening of the social dominance mechanisms. These mechanisms cannot be abolished, since in some circumstances they will continue to be adaptive; also their total abolition could hardly be accomplished without undesirable disruption of other sectors of the behavioural repertoire. What is required is, then, the capacity to dispense with dominance interactions at some times (e.g. when hunting) but not at others (e.g. when apportioning sleeping 'territory' and mates and probably also when repelling predators). We have already argued, on the basis of observations by Yerkes, Schaller, and others, that the behaviour of gorillas in the wild (and, to a lesser extent, chimpanzees) exemplifies the operation of P-factor 'withholding' or 'forbearance' in dominance transactions. It seems plausible to assume that precisely this 'forbearance' could facilitate a 'relaxation' of hierarchical social-dominance into the more egalitarian form of group structure appropriate to co-operative predation. If this were so, the transition to co-operative predation as a way of life would automatically furnish selection pressures in favour of the P-factor. (The existence of relatively high levels of P-factor in the gorilla would thus, if accepted, be further argument for the de-humanization hypothesis, see Kortlandt 1963).

Moving now to consideration of the last major development in the human phylogeny, namely the break-through to true speech from the signal-systems of instinct, it would appear that one factor, the Sensorimotor or D-factor, might be regarded as of little relevance for the *change* involved. (It would, of course, be of the greatest relevance to the several antecedent changes, as indicated above, which led up to the conditions in which speech could be evolved.) Also requisite, for there to be any advantage in the change to true speech once it had occurred, must be high levels of both C- and A-factors. The abstraction characteristic of true speech is unnecessary unless large quantities of information have to be communicated, and unless the general fund of available information can be related to present and immediate needs through the operation of A-factor appreciation of similarities and differences. The A-factor must be involved in a peculiarly intimate and direct way in the actual 'escape' of the communication system from the particularity of instinct to the abstract generality of language (see Marler 1959, 1961; Lanyon and Tavolga 1960). A simple illustration will show what is involved. We can imagine an instinctive signal-system being used by humans to communicate various types of information relating to dogs. Let us imagine that the signal-system comprises the word 'Dog' uttered in different tones of voice in different circumstances, possibly with different accompanying gestures, etc. One call 'Dog!' is directed to one's own dog, and has the function of calling the dog to one's side. (The signal achieves specificity by the dog's responding to the peculiar voice characteristics of its owner.) Another loud call 'Dog!', in a different and shriller voice, accompanied by pointing, dilation of pupils, raising of hair, signifies that a dangerous and predatory dog is approaching and that one's fellows should take cover, or grab weapons . . . Yet another call, in a softer tone, indicates to one's hunting companions that the quarry (antelopes or sheep, perhaps) are being driven by a dog within reach of the human weapons. And so on. A dozen or more easily distinguishable signals could very readily be described. The point is, that for proto-hominids still limited to the instinctive signal-system, these calls would all be *different* signals. We have assumed that all are based on what *to us* is the same word, 'dog', but of course the perception that it *is* the same word represents a radical departure from the functioning of the signal system, and is dependent on a startling exercise of A-factor abstraction and generalization. An even greater exercise in abstraction is necessary, over and above the recognition of 'dog' as the same through all the calls, before it can be used in a context-independent manner as in, say, telling a story about a dog: 'The dog stood outside the cave . . .'

At this point the contribution of P-factor in the development of language begins to become clear. The automatic association between 'Dog' uttered in some particular tone of voice and the behavioural reactions and even 'internal (emotional) tendencies' previously appropriate to that particular signal, must be broken. The break is not of course complete. Even in ourselves, most words have an emotive 'colour' to them, a purely subjective associational overlay which is to some extent additional to and/or independent of what would normally be distinguished as the 'official' connotation of the word in question. But this is quite different from what our reactions would be if the words we use were really still instinctive signals. Then, we would be unable to ask, when the work 'dog' was uttered in our hearing: 'Well, what *about* a dog? What are you saying "dog" for?' Instead, we would be automatically aroused to fear, or to thoughts of hunting, or . . . depending on the function we had come to associate with the signal 'dog'. In fact, human beings can be brought into a simulation of an instinctive signal-system even now, by the processes of stringent conditioning popularly known as 'brain washing'. After being subjected to such conditioning, certain words (or other stimuli) have lost their 'generality' or 'neutrality' for the brain-washed individual.

Thus it is the rigidity of the connection between stimulus and response which is characteristic of both instinctive systems and systems which, even though 'learned', have resulted in a loss of flexibility. If, therefore, true speech incorporates generality and flexibility to the extent that there is no necessary relationship between the physical characteristics of a word and its manner of use (i.e. its meaning) – this is to say, that words are not usually, and not necessarily, homomorphic with their meanings – it follows that the *user* of language must exercise some power of abstention, to prevent himself always using a particular word in only *one* of its possible usages, with only *one* of its possible sets of associations. What this comes down to, is that both to let a language escape from the instinctive signal-system in the first place, and also for it to retain its generality and flexibility of use, it must be subjected to the negativistic withholding influence of the P-factor of intelligence.

Hockett (1967) lists three features as being peculiar to the true speech of the hominids: 'displacement', 'productivity', and 'duality of patterning'. A fairly clear relationship may be discerned between each of these and the operation of the P-factor of intelligence. Hockett explains 'displacement' as 'being able to talk about things that are remote in space or time (or both) from where the talking goes on'. The ability to pay attention to 'remote matters' demands an ability *not* to be dominated by 'immediate matters': the withholding function of P is required. In 'productivity' also – explained as 'the capacity to say things that have never been said or heard before' – the P-factor function of breaking away from previously-standard patterns is clearly necessary. Finally, 'duality of patterning' (perhaps better as 'plurality of patterning'), the use of the same sound-units (or letters in written speech) in different permutations to give meanings which may be totally different, is also dependent upon holding-off the tendency to let similarity of sound-units dictate similarity in meaning. (Schoolboy 'howlers' often illustrate the absence of such holding-off. Stories about 'The shriek riding his horse in the desert . . .' depend for their point upon the superficial similarity in appearance, but gross difference in meaning, between the words 'shriek' and 'sheik'.) Here again, P-factor prevention of the dominance of the immediate and superficial is need for the effective use of language, and for there even to be 'language' as distinct from 'signal system'.

As always, the fact that a particular characteristic has contributed to a successful or adaptively advantageous innovation leads to its being put under positive selection pressure through the selection for the innovation in question. Thus if the evolutionary development of true speech was in fact dependent on the P-factor of intelligence, whatever advantages resulted from speech would automatically constitute selection pressures for the P-factor. And although it has been argued that the P-factor must have played a crucial role in the break-away from instinct to true speech, once speech has been achieved there must be a feed-back of positive selection upon the other factors of intelligence as well. A well-stocked memory (C-factor) can be of even greater use, to the social group as well as to the individual possessing it, if its contents can be communicated to others. By passing from an individual into a group 'repository', information can be stored and accumulated over indefinite spans of time. It should be noted that since this function is performed by the group rather than the individual, selection in its favour must also work through the group rather than the individual.

Finally in this sketch of the role of the four factors of intelligence in relation to the past evolution *Homo sapiens*, it may be suggested that although a particular factor may have been essential for the achievement of a particular level of adaptation, once that level has been achieved the selection pressure in favour of the given factor may well fall off or even be reversed. Thus although the P-factor, for example, can be represented as essential for the original achievement of speech, it does not follow from the fact that we are dependent upon speech, now, that the P-factor is necessarily under positive selection at the present time. In Chapter V it was shown that within a given social group, the emphasis of evaluation (and possibly also of genetical selection) may at some times be against the P-factor tendencies towards questioning and non-conformity. In Chapter X some of the relations between the intelligence-factors and present day trends and phenomena of education will be explored; while in Chapter IX the broader fields of interaction at the societal level will be scanned for evidence of the operation of the various factors of intelligence and of instinct. Before moving on to these major areas of investigation, it will be convenient to conclude this chapter with a brief reconnaissance of some of the possibilities for the future evolution of man, in relation to the several factors of his intelligence.

2. EVOLUTIONARY INTELLIGENCE AND HUMAN FUTURES

In relation to the evolutionary processes already discussed, in which P-factor control of social-dominance displays has been shown as a probable avenue for the independent build-up of the P-factor of intelligence, it is interesting to re-interpret various recent prescriptions for the *future* evolution of man in the terms we have been using. Leach (1968) points to the danger of being influenced by 'petty-minded people who suspect in their hearts that they are going to be overwhelmed. What is needed is greater confidence. Young people need to be shown that they are already in a position of supremacy; their problem is not to conquer . . . but to look after . . .'

'Greater confidence' – and even when we do not *feel* entirely confident, we can generate confidence, in ourselves and in our associates, by *seeming* to be confident, by putting on a demeanor of confidence and serenity. It is possible to build up a habit of calm attentiveness 'from the outside' as it were, working from the similitude to the reality. For most of us, this process would appear to be largely a matter of 'learning', of the building up of habits. It seems plausible to assume, however, that the ability of the individual to build this sort of habit must be affected by his level of P-factor endowment. The relationship between P-factor and what we ordinarily speak of as 'self-control' is undoubtedly complex, and no more than tentative general suggestions as to its nature can be put forward here. 'Self-control' in one usage connotes conscious and deliberate volition. It is notorious, however, that such deliberate attempts at self-control are often futile. On the other hand there are many individuals who achieve self-control without much effort, who do it naturally and easily – and while some achieve a similitude of 'control' simply because their instincts and inclinations are weak (cf. Chapter VI, p. 149), there are many others of whom it is implausible to suppose weakness in the drives being controlled, quite the contrary in many instances. To explain the inadequacy and even the complete absence of control in individuals who appear to have every motive and every desideratum of ability and background to achieve it, it seems necessary to postulate a deficiency in inherent capacity. This 'capacity' would, from the terms in which we have analysed the problem, have to be independent of conscious volition and all the usual kinds of motivation. It appears necessary to assume that it must be a 'central' capacity or factor, associated with the higher control centres of the cerebral cortex. The existence of such a central capacity would also account for the cases where self-control comes easily to the individual, where it does not have to be 'learned' in any obvious way – though of course this does not preclude that its efficacy could be improved with practice.

(Here, as with so many other 'exercise effects', there can be no clear-cut distinction between the usually accepted categories of 'learning' and 'maturation'. In 'learning' a new tennis stroke, for example, there is an early stage of simply learning, in an unambiguous straightforward sense, the movements that have to be performed. But this 'intention' usage of 'learning' (see Scheffler 1960; Ryle 1949) does not entail the 'success' or 'achievement' use of the word: operationally, it may turn out that one has *not* learned the stroke after this 'intention' phase of the so-called 'learning process' has been completed. In order to be able to claim that one *has* learned the stroke, in the 'success' or achievement use of 'learned', it is necessary that the first phase of learning be succeeded by a second usually prolonged phase of sustained practice. The stroke must be 'grooved in', as some tennis players say, so that it becomes automatic at the level of effector coordination. The neuromuscular integrations apparently need time and repeated practice to become 'set'; and while it is not known what changes in detail may be involved within the central nervous system, it is clear that the changes must involve central as well as peripheral adjustments, and insofar as these are of a relatively permanent nature – a stroke once properly learned can quickly be restored to high efficiency even after a lapse of many years – it seems that these long-term 'grooving in' changes should be regarded as maturational changes involving a large part of 'motor outflow structure'.

The assumption that self-control is dependent on a central factor whose strength or effectiveness can vary intrinsically between individuals and also as the result of its 'exercise' does, then, allow us to systematize what is known, what is indeed a commonplace, of human behaviour, and also what is suggested by the data so far available on the behaviour of the higher primates. The strongest argument for such self-control being a function of an inherent factor or property of the central nervous system, i.e. that self-control is not *merely* to be learned by anyone so inclined, is that unless it has a definite genotypic basis it is unlikely that it could have been developed, to the extent observed in at least some individuals, in the course of evolution. In short, the argument that self-control is solely a learned or socially-induced type of behaviour is liable to involve Lamarckian presuppositions; and would be open to the various weaknesses of, for example, the 'blending inheritance' theories (see discussion in Huxley 1942). But provided that there can be varying levels of actualization of an inherent potentiality for self-control, it is possible and indeed likely that a form of pseudo-Lamarckian evolution will occur, the phenotypic behaviour (i.e. actual behaviour) swinging quickly to a new adaptive 'peak' (see Lewontin 1965) as a result of 'learning', socially-induced or ideationallyinduced habit formation, etc., this rapid phenotypic change being followed much more slowly by a genetic change which might (but need not) result in genetic 'fixing' and stabilization of the phenotypic adaptation (Waddington 1957). If for example it did become adaptively necessary to suppress certain instinctive tendencies – notably perhaps those towards aggression – this might be achieved in the short run by various familiar means, coercion, indoctrination, education, the setting up of ideals, and so on. To a considerable extent intra-national aggression has already been controlled by these means. Inter-national aggression has not yet been so satisfactorily controlled; and part of the difficulty may be due to inadequacies in the control of basic inter-personal aggression within national societies. In other words, international wars may occur simply as 'spill over' activities closely analogous to or even homologous with the 'vacuum activity' (Lorenz 1935; Tinbergen 1951; Hinde 1966: 222–3) observed with the simpler expressions of instinct; and this may be because ordinary interpersonal aggression is merely 'bottled up', repressed in the Freudian sense rather than suppressed, *within* each national society. It seems probable that many assumptions of social theory involve the presupposition that if external stimuli for aggression are absent, there will be generated no aggressive motivation. The demonstration that many instinctive drives are endogenous (see Tinbergen 1951, Hinde 1966) casts doubt on this. Possibly we do have a tendency towards long-term build-up of aggressive motivation (Lorenz 1966). This theoretical possibility, interesting and important enough in its own way, is however of little immediate practical significance, since it seems unlikely that we will in the foreseeable future eliminate the myriad incidents which do in fact stimulate aggression in everyday life (this *contra* some of the criticisms of Lorenz and others, in Montagu 1968). The problem is that aggressive drives are present, at some level or other, most of the time in most individuals. The question of the extent to which agonistic motivation is endogenous, as against the extent to which it is stimulated by all the petty pinpricks of even the most salubrious existence, is an hypothetical and abstract one. The possibility of controlling aggressiveness by eliminating its causation, either way, is not practicable. It remains to enquire what *else* can be done.

It has been implicit in much of the foregoing discussion that P-factor control can contribute directly and immediately to the non-performance of overt acts of aggression. But it can contribute to a lessening of aggressiveness also in other, more subtle, and in the long run perhaps more efficacious ways. For example, by allowing an individual to remain in situations where stimuli to aggression are present, without responding to them, the P-factor would appear to allow habituation effects to lessen responsiveness to subsequent stimuli of the same or similar type. Again, the withholding of overt aggressive behaviour can provide opportunity for rational appraisal of a situation and for a strategy of constructive long-term response to be worked out. It is not being suggested that the mere abstention from overt agonistic activity will always have this effect – far from it. As already indicated, if the abstention involves nothing more than a 'bottling up' of resentment, there is a need to provide some sort of outlet, preferably of a harmless nature. Probably the agonistic displays of the gorilla and other hominoids (Schaller 1964: 210–15) have this safety-value function among others. As Schaller remarks: 'Man behaves remarkably like a gorilla in conflicting situations [presumably "conflict situations"]. A marital squabble, for example, in which neither person cares to attack or retreat, may end with shouting, thrown objects, slamming doors, furniture pounded and kicked – all means of reducing tension' (p. 215).

These are typical 'redirection activities', as Lorenz (1966), Tinbergen (1952) and others have pointed out. It must be emphasized that even if nothing more is achieved than redirection of the aggressive behaviour, this is still of great adaptive value. Indeed it is possible to use such redirected outbursts with deliberate purpose as tension reducers. The parent who, when roused from sleep in the cold hours of the night by a crying child, leaps with violent curses from his bed and wrenches the bedroom doors almost off their hinges, may thereby be enabled to deal with the child calmly and sympathetically (relatively speaking) when he does arrive at its bedside. By contrast, the parent who practices compulsive total repression of his violent emotions may eventually 'fly off the handle' and do the child real damage.

Intelligently directed 'sublimation' of aggressive impulses (supplemented perhaps by harmless 'redirected' or symbolic aggression as a safety valve) seems the most useful adaptation. The term 'sublimation' is intended to suggest, not the Victorian assumption that if we ignore an unwelcome fact of physiology, it will cease to exist, but that the *expression* of any basic behavioural tendency is open to at least some degree of management and long- or short-term control. Some of those who have attacked the books by Lorenz (1966) and others on instinctive behaviour (including, in particular, aggressive behaviour) in humans seem to assume that, if a behaviour pattern is susceptible in the least degree to modification by learning, then it cannot be instinctive and it must be completely modifiable by learning. The latter assumption would appear to involve the logical fallacy of composition; while the former presupposes that 'instinctive' and 'learned' are absolute and totally exclusive categories. Such a presupposition is quite unwarranted. On the contrary, a great deal of clearly instinctive behaviour is modifiable, and indeed depends upon various 'learnings' for its adaptiveness (see Chapters II, III, and XI); and, as Crook (1968) points out in a cogent discussion, the fundamental desideratum is for vastly greater understanding of *all* the mechanisms of behaviour, be they instinctual or learned, so that cultural patterns especially in child-rearing and education may be emended, where necessary, to promote the types of behaviour necessary for long-term adaptiveness. 'It is to be expected,' he says (p. 155), 'that the tendency to show aggression may be modified by . . . socialization procedures. Much depends upon the nature of parental "permissiveness" and the sort of rules used in family control of aggression.'

It is highly significant, from the point of view of the present work, that in discussing past behavioural changes in the human phylogeny, Crook should make the following statement: 'Certainly the necessity for increased control of individual behaviour in relation to hunting and role playing in co-operative social life will have entailed the marked development of *the ability to delay responding*, to interiorize and reflect on alternative causes [presumably "courses"] of action, and to cope with motivational frustration' (1968: 171; my emphasis added).

This seems a particularly clear acknowledgement of the importance of the P-factor 'withholding' function. From the context in which the statement is made, moreover, it appears that an 'inherited trait' is visualized – presumably something like the 'basic capacity for withholding' which is argued in Chapter XI as being genotypically-based.

The ability to withhold the instinctive response, notably but not exclusively an aggressive one, has been significant in the evolutionary past and continues to be so under the present conditions of human society. The individual who, when thwarted by unfavourable circumstances or the dictates of higher authority in a cherished project in his work, can transpose his activities into, say, a theoretical exploration of the causal factors of the thwarting, has in a sense been able to outflank his restrictions and move into a realm where his freedom to achieve is limited only by his own nature not by any extraneous prohibitions. Freedom to prosecute some worthwhile activity makes a man happy and healthy. Thus he can tolerate frustration in some of his other activities without becoming refractory and bitter, without becoming an infection-site for destructiveness. The results of his private projects may themselves be of benefit to the social group; but the primary value of 'sublimation' is in enhancing the quality of life of the individual, and reducing the amount of agonistic behaviour within the social group.

There seems to be a case, over and above this, however, for incorporating quasi-sublimation devices among our social mechanisms. Legal and political institutions already to some extent embody such devices – what is needed is an extension of them, so that instinctive drives, and notably the aggressive ones, can be harnessed to socially useful purposes. If this can be done (and of course it already is done to a considerable extent), the individual will have scope for tension-reducing expression of his aggressive drives of various sorts, while society benefits. Again, the shadowy judo-like principle, recurrent in biology at all levels, seems applicable, that when confronted with an array of potentially destructive forces the thing to do is to use some of them, swing them round so that they work against the others. Already-existing social mechanisms for the control and exploitation of aggression and social-dominance drives are, for example, the procedural rules for the conduct of meetings. These rules have evolved on the two assumptions:

(a) that many of the people attending the meeting will have conflicting aims and policies and are likely to be antagonistic to each other on a purely personal basis; and

(b) that an overall benefit can nevertheless result if the conflicts and antagonisms are to a sufficient degree resolved.

This second assumption comes down to this, that the business (whatever it may be) is worth holding a meeting about.

Many of the rules can be seen quite clearly as devices for allowing expression to, yet controlling and usually minimizing the effects of, the aggressive tendencies of the individuals present.

Mutually antagonistic individuals, for example, are not allowed to address each other directly, nor they allowed to mention each other by name. If they do not directly and physically face each other the effects of their agonistic displays are reduced – though of course 'facing' encounters cannot always be prevented, especially if the individuals comprise two opposing parties ranged facing each other across the debating chamber as in the British and most Commonwealth Parliaments. The Chairman, as his very title suggests, remains seated in his chair, a non-bellicose situation and posture, and this contributes to the lowering of instinct-tensions. The general aim of the procedural rules for the conduct of meetings is to allow some expression to the instinctive drives which form perhaps the bulk of the motive force, but to mitigate the effects of the instinctive 'displays' so that rational considerations can exert a determinative effect. The driving power for action comes largely from the instincts and emotions, but it must be controlled by intelligence. The rules of procedure for meetings have been designed by social selection to play down instinct and play up intelligence.

Another area where aggression might usefully be allowed greater scope than at present is in public debate through the mass media which have become so influential. It may be necessary to overhaul some of our legal machinery, for example the laws relating to libel and slander, to permit more overt aggression through socially-useful activities. There is enormous diversity, on the international scale, in the extent to which members of the 'Establishment' can be called to account for their public actions and subjected to criticism. There is an optimal range, varying with the type of public office and function in question, for the degree of direct criticism which should be allowed. Too little public criticism leads to authoritarian and eventually totalitarian government; but too much can lead to inertia and paralysis. The excesses of subservience and of violent 'social protest' should both be avoided, and there seems to be no reason in principle why they should not both be avoided. What is required is adequate feedback mechanisms so that Establishments can be kept in reasonable accord with the realities with which they have to deal. One of these realities is human aggressiveness. Recognition of this has often led to pessimism. But intelligence is just as much a reality as is aggressiveness; and through the use of intelligence, and in particular through the intelligent utilization of P-factor control, it may be possible for us to achieve a new synthesis of psychological adaptation, a new level of cultural evolution. Some of the possibilities for this in terms of education will be discussed in Chapter X. The key assumption in all this must be, that just as intelligence has emerged in the past to exert a decisive effect upon human evolution, so it can continue in the future, *given the right conditions, the appropriate mechanisms*, to lead the human species past the blind alleys of instinctual specialization and the death-trap of behavioural disintegration.

VIII

Phylogeny and Ontogeny

This book has been concerned with the evolution and phylogeny of intelligence. Its ontogeny, which is the complementary field of study (see Garstang 1922), has been almost entirely neglected. There are many reasons for this. In the first place, there is a considerable danger of 'recapitulationary' assumptions creeping in unnoticed, and distorting one side of the argument or the other; and, since ontogeny is open to direct empirical investigation in a way which is impossible with regard to phylogeny, it seemed best to postpone a comparison between findings in the two areas until the evolutionary/phylogenetic picture had been explored in some detail. Once this has been done, the 'factually establishable' ontogenetic side is less likely to overpower the more inferential case presented on the side of evolution. Since the skeleton of the evolutionary argument has by this time been presented and some of its implications explored – this exploration being in the nature of a 'coherence test' (see Stebbing 1933) of the truth of the theory – it will be appropriate at this point to attempt a brief comparison between the phylogenetic and the ontogenetic views of intelligence.

The pre-eminent name in studies of the ontogeny of intelligence in humans is undoubtedly that of Piaget. What appears to be an excellent account of his views has just become available: the book *Piaget and Knowledge* by H. G. Furth (1969). (Interestingly enough Chapter II of that work, though entitled 'Knowledge in Evolution', deals with the relationship between instinct and intelligence in the course of human evolution, and a sketch is attempted of the development of intelligence from a basis of Lorenzian instincts. That chapter is thus an exact converse of this present one: within a book on the ontogeny of intelligence it touches upon the evolutionary-phylogenetical problem while this chapter looks across in the opposite direction.)

1. GENERAL SIMILARITY WITH PIAGET'S THEORIES

Furth emphasizes the complementarity between ethological studies and Piaget's work on the ontogeny of intelligence in humans. 'The comparative observations of ethologists regarding the spontaneous knowing behaviour of animals . . . seem to be a . . . direct parallel to Piaget's own sampling of developmental observations in humans' (p. 185). (An open implication of this is that ethological investigations of instinctual behaviour, not on infra-human animals but on humans themselves, should be extremely fruitful and are in fact necessary if a proper understanding of human behaviour itself is to be attained.)

Piaget's own view of the relationship between instinct and intelligence is worth quoting at length.[1]

The basic fact of the breakup of instinct, in other words, of the nearly total disappearance in anthropoids and in man of a cognitive organization that was predominate during the entire evolution of animal behaviour is itself highly significant. This is not, as one says quite commonly, because a new mode of knowledge, namely, intelligence considered as a whole, replaces the now extinguished mode. Much more profoundly, significance lies in the fact that a form of knowledge until now nearly organic extends into new forms of regulations. While this new mode of knowledge takes the place of the former, it does not properly speaking replace instinct. Rather, the change involves the dissociation and utilization of its components in two complementary directions (p. 199).

This, in terms of the theory argued in this book, would involve the utilization of the sensorimotor D-factor which had been under positive selection since the earliest beginnings of metazoan (or even protozoan) organization, along with the 'memory' C-factor and 'abstracting' A-factor which had been developed with the 'higher' levels of instinctive and supra-instinctual type of behaviour. The crucial issue is, whether or not the existence and role of the P-factor is recognized. In a very real sense it can be regarded not so much as a 'component' within instinctive behaviour as a definitely anti-instinctive factor. Certainly the P-factor must have played a decisive part in the 'breakthrough' to

[1] From Piaget, J., *Biologie et connaissance* (1967), translated by H. G. Furth in *Piaget and Knowledge* (1969).

intelligence in human evolution, in breaking the quasi-mechanical link between stimulus situation and the previously-standard activity and, notably, in allowing this to happen within the lifetime of the individual animal. But while it is necessary to appreciate the importance of this 'negating' function, as was argued in Chapter III, it is also necessary to recognize, as was pointed out in Chapter VI, that the operation of some instincts, e.g. exploration, also demands a mechanism similar to or identical with the P-factor, for withholding the other 'competing' instincts. Whether in this case one regards the 'withholding factor' as part of the exploratory instinct or as part of the system for allocating priorities between the instincts is perhaps of little importance. The latter view would have the theoretical advantage of placing the proto-P-factor 'centrally'; whereas it if were seen as initially being 'part of' a particular instinct, the problem would then arise of allowing for its subsequent 'extension' to cover the operation of all the instincts. (Here again, it might be regarded as doubtful whether the P-factor does have an equal 'jurisdiction' over all the instincts – but there seems little point in pursuing this issue, in the present state of knowledge in this field.) That the proto-P-factor was initially tied to exploratory activity seems to accord better with what can be inferred from observations on such activity in lower animals, e.g. insects (see Chapter VI).

What, then, is Piaget's position with regard to the P-factor? Again, it will be appropriate to allow a lengthy quotation to speak for itself:

What disappears with the breakup of instinct is the hereditary programming in favor of two kinds of new cognitive self-regulations, which are flexible and constructive. One might say, this is a replacement, indeed a total change. But one forgets two essential factors. Instincts do not consist exclusively of hereditary programming. As Viaud expresses it so well, such a concept views instinct at its extreme limit. On the one hand, instinct derives its programming and particularly its "logic" from an organized functioning which is implicit in the most general forms of biological organization. On the other hand, instinct extends this programming in individual or phenotypic actions that involve a considerable margin of accommodation and even of assimilation that is partly learned and in certain cases quasi-intelligent.

With the dissolution of instinctual behavior, what disappears is only the central or middle part, i.e. the programmed regulations, while the other two realities remain: the sources of organization and the results of individual or phenotypic adjustment. Intelligence inherits therefore what belongs to instinct while rejecting the method of programmed regulation in favor of constructive self-regulation. That which intelligence retains makes it possible to branch out in the two complementary directions, interiorization towards the inner sources and exteriorization towards learned or even experimentally controlled adjustments (Furth 1969, p. 199).

The proper interpretation of some of the central terms in the above is not altogether obvious to inspection. What are the 'sources of organization'? That 'the results of individual or phenotypic adjustment' are behaviour patterns that have been learned by the individual seems clear enough – but what exactly are the 'programmed regulations' which were present in instinctive behaviour but which have been 'lost' in intelligent behaviour?

It seems possible that the 'sources of organization' might be equatable with the four factors of intelligence *plus* the instinctual motivational systems whose operation they jointly control and direct. The 'programmed regulations' might then be regarded as being the inbuilt sensory and motor mechanisms providing direct coupling between 'total stimulus situation' and the species-specific standard response to it – phylogenetically adapted mechanisms of wider scope than, and including, the 'innate releasing mechanisms' (Tinbergen 1951). Further investigation of some of these possibilities will be attempted in Chapter XI, especially in relation to recent theoretical explorations by Lorenz (1965). For the moment it will be convenient to refer again to Piaget's account of the transfer of control from instinct to intelligence, with a view to assessing the adequacy of my preliminary and tentative interpretations of it.

'Interiorization' may be equated, it appears, with what I have called 'vicarious imaginative exploration of possible courses of action' (see Fig. 8, p. 86). 'Exteriorization' might involve either attempts or trials of courses of action, or playful exploration of behavioural capacities in relation to the environment. This last would be conducive to the adaptiveness of behaviour only at second remove, via the 'interiorization' of its results. This could lead to large jumps in adaptive innovation, in proportion as the 'playful exploration' departed from previously standard behaviour patterns; but of course all innovations would ultimately have to be put to the test of 'exteriorization'. As Piaget says:

A preliminary condition for this double progress is of course the construction of a new mode of regulations. These regulations, from now on flexible and no longer programmed, start by the usual interplay of corrections based on the results of actions and of anticipations. Tied to the construction of schemes of assimilation and the coordination of schemes, the regulations tend by a combination of proactive and retroactive effects in the direction of the operations themselves. In this manner they become precorrective rather than corrective regulations. Moreover, inverse operations ensure a complete and not merely approximate reversibility.

These novel regulations are a differentiated organ for deductive verification as well as for construction. Thanks to them, intelligence manifests itself simultaneously in the two above-mentioned directions of reflective interiorization and experimental exteriorization (Furth 1969, pp. 199–200).

If 'experimental exteriorization' were to occur by itself, in the absence of 'reflective interiorization', it would have to be equated with pure 'trial and error learning'. But as Thorpe (1963) says, it is unlikely that this often occurs in higher animals. A degree of 'interiorization' must then be assumed to accompany or precede 'exteriorization', so that it may be to as great an extent as possible 'precorrective'. But what must be involved for this to be the case?

'Although the cerebral nervous system as well as intelligence insofar as it is a capacity to learn and to discover is hereditary, the activity that has to be accomplished is henceforth phenotypic,' says Piaget (p. 200). 'As the work of intelligence consists in going deeper into the sources and in extending the explorations it makes new constructions of two types. One type is the operational schemes that derive from reflecting abstraction and [which] focus on the necessary conditions of general coordinations of action; the other type assimilates experiential data to these operational schemes.' I suggest that Piaget's first type of 'construction' is to be equated with the C-factor of evolutionary intelligence, the dynamic memory-store or 'central integrative field' as Halstead calls it. This memory-store is both furnished and continually reorganized and sorted through the operation of the A-factor, whose activities of abstracting and generalizing, establishing similarities and differences, are in one sense usefully regarded as essential activities of the C-factor itself: the two factors are certainly not to be regarded as separable. Nevertheless their functions are logically distinguishable, and the C-factor is best regarded, metaphorically, as a store of information both particulate and in the form of organized schemata (cf. Chapter IV, pp. 70–3). These schemata would be comprised largely by 'cause and effect' data, including information on the individual's own adaptive capabilities. Compared to these several schemata would be a continuing stream of information about 'present circumstances'. It seems useful, instead of leaving this inflow described merely as an 'assimilation of experiential data', to distinguish two different factors or factor groups which affect its assimilation in different ways. Thus the four-factor theory would appear to represent an advance upon Piaget's less differentiated view, in that both A- and D-factors affect the 'assimilation' of present situation to past experience. As argued in Chapter VI (p. 151), the different balances between the factors, their different strengths relative to one another, could be expected to have very different effects in terms of behavioural or 'operational' capacity. These differences would be found in the type or modality of operational effectiveness, rather than in its 'absolute' level. For example, two painters of identical overall intelligence who differed only in the relative balance of A- and D-factors, could be expected to have different work-patterns in their painting activities and to produce finished articles of dissimilar style and treatment. One with relatively high A- and low D-factor would be likely to produce balanced, well-composed pictures with relatively little subtlety of line or colour. He would work methodically, with few re-workings, erasures, or paintings-over. He would excel in watercolour rather than oils. The painter with low A and high D, on the other hand, would probably utilize sensory feed-back a great deal. His pictures would tend to grow more slowly, perhaps more messily (at least as regards his studio if not the painting itself); and they would achieve strength through colour harmonies and line in detail, rather than through general features of composition. Probably this painter's basic composition schemata would be few and simple. He would work through the variations on a theme, not variations between themes. He would be more effective in oils, less so in watercolour.

The functions elucidated for the A-, C- and D-factors of evolutionary intelligence are, then, compatible with the views of

Piaget; and can indeed be said to provide enhanced discriminatory and explanatory power, compared to his views as expressed in the passages quoted from Furth. But what about the factor newly contributed by the evolutionary argument of this book and by Halstead's (1947) findings, namely the P-factor? This occupies a central place in the present theory. It has been argued (see Chapters III, V) as the *sine qua non* for the breakthrough to intelligence. It would be odd indeed if Piaget's researches had revealed no hint of it.

A possible explanation for the overlooking of the P-factor, along with various other phenomena relating to biological and evolutionary considerations, is furnished by Piaget himself. The operation of intelligence, he says, involves 'a beginning again from zero, since the innate programming of . . . instinct has disappeared . . . Intelligence provides . . . convergent reconstructions with further evolving. In the case of human knowledge this reconstruction appears complete to such a degree that hardly one theoretician of . . . knowledge has thought to search for an explanation [of it] in the indispensible framework of biological organization' (p. 200). In other words, P-factor operation is *so* basic that it escapes notice!

There are nevertheless a number of clear hints at P-factor operation within Piaget's findings.

Referring back to our earlier discussions of 'interiorization' and 'exteriorization', it was argued that *pure* 'exteriorization' would be equivalent to pure 'trial and error learning', and that this would seldom if ever occur in the higher animals and especially man. If in a given situation previously 'interiorized' knowledge is to be able to give rise to enhanced adaptiveness – if, in short, it is to be 'precorrective' as Piaget says – then it will normally be desirable that action should not commence immediately, i.e. prematurely. (Emergency action of course has the opposite requirement, namely that of immediacy, urgency, speed – but even here, it is desirable that emergencies should have been prepared for. . . .) In a non-emergency problem situation the desirable course is to 'pause and consider', to 'look before you leap', to assess the situation and attempt, in imagination, to foresee the likely outcomes of possible courses of action. A few moments consideration of the situation are usually necessary before one's past experience can effectively be deployed for successful action. If action is to be deferred while C-, A-, and D-factors operate to achieve an optimal use of past experience (i.e. 'knowledge', in Piaget's terminology), it is clear that the individual must possess the ability to defer it and to keep his mind 'open' as long as possible. Note that it is not merely a question of deferring overt action. That in itself would be of little account, if in the meantime the individual had seized upon the first action-strategy that entered his head and was merely waiting for the chance to put it into effect. The delay in 'closure' in the face of the demand for action, is the important adaptive feature.

The function of P-factor in giving time to 'pause and consider' could, it seems, be interpreted in Piaget's terms as providing an opportunity for 'accommodation'. Furth states (1969: xi) that 'Accommodation is defined as the application of a knowing structure to a particular instance regardless of whether it issues in a new structure.' If we roughly equate 'knowing structure' with C-factor, and 'application' as involving A- and D-factors, then what is required (unless we are prepared to assume an *instantaneous* 'accommodation') is P-factor 'withholding' to allow whatever time may be necessary for 'accommodation' to take place.

There are stronger indications of P- factor than this in Furth's book. 'Within practical intelligence [i.e. intelligence concerned with or manifested through overt action] one can observe an evolutionary development from specific and immediate behavior reactions (e.g. instincts) to more general behavior coordinations. This tendency indicates a *loosening* of the tie between the form of coordinations and the content of its external manifestations; however, a true *dissociation* is realized only in operational intelligence' (p. 246, my italics). Here we see a pretty clear acknowledgement of the importance of the 'negativistic' functioning of P-factor which has been mentioned so often in this book.

The basic similarities between Piaget's ontogenetic and my own evolutionary or phylogenetic theories of intelligence are well brought out by comparison of my diagram of the operation of 4-factor intelligence (Fig. 8, p. 86) with Furth's 'Diagram of Piaget's theory of knowing' (Furth 1969: 75), reproduced here as Fig. 14). Here again, his 'Inner Structure' seems to equate almost exactly with my C-factor, the 'central integrative field'. In both diagrams the sensory-motor intercourse between organism and environment (involving D-factor) is mediated and monitored by A-factor. What are in effect feedback loops to the

G

right of the Piaget diagram, involving a running correction of motor output along with simultaneous alteration in the 'knowledge' or 'inner structure (= C-factor 'memory store'), are operated by an interaction of A- and D-factors. Efficiency in such correction may be due either to unusual acuity in perception and/or proprioception (high D-factor) or to high development of critical/comparative ability (high A). In some senses, then, the Piaget diagram amplifies and gives greater detail to some of the mechanisms involved in intelligent behaviour than does my own – though in this respect and in the one now to be mentioned they are in no way incompatible. (My hypotheses concerning operation of the four intelligence-factors in the actuation of exploratory behaviour (Chapter VI, pp. 151) may perhaps be regarded as giving an even more detailed account.)

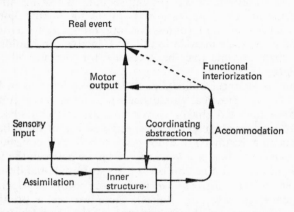

Figure 14. Diagram of Piaget's theory of knowledge (*after Furth 1969*).

The main omission from the Piaget schema, then, is the P-factor. It might be said that, in some sense, it is the *time* dimension which is omitted. It is easy for confusion to arise at this point. The most fundamental and essential feature of P-factor operation is that it delays cognitive/imaginative 'closure' on any one concept or course of action; it prolongs the period of open scanning of possible courses of action, possible relevancies. There are difficulties in thinking of this scanning as a 'process', as Ryle (1949) has pointed out, especially if it is assumed to be a 'process' which necessarily takes an appreciable period to be completed. But these difficulties can be avoided. The most basic feature of P-factor operation is describable in logical terms simply as the prevention of 'closure', to enable scanning of possibilities to be as free and as wide-ranging as possible. It must be recognized also that this scanning – or whatever it is that happens, consciously or unconsciously, whether called 'process' or called by some other name – does often appear to take time. Sometimes, perhaps, a new idea seems to appear instantaneously, and often spontaneously. Equally or more often, however, our common experience is that we 'work out' our new ideas, new behaviour strategies, over a period of time. This is not to say that we do the 'working out' *consciously* over an appreciable period, though this may sometimes be the case: but the least that can and must be said is that there is usually a time interval, sometimes a long one, between a problem's being perceived and a solution to it being found (cf. Chapter III, p. 65). The question of what 'processes' go on during this time is, at the neurophysiological level, still largely an open one. In terms of the four-factor model, the A-factor mediates a matching-up of the significant features of the given situation with whatever relevant items can be found within the C-factor 'memory store'. The two-way flow of information involves the D-factor; and the channels are held open by the operation of the P-factor.

It is this holding open of the information channels, this delay (whether appreciated as such or not) in 'closure', which Piaget appears to have taken for granted and hence neglected.

2. SEQUENTIAL APPEARANCE OF INTELLIGENCE-FACTORS IN PHYLOGENY AND ONTOGENY

Since it may by this stage be safe to compare the ontogenetic with the evolutionary accounts, it is interesting to note that there is, it seems, a rough general recapitulation of phylogeny in the ontogeny of intelligence. A thumbnail sketch of the phylogeny of intelligence (cf. Chapter VII) would run as follows:

The first factor eventually to be incorporated in intelligence, is the D- or Sensorimotor factor. This or its analogue must have appeared right at the beginning of animal life.

Second to appear were the C- and A-factors, presumably in conjunction. These are the necessary prerequisites for learning: any learned modifications of or elements within instinctive patterns above the level of the conditioned reflex must be dependent upon these factors. Their relative importance varies from group to group, but presumably not until the rise of the mammals were they able to assume a major and determinative as distinct from an auxilliary role in behaviour.

The P-factor was last to appear. In the higher mammals, notably in the anthropoidea, it becomes clearly demonstrable. Although individual great apes appear to have high P-factor endowment (individual variation in this seems to be very marked), it is difficult even with them to tie down the precise adaptive value accruing from the withholding of response thus made possible. Possibly the de-huminization hypothesis may be able, when elaborated in detail, to illuminate the issue. Whatever may be the situation with regard to the great apes, however, it seems clear that the P-factor has had, and continues to have, an immense though subtle effect on human behaviour and human evolution (cf. Chapters V and VII).

The phylogenetic sequence, then is D, followed by A and C more or less in conjunction, and lastly P.

In ontogeny, Furth (1969: 29–33) outlines three stages as distinguished by Piaget.

The first stage is called that of 'Sensory-motor operations'. It typically occupies about the first eighteen months of life. 'It is characterized by the progressive formation of the scheme of the permanent object and by the sensory-motor structuration of one's immediate spatial surroundings' (p. 29). The heavy emphasis on D- factor is obvious.

The second stage is that of 'Concrete thinking operations'. It lasts 'from the middle of the second year until the eleventh or twelfth year. It is characterized by a long process of elaboration of mental operations. The process is completed by about the age of seven and is then followed by an equally long process of structuration.' In other words, during the first part of this stage new types of organization are being attempted and mastered; while the latter part consists of a period of largely quantitative extension of these 'organizations' to cover larger and larger sweeps of experience.

'During their elaboration', says Furth (p. 30), 'concrete thought processes are irreversible. We observe how they gradually become reversible. With reversibility, they form a system of concrete operations.'

That this is an account of the coming into operation of the C- and A-factors is not superficially apparent. Some re-interpretation using the terminology of four-factor intelligence must therefore be offered.

On the face of it, 'concrete thought processes' is a contradiction in terms. The processes themselves can hardly be 'concrete' – it must be the 'items thought about' which are concrete. If 'concrete' were interpreted as 'concrete *and particular*', the 'items' of thought might then be equated with the particular items, 'this branch', 'that rock', which were argued in Chapter III as forming the primary 'raw material' of the C-factor memory store. What then could Piaget and Furth's 'reversibility' stand for? Why should 'reversibility' allow for systematization, and the lack of it presumably prevent such systematization? Here again it appears that an intelligible account can be given if we add A-factor operation to allow abstraction and generalization upon the particularities of 'concrete' first-order experience. From 'this', 'that', we generalize to 'this sort', 'that sort'; and once a level of abstraction and generality is reached, it is possible to talk of *systemic* relationships. At the 'concrete' and particular level we are limited to statements such as: '*This* is here, *that* is there. . . .' At the more abstract and general level we can say 'This rock is to the left of the fallen branch, and the bush is to the right of it.' From this we can illustrate what may be meant by 'reversibility' by saying, for example: 'The branch is to the right of the rock and to the left of the bush.' It should be noted that the use of language itself involves a considerable degree of abstraction and generality (cf. Chapter VII). The use of words like 'rock', 'branch', etc. involve the subsumption of *this* particular under the appropriate general term. But there is an appreciable jump in abstraction and hence freedom of association when relational concepts come into use. Furth talks of 'two forms of reversibility: (a) negation . . . in which a perceived change in form is cancelled by its corresponding negative thought operation; and (b) reciprocity, as expressed in the child's discovery that 'being a foreigner' is a reciprocal relationship, or that left-right, before-behind spatial relationships are relative' (pp. 30–1). Furth's 'reciprocity' is exactly the type of abstract and general relationship postulated earlier as dependent on A-factor operation.

It seems not unreasonable, then, to accept as a working assumption that the second stage of individual development involves the appearance of the C- and A-factors. Development is gradual, and is obviously dependent on the interaction of both factors. 'We can follow the genesis of thought processes which –

at about seven years of age – issues in the elementary logico-mathematical thought structures. Nevertheless, it still requires years before these structures are brought to bear on all possible concrete contents. It can be shown, for example, that the principle of invariance (constancy, conservation) is applied to the quantity of matter earlier than to weight, and to volume still later. In every case, as earlier schemes are integrated into later ones, they are altered in the process' (p. 31). That is to say, the mental schemata due to A- and C-factors affect what 'concrete' particulars are to be assimilated – but the assimilation of particulars also affects the schemata. This seems to be the nature of the 'equilibration' which is important in the systematization of knowledge at this stage. The schemata or 'thought structures' are subjected both to correspondence tests of truth through the D-factor and direct sensory and pragmatic experience, and to coherence tests for internal consistency and compatibility through the operation of the A-factor. 'Equilibrium within this system is attained at about 11 or 12 years of age. This operational structure, in turn, forms the basis of the development of formal thinking operations' (p. 31).

The third ontogenetic stage is entitled by Piaget and Furth that of 'Formal Thinking Operations'. It 'is characterized by the development of formal, abstract thought operations.' The word 'abstract' in this quotation would seem to point to the A-factor which, as we have argued above, had already assumed a major role in the previous stage (II) of intellectual development. There might appear, then, to be no new factor involved at Stage III, beyond perhaps a strengthening of the relative influence of the A-factor. If this were so, no account would have been given of the appearance of the P-factor in ontogeny – yet there are grounds, as we have seen, for saying that P-factor operation is implicit in various other theoretical formulations put forward by Piaget and Furth. Is there a discrepancy here, between their general account of intellectual functioning and their account of the ontogenetic development of intelligence? Or can the developmental account be modified to allow for P-factor intrusion?

Briefly, it seems that there is no reason to conclude that a discrepancy exists; and 'modification' of the developmental account would seem to involve nothing more radical than making explicit, in appropriate terms, what is already implicit in Furth's statement of Piaget's findings.

'In a rich cultural environment', says Furth (p. 31), 'these

(Formal Thinking) operations come to form a stable system of thought structures at about 14 or 15 years of age. . . . The adolescent is capable of forming hypotheses and of deducing possible consequences from them. This hypothetico-deductive level of thought expresses itself in linguistic formulations containing propositions and logical constructions . . . It is also evident in the way in which experiments are carried out, and proofs provided.

In the experiment on combinatorial logic, the child is presented with five bottles of colourless liquid. The first, third, and fifth bottles, combined together, will produce a brownish colour; the fourth contains a colour-reducing solution, and the second bottle is neutral. The child's problem is to find out how to produce a coloured solution. The adolescent in this third stage of development gradually discovers the combinatorial method. This method consists in *the construction of a table of all the possible combinations* and of determining the effectiveness or the ineffectiveness of each factor' (p. 32).

The significance of the italicized portion of the above paragraph (my italics) is that, in order to solve the problem, the adolescent being tested must in the initial phase of the experiment push himself right away from the bottles of liquid. He must for the moment completely ignore the physical situation. Probably it would be helpful if he were to do this physically, by going off to a different part of the room and sitting down with pencil and paper to draw up a table of possible combinations. After the scheme of action has been drawn up (in the experimenter's mind if not on paper), it is necessary to return and actually carry out the physical combinings of samples from the different bottles. But to do this prematurely, in the absence of a proper programme of operations, is almost certain to be inefficient.

The odds against random experimentation coming up with the right answer are calculable, and high. So the essential first step towards solving the problem is to dissociate entirely from its physical setting, pay no attention to the bottles of liquid, and instead work out the abstract pattern of combinations which will later be tested physically. In doing this, it is clear that the A-factor must play a major part. Before the A-factor can come into full operation at the abstract level, however, the physical situation has to be disregarded. This means more than just turning away from and not looking at the row of bottles. A mental reorientation must also take place (could take place, in fact, even without a physical dissociation): the bottles must be got rid of, temporarily at least, even from the conceptual schema, even from the experimenter's imagination. This act of lifting away from the narrowly empirical, the observable, the physical and concrete, must be ascribed to the operation of the P-factor.

Once the dominance of the immediate and observable situation has been broken by the P-factor, the A-factor is free to operate (in conjunction of course with the other factors of intelligence). But the initial act of 'abstraction' is a P-factor function. It is the same negative-seeming, dissociating, bond-breaking function which we have had to attribute to the P-factor in other contexts (see especially Chapters III and V). 'Abstraction' here must be given the meaning of 'not paying attention to'. Of the various meanings given for 'abstract' and 'abstraction' in the Shorter Oxford Dictionary, some are appropriate to describe A-factor operation, while others approximate to P-factor descriptions. 'The act of separating *in thought*' and 'The *result* of abstracting; a mere idea; something visionary' and 'Seclusion from things of sense' are given for 'abstraction' – and these all tend towards the A-factor. One meaning of the verb 'abstract' also belongs in the A-factor sub-family: 'To separate *in* mental conception; to consider apart from the concrete.' All of these meanings assume that the 'separation from the concrete' *has already occurred*. (My italics have been added to emphasize this aspect.) We can turn to another sub-family of meanings, however, in which the emphasis tends to be on the *act of separating* from the concrete. This we may take as the P-factor sub-family. It includes, for the verb 'to abstract': 'To *draw off*, disengage *from*', and 'To *withdraw oneself*, to retire from'; and for 'abstracted': 'Drawn off; separate apart from', 'Withdrawn from the contemplation of present objects', and 'Separated from the concrete'.

Thus Piaget and Furth's use of the word 'abstract' leads not only to the A-factor but also and most importantly in the present context, to the P-factor. The term 'formal' itself, used in the logical sense of form as contrasted with content or matter, serves to indicate the separation from the concrete and immediate. And Furth quite explicitly contrasts the adolescent who has achieved the Stage (III) of 'Formal Thinking Operations' with 'the child in Stage II, whose thought is still bound to the concrete here and now'. Clearly, the major feature of Stage III is that of being *not* 'bound to the concrete here and now' and equally clearly, this 'being not bound' at the behavioural level must be dependent upon some newly-developed causal factor which, as I have argued, must be the P-factor of evolutionary intelligence.

Thus a modicum of straightforward re-interpretation of Piaget's findings on the ontogeny of intelligence enables us to discern at least a broad similarity between phylogeny and ontogeny. In very general terms ontogeny does recapitulate phylogeny. The same factors appear, and in the same order. This does not, of course, prove that the evolutionary theory is true. But the general agreement between ontogeny and phylogeny does add substantially to the support for the evolutionary theory. Huxley (1963: 23) touches upon the validity of recapitulationary argument:

If we regard recapitulation as a theory implying that phylogeny in some way causes ontogeny, that adult ancestral stages are automatically pressed back into the individual development of later descendants, it is untrue. But if we use it as in a purely descriptive sense, implying no more than that ancestral plans of structure may be retained in development, and so may shed light on evolution and even reveal unexpected relationships, then it is a legitimate and useful term.

Sometimes the revelation is spectacular. Sea-squirts (Ascidians) are fixed sessile filter-feeders, with two siphons for taking in and expelling water. They lack nerve-cord and sense-organs, and were for a time classed with clams and similar molluscs. Then, in the 1870's, the Russian zoologist Kovalevsky discovered their larva – a little free-swimming creature with tail and notochord-rod (precursor of the backbone), a nerve-cord along the back, primitive hollow brain-vesicle, eye-spots, gill-clefts; its ground-plan is revealed as similar to that of an embryonic vertebrate, but with no resemblance to the ground-plan of any other group (Huxley, p. 23).

The Chordate affinities of the Ascidians are now generally accepted as factual (Kerkut 1959). As Huxley points out, these affinities can be argued for only on the basis of recapitulationary assumptions. The evidence is all indirect – the conclusion that the affinity is real depends on inference rather than observation, on a coherence rather than a correspondence test of its truth. It would appear then, that at least some weight may be accorded the recapitulationary findings of the present argument.

3. IMPLICATIONS OF A LONG 'SOCIAL DEPENDENCY' PERIOD IN JUVENILES

The parallel between ontogeny and phylogeny could never hold at the level of detail in the case of intelligence, since the originally instinctive adult animals which over many generations became intelligent had to preserve the adaptiveness of their behaviour, and hence their instinct systems, until intelligence had developed sufficiently to be able to replace instinct in the control of behaviour. (There are grounds for querying whether this has even yet fully occurred). The human infant and child, on the other hand, due to their long period of dependence on the parents – the early part of it involving an absolute dependence on the mother – have little need of the complete instinct-system found in 'instinctive' adult animals. Instincts are, of course, still present and vitally important: those of feeding, breathing and sleeping are noteworthy (see Spurway and Haldane 1953 on breathing as an instinctive activity). But a great part of the eventual behavioural repertoire of the adult is normally made up of 'learned' or experience-synthesized patterns (Ewer 1957). That these 'learned' patterns are dependent upon an instinctive basis, for motivation and for differentiation between propensities to learn some activities rather than others, will be argued in a later chapter. But it is important to notice at this time that it is the relative absence of a full instinct-system in infancy and early youth which allows the human species to rely so heavily for its adult behaviour upon learned patterns originally deriving from creative intelligence (see Chapter IX). There is no need for an elaborate system of built-in reactions in the human infant, since its mother cares for it – this means that there is no elaborate system to demolish later, hence at certain stages in its development the human child has open to it possibilities for a maximal development of intelligence. The actualization of these possibilities is dependent, as Furth (1969: 31) implies, upon the availability of 'a rich cultural environment' and upon other factors; and unfortunately in many cases the right environmental factors are lacking. The potentialities for intelligence tend to atrophy in such cases, and instead of going into adulthood with the resource of adaptive variability and of an *increasing* capacity for adaptation, the lives of many individuals show nothing more than an increasing stereotypy, a rigidity of custom and habit fully as mechanical as the most fixed patterns of instinct.

The absence of a full instinct-system in the human child would appear to have its penalties as well as its advantages. There is little in the way of built-in behaviour for the P-factor to operate upon. On the assumption that any phenotypic feature needs 'exercise' for its proper development in ontogeny, there might be some danger of the P-factor being under-developed in some human individuals – that is, of their failing to actualize their genotypic P-factor potentialities. Since the P-factor confers the power of abstention from action, the power to pause and consider, it seems possible that it, or rather its absence, may play a significant part in criminality. A relative under-development of P-factor might account for the susceptibility to temptation, the 'easily-led' nature of some individuals, even those of otherwise high intelligence. It seems possible that this 'weakness' might to some extent be remediable, through increased opportunity for the exercise of P-factor 'withholding' during childhood. The merest sketch of a nurturing policy to maximize the actualization of the 'natural' (i.e. genotypic) endowment of P-factor can be attempted here. Nevertheless it seems worth hazarding some quick indications.

Although the human child has little in the way of an instinct-system upon which to exercise his developing P-factor, it is possible to provide a substitute for instinctive patterns in the form of 'learned' habits. These are normally acquired mainly from the parents, partly from school, partly from other sources. As was suggested in Chapter V, flexibility of action and thought in relation to acquired regularities seems to demand the P-factor 'negating' function just as it does in relation to the inherited regularities of instinct. So the developing individual may use these as the 'opposing force' against which to exercise P-factor withholding. In one way, however, 'learned' patterns may be inferior to inherited. Severe and intractable problems may be generated by the attempt to break them. The reason for this is, that since the 'learned' patterns would have been acquired largely from the parents and other adults with whom the child comes into relatively intimate contact, the attempt to break the old patterns and synthesize new ones is likely to alienate and even antagonise these powerful elements – the adults – in the child's immediate environment. Few adults are happy or are even able to allow the behaviour patterns *they* have inculcated to be changed for something different. All too often, any change is automatically assumed to be a change for the bad. (For this not to be so is dependent on high P-factor development in the parents – hence a tendency for 'behavioural inheritance', balanced and 'creative' parents tending to produce balanced and 'creative' children.)

Thus the optimum conditions for the upbringing of children, for P-factor development, would be in families and schools where a firm structure of rules was imposed but where departure from the rules was not harshly penalized either physically or emotionally. 'Harsh' here is relative – the real point is that the adult must not use the child as a means to his own ego-aggrandisement. The adult must not interpret rule-breaking as an attack upon himself, to be a pretext for retaliatory *attack* under the guise of punishment. In other words, he must control his agonistic instinctive tendencies, as argued in Chapters VII and IX. But if the child is to 'learn' self-control, or, to put the issue in a non-tendentious form, if he is to *develop* the ability to be independent, self-controlled, autonomous, and intelligent, he must be given opportunities for both intellectual and social divergence, non-conformity. Dissent becomes violent when social controls are too rigid; but if they are not rigid enough, there is nothing to dissent *from*. Home and school need to provide patterns strong enough that the developing individual has something to diverge from, something against which he can if need be rebel. Over-permissiveness is like a vacuum or a swamp: the individual can struggle in it, but his struggles don't lead to anything. Specifically, he has nothing upon which to exercise his P-factor endowment; and this, as a result, is likely to become atrophied or distorted.

Following on from the foregoing line of argument, it appears possible that the age at which the transition to 'Formal Thinking Operations' occurs may be explainable in terms of considerations relating to the appearance of instinctive components in the behaviour of the developing human. It is recognized that whatever instinctive drives may be present during childhood are likely to be conducive to 'learning' from parents, parent-surrogates, elder siblings, etc.; hence that there is little likelihood of major conflict between the individual's endogenous 'drives' and the social pressures impingeing upon him. In general, the behavioural tendencies all 'flow the same way'. At puberty, however, various tendencies connected directly or indirectly with reproduction begin to manifest themselves. There arises the possibility of conflict between the developing sex and sex-connected drives and the pressures and expectations of the social

group. There is thus a relatively sudden increase in the potential
need for the control of behaviour, a need for *self*-control to
obviate or keep within bounds conflict between the adolescent
individual and his social group. There is, therefore, a corresponding sudden increase in the function required of the P-factor
of intelligence, in performing its phylogenetic role of controlling
the expression of instinctive tendencies. The instinctive drives
connected with reproduction come in at this stage, and generate
a crisis for the individual in terms of his relationship with his
social group. For the first time he finds himself with wants which
may be substantially (rather than trivially, in minor issues only)
at variance with the interests of the dominant members of the
social group. In a primitive human society, presumably even
more in a proto-human society, the post-pubertal male is in real
danger if he allows too much or indiscreet expression to his
inclination to indulge in mating behaviour with the females of
the group (even if only at the appetitive level). He may arouse
dangerous antagonism in the dominant or senior males (cf. below,
Appendix; also van Lawick-Goodall 1971: 158 ff.). Here as in so
many other situations the individual can hardly afford to leave
too much to 'learning by experience' – such learning can be
dangerous, even fatal. Even for the adolescent female the dangers
of too-early involvement in the sex-life of the adults are substantial. These dangers are not restricted to the physical hazards
of mating and pregnancy, substantial though these undoubtedly
are. But in addition there are dangers in becoming embroiled in
agonistic encounters with full adults. The adolescent of either
sex is likely to get hurt in a 'punch up' among the adults. Adult
males are particularly dangerous, due to their usually greater
size and strength; and on the basis of a differential effect of their
influence on the adolescent male and the adolescent female it is
possible to advance an hypothesis to explain alleged differences
in intellectual functioning between male and female.

The adolescent male, in his relationships with the senior males
of the social group, is, as we have argued above, under pressure
to control, withhold the expression of, the various instinctive
drives connected with reproduction. Smaller and weaker than the
adults, even flight probably will not save him if he seriously
antagonizes one of the seniors. Endogenous control – strengthened
and reinforced by 'learning', no doubt – is required, and the
mechanism of such control must as we have argued involve the
P-factor of intelligence. Therefore P-factor control must have

been under strong positive evolutionary selection-pressure in males. For the adolescent female, on the other hand, the situation is rather different. She too will be under pressure to conceal her reproductive inclinations, but only up to a point. If the interest of an adult male has been seriously aroused, it is *dis*inclination to give expression to the reproductive drives which is dangerous. The adolescent female in that situation is better to acquiesce in the reproductive or sex demands of the male. (There seem to be grounds for believing that the adolescent female may be protected in such a situation more by her reproductive physiology than by endogenous central control over the expression of instinctive drives.) It appears likely, then, that the P-factor may have been under somewhat less, and different, selection pressure in females than in males; and if this were indeed the case, it might explain an alleged tendency for women to be less abstract in their thinking, more immediate and 'intuitive'. Since abstract or 'formal' thinking appears to be dependent on P-factor, the general picture that emerges is that men, having a greater tendency towards P-factor, could be expected to be more 'formal' in their thinking than women, and to furnish a greater propertion of substantial innovators (cf. Chapter V); while women might be expected to be more emotional and impulsive, and to depend more on the D-factor in their intellectual creativity. (Support for this prediction of sex differences in cognitive style, at least as regards P-factor expressed in Piagetian 'formal thinking operations', is offered by findings of Field and Cropley 1969. It is noteworthy in the present context that, as they point out, research on children younger than the sixteen to eighteen years of their own subjects has revealed no significant sex difference.)

The comparison between the ontogeny of intelligence and its phylogeny can provide further illumination, then for such diverse issues as sex differences in intellectual 'style', the development of creativity in the individual, and the fostering of 'social responsibility' and self-control. In all of these essentially developmental issues the P-factor can be seen to assume a central role. In particular, it can be seen that the *exercise* of P-factor control is important. The full development at the phenotypic level of whatever P-factor 'potentiality' is available at the genotypic level is dependent, as with all similar actualizations of potentiality, upon the exercise of the function during the course of its development. In the simplest possible terms, the P-factor needs exercise. If too much social control of behaviour is imposed during childhood, the P-factor gets inadequate exercise, hence its development is likely to be sub-optimal. Similarly if there is too little imposition of social pressure. Either way, the P-factor level in the adult is likely to be below its potential. Two consequences of this are a lesser potentiality for creativity and/or an increased propensity towards anti-social behaviour.

Further support for the hypotheses advanced in this chapter (also in Chapters V and VII), is provided by various findings on the characteristics of proven creative scientists. Consider, for example, a passage from the section 'Reflections of the Conference Participants and the Editors' in the Utah Conferences on the Identification of Creative Scientific Talent (Taylor and Barron 1963: 385-6). I have emphasized the sections which seem especially suggestive of P-factor operation:

A highly consistent picture of the productive scientist has emerged from the researches of Roe, McClelland, Barron, Saunders, MacCurdy, Knapp, and Cattell, though the methods employed in these researches were highly varied, ranging from clinical interviews and projective techniques through empirically developed biographical inventories to factor-based tests. This or that investigator may use slightly different terms, depending upon his theoretical preferences or biases, but so consistent is the common core of observation that little is needed in the way of translation from one terminology to another. In what follows we shall try to abstract from these descriptions a single unified delineation of the productive scientist by listing the traits which are found in study after study.

1. A high degree of autonomy, self-sufficiency, self-direction.
2. A preference for mental manipulations involving things rather than people: a somewhat *distant or detached* attitude in interpersonal relations, and a preference for intellectually challenging situations rather than socially challenging ones.
3. High ego strength and *emotional stability*.
4. A liking for method, precision, exactness.
5. A preference for such defense mechanisms as *repression and isolation in dealing with affect and instinctual energies*.
6. A high degree of personal dominance but a dislike of personally toned controversy.
7. *A high degree of control of impulse, amounting almost to overcontrol: relatively little* talkativeness, gregariousness, *impulsiveness*.

8. A liking for *abstract thinking*, with *considerable tolerance of cognitive ambiguity*.
9. Marked *independence of judgment, rejection of group pressures* toward conformity in thinking.
10. Superior general intelligence.
11. An early, very broad interest in intellectual activities.
12. A drive toward comprehensiveness and elegance in explanation.
13. A special interest in the kind of 'wagering' which involves pitting oneself against uncertain circumstances in which one's own effort can be the deciding factor.

Some of these traits are descriptive of productive scientists in general, while others are especially pertinent to the appearance of 'originality specifically' in the scientist who is productive. As we have seen, productivity need not require originality, although what we have been calling creativity involves both: the creative scientist produces a high volume of unusual ideas which work effectively.

Roe (1964, pp. 68–70) who also quotes this passage, remarks of it: 'I can give you no more authoritative statement on the motivational and personality characteristics of scientists than that.'

There are other extant lines of argument suggestive of P-factor withholding of instinctive activities as a prerequisite for intellectual creativity. McClelland (1956) has shown that 'Radical Protestantism' has produced a disproportionately large number of significantly creative scientists; and he has gone on to suggest (1962) various hypotheses relating to empirical findings from personality tests which might explain the association between radical Protestantism and scientific creativity. He examines and rejects – while recognizing the considerable weight of positive evidence – the thesis that sexual repression might result in the redirection of energy into scientific creativity. He cites Roe's (1953) evidence that 'young scientists are typically not very interested in girls, date for the first time late in college, marry the first girl they date, and therefore appear to show a rather low level of heterosexual drive'. This might mean no more than that they had low reproductive drives in the first place, but it could also mean that the scientists in question had already repressed and/or redirected their reproductive drives before they came (in their late teens or early twenties) to be classifiable as 'scientists'.

In addition to the peculiarly sexual hypothesis, McClelland indicates a further possibility:

There is another problem peculiar to the male sex, and perhaps peculiarly so to males in radical Protestant households: the problem of aggression. A key characteristic of radical Protestantism is its emphasis on asceticism, on the necessity for curbing impulses early in life . . . Severe frustration of all such impulses should produce strong instigation to aggression in children. Yet the direct expression of aggression is one of the impulses most severely controlled in such families. So a conflict should often arise in the children of such families between the strong impulse to aggression and an equally strong fear of expressing it. The problem should be even more acute and more prevalent among boys than girls, since for boys controlling aggression is more of a problem because of their innately greater strength and destructive power.

The relevance of this passage to the argument of pp. 204–7 above must be clear – and one of the problems facing McClelland, that of explaining individual differences in creativity within uniformities of environment and of I.Q.-type intelligence, can be overcome with the present theory on the assumption of individual differences in P-factor potentiality. The genetic basis of P-factor is expected to vary (see Chapter I, pp. 32–3], hence there is bound to be variation in effective (i.e. phenotypic) P-factor functioning over and above, and independent of, environmentally-induced variations.

Since much of this 'environmentally-induced variation' is in fact socially-induced – and since many interactions between individual and society have been touched upon in relation to mechanisms of instinct and intelligence – it seems appropriate that the next chapter should be devoted to a more extensive exploration of the functioning of instinct and intelligence within societies.

IX

Instinct and Intelligence in Human Societies[1]

1. THE NEED FOR NEW IDEAS

It has become a commonplace of present-day discussion that every society must 'develop' if it is to retain its identity, independence, and prosperity. The term 'development' is ambiguous, and in some of its usages seems little different from various earlier but now discredited catchwords like 'progress'. Perhaps it is a sign of intellectual 'progress', however, that we do not now assume that progress and development are inevitable. They are dependent upon effort and planning. Conditions have to be manipulated before the desired 'development' will take place. Sometimes the 'developments' which do eventuate are unexpected and unwelcome. This leads to the notion of long-term planning for development, so that unpleasant surprises may be minimized. Basic to all arguments, however, are the assumptions that

---

[1] It should be emphasized that I come to this topic mainly as a biologist. I have only recently become aware (Ronald Fletcher, pers. comm. 9/1/72) of the great body of sociological theory, going back a century or more, in which the relationships between social structure and interactions, and the mechanisms of behaviour involving the individual, have been explored and discussed. As Fletcher (1968 : xviii) points out, '. . . all the major sociologists have wished to incorporate psychology in their conceptual schemes for the analysis, description, comparison, and explanation of human societies'. But 'where are sociologists to go for the "reliable" pyschology they need?' As Fletcher and others have shown (e.g. Palermo 1971), psychology has not yet achieved a unified and coherent theoretical basis: '. . . it offers a multitude of different approaches, many of which appear to bear scarcely any reference to those personal and social aspects of behaviour . . . with which sociology is concerned' (Fletcher 1968: xviii).

In the light of this, my relative ignorance of the sociological literature, especially in its historical aspects, may not be a complete handicap. True, I cannot relate my argument positively or negatively to prior literature. But if man's phylogeny means anything, it should mean that a soundly-based development of some sort is desirable, and that it is dependent, ultimately, on individual intelligence and creativity.

It seems unlikely that individual creativity could ever be entirely suppressed, no matter what the condition of the society. It is clear, on the other hand, that the form of expression which will be taken by the creative individual is very much susceptible to societal pressures. In some circumstances the individual may be encouraged to propose social and political innovations; in others this may be frowned upon, but musical and literary innovation may be welcomed and rewarded. The bare *'having'* of new ideas by the individual would, however, in terms of the mechanisms of creativity proposed in Chapter V be largely independent of particular societal conditions. This is a consequence of the nature of P-factor contrariness: a measure of adversity and antagonism is likely to stimulate rather than inhibit originality in thinking. (Perhaps the worst situation for the individual creative is a regime of fat complacency). It would appear, then, that new ideas are always available at the individual level. There will always be individuals who have new ideas, even if for one reason or another they do not tell anyone else about them. So we must assume that every society has at all times a reservoir of ideas for innovations in all or most fields – the decisive issue for the wellbeing of the society is whether or not the new ideas can be given a hearing with a view to their being implemented.

---

biological theory could be expected to have relevance within the human context.

As Tinbergen (1969: ix) argues, '. . . we should try much harder to understand the state of adaptedness and the process of evolutionary adaption, if for no other reason than that this has led to the structuring of our behaviour'. While agreeing with Crook (1970) that facile extrapolations from animal to man can be misleading and pernicious, I cannot agree that the animal-human gap is as great as he seems to imply. I cannot agree that we are irrevocably limited to 'comparison of explanatory models' across the animal-human boundary fence. As an evolutionary and field biologist, and as a theoretician, I feel bound to believe that a *single* theory or theoretical matrix, encompassing both the human and the infra-human animals, must be a possibility and must be the goal to which we strive. Otherwise, the inclusion of man in the Animal Kingdom means nothing. So, while it is needful to appreciate the differences as well as the similarities between human and animal behaviour, human and animal societies – and while straight extrapolation at the level of superficial detail, either way, is bound to be hazardous – it does seem to me that, especially at the deeper, more general levels of theorizing, zoology-ethology can offer the substance of unification to psychology, sociology, and the other social sciences.

2. MECHANISMS OF DISSEMINATION AND APPRAISAL

(a) Theoretical Possibilities

The step from idea to implementation is by far the biggest and most important one, and the most difficult. Its occurrence may necessitate reference to a measure of 'rational persuasion'; but rational persuasion by itself would scarcely suffice to obtain the extent of effective support necessary for implementation. This problem, of obtaining *enough* support, would be particularly acute with regard to any proposal whose implementation needed the support of large numbers – it must be particularly acute where decisions are made on a basis of democratic universal suffrage. Taking as a first assumption that rational persuasion by itself cannot be a sufficient means of obtaining support in at least a proportion of cases – that is to say, that the support of a large section of society is not often likely to be gained on the basis of intelligence alone, involving as it does the independent thinking-through of an issue by each member of the community – it remains to ask whether there may be other mechanisms which can supplement intelligence and provide an acceptable explanation of the actual process of idea-assimilation.

It is going to be argued that there are such mechanisms, and that an examination of their nature can provide enhanced understanding also of the processes involved in intra-societal interactions.

Malleson (1961), in his paper on 'Instinct and History', makes a case for the study of instinctual transactions in human societies: Those instincts, like hunger, thirst and sex, which have a clear consummatory component, can be studied by the methods of physiology, psychology, and ethology. One presumes, however, that there are other instincts, in particular like those concerned with dominance-submission, with group formation, and with territorialism, which have no readily recognizable consummation. It is suggested here that these may be better studied in man in the broader fields of the social anthropologist and the historian. Their manifestation is necessarily in the interrelationships between individuals; in man these interrelationships are what constitute society. In the form and the changes of society as a whole, rather than in the behaviour of individuals, instinctive elements may be clearer; on the broader canvas individual quirks may be less misleading . . . (p. 117)

We take it that each species has its endowment of separate instincts. However great the manifest plasticity, the conative-affective core of each presumably remains discrete, and in principle if not yet in practice, recognizable. For our purposes . . . we can hardly hope to recognize the influence of single instincts on a developed sequence of human behaviour. We have to contend with the fact that much behaviour, even of lower animals, is the outcome of a synchronous interplay of more than one instinct . . . It seems best to by-pass the question [of the identification of specific instincts in man] altogether, and simply refer to 'instinctive proclivities' to behave in such and such a manner, not concerning ourselves as to whether one or more separate instincts are involved (p. 118).

On the basis of comparisons between the behaviour of other higher primates and man, Malleson lists several types of behaviour for which instinctive proclivities in man might be expected:

(i) To manifest dominance-submission behaviour, and to respond to what is felt as a challenge by dominating aggressive behaviour.
(ii) To respond submissively to individuals felt to be dominant, thereby establishing in effect a dominance scale.
(iii) To engage in amicable behaviour even though, at other times, dominance-submission behaviour prevailed between the same individuals.
(iv) To form groups which encompass both dominance-submission behaviour with a dominance scale, and amicable behaviour.
(v) To cease intragroup conflict and engage in group-cohesive outwardly directed and aggressive behaviour in response to an event perceived as a challenge from outside the group.

This last point, concerning the group response to an outside challenge, will be taken up later. The first four behavioural proclivities are involved in the process of gaining support for a new idea. If we assume that the stable state of the human group is one of amicability, with a minimum of overt dominance-submission contests, it follows that fixity of the dominance scale is also the stable state. Thus the dominance scale tends to remain fixed, allowing interpersonal transactions to be conducted on a basis of amicability, unless something within or outside the group causes a disturbance of the 'peck order'.

If we adopt and adapt in this context, something similar to

Getzels' and Jackson's (1962) distinction between 'high creative' and 'high I.Q.' individuals, and if we assume that any reasonably large social group will contain some 'high I.Q.' individuals who will be able to understand and see the value in a new idea even though they are unlikely themselves to produce new ideas, we can allow the first step in the dissemination of the new idea to involve only a mechanism of intelligence. If the general social context is such that intelligent individuals tend to achieve social dominance, i.e. high positions in the peck order, at least as easily or more easily than non-intelligent individuals – and this assumes that intelligence is rewarded and/or has survival value in the society – then as soon as some of the individuals who are both intelligent (i.e. of the 'high I.Q.' type) and socially dominant have perceived the worth of the new idea, they can push it off into wider circulation on a basis of the dominance mechanisms. The spreading-power of an idea can be equated with a complex product of its utility-cum-intelligibility. If the idea can start off from a very intelligent person who is high on a dominance scale, then:

(a) Other intelligent individuals will pay attention to a new idea which is supported by a dominant individual, and some of them may come to accept and support it themselves as a result of intelligent appraisal.

(b) It can spread through the instinctive dominance-submission mechanisms: a low-ranked individual supports an idea, irrespective of whether or not he understands it, just because it is supported by the dominant individual(s) above him.

The idea can spread by a third mechanism involving both instinct and intelligence. If intelligence does have survival-reward value (as postulated), the intelligent individuals high on the dominance scales will keep watch on the spread of ideas among the people below them. They have to do this in self-defence, to maintain their own high position – because if a subordinate gets hold of a new idea which does turn out to be valuable, the subordinate is likely to climb up and reach a dominant position while the previously dominant individual is forced down below him. It is partly through the operation of this mechanism that intelligent individuals come to occupy socially dominant positions in the first place (see Packard 1957, 1959). Thus the instinctive drive to achieve or retain a position of dominance, plus an intelligent understanding of the value and functioning of new ideas within a society, can generate in the individual a motivation for 'open-mindedness' even if only in the somewhat exploitive sense of being on watch for ideological 'prey', 'competitors', or 'predators'. The main point is, that an understanding of the potential value of new ideas tends to increase receptivity towards them.

If on the other hand the dominant individuals within a social group are either unintelligent, or unappreciative of the value of new ideas, they will tend to be antagonistic or contemptuous of ideas held by individuals whom they feel to be lower in the hierarchy than themselves. New ideas originating low in the social-dominance hierarchy will tend, therefore, to spread 'upward' only very slowly; and even their 'lateral' spread is likely to be hindered by the influence of antagonistic 'dominants', especially in a society in which instinctual dominance-mechanisms tend to predominate over the individual exercise of intelligence. There is a further possibility, that dominant individuals may actively try to extirpate ideas at variance with their own, as a result of feeling threatened by them (cf. Chapter V, p. 98).

The value placed on new ideas, and the speed with which they are spread through and assimilated by the society, will tend to be increased in situations where the fifth 'instinctive proclivity' mentioned by Malleson is actuated. The balance of effect between the cessation of intragroup conflict and the intensity of outwardly directed aggressive behaviour must vary with the nature of the society and of the 'event perceived as a challenge from outside the group'. A mild to moderate challenge should in general facilitate receptivity and implementation through the diminution or cessation of intragroup conflict allowing more play to intelligent appraisal while the motivation to look for new devices and strategies would be heightened by the desire to surmount the challenge. An over-severe challenge, on the other hand, would tend towards emphasis on a direct response. Aggressive motivation would tend to block out the individual intelligence of the members of the society, thus nullifying the good effect of increased amity within the group. The structure of the society, and particularly the characteristics of its leaders at the time of the challenge, would also exert a strong effect. Leaders usually appear to have the instinctual side of their behaviour strongly developed, especially, as would be expected, the types of behaviour relating to social dominance. Intelligence is present to a much more variable degree, and may be utilized in different ways depending on the modality of operation of the individual. Some employ their intelligence directly in the manipulation of personalities, with less emphasis for the substantive policies involved. Neville Chamberlain and Eisenhower might be examples here. Others, like Roosevelt and Churchill, for example, appear to have concentrated on policies, sometimes with little regard for personalities – indeed Churchill's lengthy 'out of office' periods seem to have been due largely to his neglect of personality-manipulation and the need to placate opponents. It seems that an individual who is predominantly intelligent, even though his instinctive drives may also be strong, is likely to achieve political leadership only as the result of accident. The Presidency of Abraham Lincoln, for example, is generally acknowledged to be of significance on a world scale, because of the far-sighted humanitarianism which he manifested – yet he achieved office to some extent as a result of political accident, and due to unusual circumstances in which his intelligence could be effective despite any shortcomings on the instinctual side. (The imbalance between instinct and intelligence, unusual for a politician, is attested by reactions to his Gettysburg Address. The immediate reports of the ceremony gave most emphasis to the oration by Edward Everett, while Lincoln's Address – despite the fact that he was President – received scant attention. It was only later, when the form and substance rather than the manner of delivery became the basis of judgement, that it became recognized as a supreme statement of democratic ideals, indeed of the ideals of humanity in general.)

The reception and implementation of new ideas emanating from the creative minority is likely to be improved, then, in proportion to the intelligence and the degree of dominance of the individual leaders. This will hold in any sphere of activity, not just the political. It follows that the long-term wellbeing of any corporate enterprise – a nation, a society, a commercial or business company, a civilization – is likely to be enhanced by mechanisms which tend to select for positions of leadership those individuals who embody a desirable combination of intelligence and instinct, motivation and personal capacities. Just what the desirable combination should comprise will be investigated briefly in a later section – before this, it will be convenient to attempt an assessment of the mechanisms which operate in the transition from an idea being accepted by some of the dominant individuals of a society to its being implemented in the institutions of the 'mass' of the society. So far in this Section we have been arguing in terms of abstract probabilities based implicitly on the assumption of a behavioural continuity, indeed a strong similarity, between human and infra-human (mainly ape) behaviour. It is now time to look within human behaviour itself for direct evidence of the mechanisms and propensities which are actually operant in social interactions.

(b) Direct Evidence

Is there direct evidence of mechanisms, basically instinctual though perhaps including 'learned' modifications and elaborations, through which the individuals forming the 'mass' of a society can be brought to support some new policy or new institution? In framing the question in this way I am for the moment excluding from consideration the mechanisms of 'rational persuasion', not because they are assumed to be non-operative or unimportant, but because the possibility of the relevant social transactions being on an instinctive basis is one which has been neglected up until very recently. This possibility is, therefore, the one most in need of sympathetic scrutiny. I am assuming throughout that *some* support for innovation does result from 'rational persuasion'; and am simply choosing to concentrate upon the other type of mechanism. The question of the relative importance of the two main types of mechanism is probably best left in abeyance until much more is known about this aspect of human behaviour.

It may be convenient to initiate this phase of the inquiry by remarking that the mechanisms described by Packard in *The Hidden Persuaders* seem to involve, in most instances, a mixture of rational persuasion with the elicitation of instinctual responses. As he remarks (1962: 37) in discussing the new scientifically-based approach to advertising:

People's subsurface desires, needs, and drives were probed in order to find their points of vulnerability.

One . . . project of note was a psychiatric study of women's menstrual cycle and the emotional states which go with each stage of the cycle. The aim of the study . . . was to learn how advertising appeals could be effectively pitched to women at various stages in their cycle. At one phase (high) the woman is narcissistic, giving, loving, and outgoing. At a lower phase she is likely to need and want attention and affection given to her and have everything done for her.

For example a single ad for a ready cake mix might appeal to one woman, then in her creative mood, to try something new; then at the same time appeal to another woman whose opposite emotional needs at the moment will be best satisfied by a cake mix promising 'no work, no fuss, no bother'.

It might be objected that while an appeal to instinctual/affective mechanisms in women might be effective enough in selling cake mixes, the 'selling' of new ideas to men – in whom the obvious physiological cycle of reproductive behaviour is apparently absent – cannot be based on instinctual mechanisms. The assumption that men have no instinctual mechanisms to be played upon, by commercial advertising or by any other agency, is of course quite mistaken. Male instinctual mechanisms are different from the female ones, but they are just as obvious in their own way. Packard quotes (p. 15) an advertising executive as pointing out that 'We no longer buy oranges, we buy vitality. We do not buy just an automobile, we buy prestige'. And as has been emphasized many times, a pretty girl in a swimsuit will sell almost anything!

The mechanisms involved here are not simple, nor are they of the unvarying 'press button' type. The denial that human reactions can be instinctual at all seems often to be based on the assumption that *only* reactions which are elicited by simple and obvious physical stimuli, and which are grossly reflexive and automative, can count as instinctual. Few instinctive reactions even of, say, arthropods, are of this type. The appeal of a pretty girl on an advertisement for whisky is not *simply* a sexual one. The sexual mechanism is itself likely to be linked with mechanisms of social dominance. Since the possession of a female regarded as desirable by the other males of his social group confers status upon him, the vicarious possession suggested by the advertisement puts a man in fantasy near the top of the 'peck order'; and it is the vicarious feeling of satisfaction elicited by this which 'rubs off' onto the whisky and causes the man to buy that particular brand. The connotations of 'vitality' and 'prestige' in terms of the dominance scale are clear enough.

The most powerful instinctual mechanisms operating among the males of the human species, especially in any activity even remotely social in character, seem to be those pertaining to the dominance-submissiveness dimension. The social promulgation of new ideas is likely, then, to be dependent on dominance mechanisms. Suggestive analogues to human transactions and relationships have in recent years been discovered in the behaviour of other primates. In discussing the relationship between the males in the dominance hierarchy of a monkey colony, the Russells (1961: 143–4) describe the behaviour and demeanor of the top-ranking male (which they designate 'the overlord'):

... the overlord ... has complete freedom of movement within the colony, and perfect licence to pursue his own appetitive behaviour of any kind. He is under no kind of conflict or stress. Further, dominance in the extreme case implies the elimination not only of fighting but even of threat. A dominant animal must be ready to threaten at need ... but the occurrence of threat (always a blend of rage and fear, however little of the latter ...) is itself a sign of imperfect dominance. If dominance is complete, subordinates will never provoke the overlord to threaten ... So the mark of a completely dominant animal is a virtual absence of fear and rage, as the mark of a completely successful dominance system is a virtual absence of any trace of overt conflict between animals ...

Hence the successfully dominant animal has a characteristic appearance and stance ... He is in perfect health, which shows itself conspicuously in a perfectly groomed coat, the external mark of complete bodily comfort ... This health is a result of complete freedom to execute routine behaviour, and complete absence of stress. His posture, in motion or repose, is quite relaxed, and his movements leisurely and deliberate. He displays what we may call complete *self-assurance* ... A successful overlord, whom it would be dangerous to challenge, is recognized by his health, well-groomed look and relaxed, confident posture and movements.

The similarity to human behavioural status signals is striking. The features mentioned above may be seen in socially dominant humans, exaggerated sometimes to the point of caricature in members of a long-established 'dominant minority' or aristocracy. The drawling speech and emotionless demeanor of the 'upper-class Englishman' stereotype may be instanced – the parody has point only because there is a reality behind it. The characteristic behaviour-patterns have been ritualized and 'fossilized' into a rigidity equal to that of the most mechanical of instinctive patterns.

The 'releaser signals' of human instinct-systems, and in particular the dominance value of an appearance of self-confidence, have achieved greater importance since the spread of television and its use by politicians, educators, and advertisers. The famous Kennedy-Nixon TV confrontation in the presidential election of 1960 is a useful example of the difference in 'impact' which can be generated between a person who worked well with, and one who worked at that time against, the instinctual predilections of the audience. Nixon wore a light-coloured suit which tended to lose itself – and its wearer – against the background, while Kennedy wore a dark one which made him clearly visible. The most important difference between the two men, however, was in posture. Nixon sat tensely upright, crossed his legs awkwardly or put the soles of his feet on the floor as though ready to jump up any moment. Kennedy, by contrast, lounged back in his chair and crossed his legs comfortably, his attitude was one of calm attention, alert enough to remove any suggestion of lethargy or indifference. (Over-compensation, an *exaggerated* calmness, suggests the opposite of the behaviour exhibited; and as Tinbergen pointed out long ago (1951), yawning in humans may be indicative of high escape motivation, i.e. fear.) Thus Kennedy was projecting the social signals of dominance, Nixon those of the self-doubting challenger: the expected effects on their respective electoral support were noted by many commentators. Besides the difference in posture, it seems not entirely fanciful to ascribe some efficacy, in terms of instinctual reactions, to the difference in visibility between the two candidates on the TV screen. Nixon was 'camouflaged', 'hiding in the background', an ambiguous position between threat and flight which suggests a predatory function and/or fearfulness. The fact that Kennedy allowed himself to be clearly visible, indeed conspicuous, could again be read as a signalling of self-confidence.

The TV interview with Kennedy, at the time when in-coming electoral results first indicated decisively that he would be President, showed very strikingly the lack of overt emotion of the dominant individual. Watching and listening to Kennedy as he spoke to the nation that had just made him one of the most powerful individuals on the face of the earth, the impression was so strong as to be disturbing *either* that the position meant very little to John F. Kennedy *or* that he had been certain all along of being elected. His general demeanor was an extreme example of the 'impassivity display' postulated in Chapter VI

A great deal of popular support for ideas new and old results from the operation of dominance mechanisms. Note that 'support'

in this context does not presuppose 'understanding': of the sup-
porters of 'democracy', for example, what proportion understand
the real grounds for saying that a particular form of government
is or is not democratic? Many can trot out catch-phases like
'majority rule', 'government of the people by the people for the
people', and the like; but these phrases are themselves of almost
infinite ambiguity – that is why they are so useful to politicians –
and people support them mainly on non-rational grounds,
through what is called the emotive meaning of the terms. Thus
the word 'democracy' is used of their own forms of government
by both Russians and Americans. The connotations are similar
in the two usages, in that each group has a strong pro-attitude
(Nowell-Smith 1954) towards its own form of government; but
the denotations are very different. The denotations are in fact
the two opposed forms of government of the two nations.

It seems plausible to assume that the attitudinal or emotional
orientations, which in many instances are quite independent of
rational justification, must have been set up as a result of instinctual mechanisms. The intimate connection between emotion
and instinct has already been stressed; and although there is
considerable latitude in the objects or ideas which can be associated with particular emotions, just as there is great though not
unlimited latitude in the objects to which young goslings or other
animals can become imprinted, the very fact that there is an
emotion 'available' for association with objects or ideas is itself
the strongest possible argument for the existence and operation
of mechanisms which are basically instinctive.

It is noteworthy, finally, that the emotions played upon most
often by commercial advertising, and by all the opinion-moulding
agencies of politics, education, and social pressures in general,
are just those corresponding to the well-recognized instincts in
our nearest (as well as our more distant) phyletic relatives.
Whether one shies off using the term 'instinct' and talks only
of 'emotion', whether one makes an implicit affirmation of
phyletic continuity by calling human mechanisms 'instinctive', or
whether one hedges by calling them 'instinctoid' – the *fact* of
continuity of behavioural mechanism between ourselves and our
relatives is the only point of importance. And it is necessary to
insist on its recognition. To assert the continuity of behavioural
mechanisms between man and the 'lower' animals is not necessarily to become involved in the fallacious reductionism of saying
that human behaviour is *nothing but* animal behaviour. To assume

an absence of such continuity, on the other hand, is to brand oneself a pre-Darwinian.

There are grounds, then, for asserting that instinctual mechanisms do play an important part in group activities, among the infra-human animals, notably in aspects of our commercial and political life. The investigation of the instinctive aspects of human behaviour *as instinctive* has of course hardly begun (but see Bowlby 1951, 1952, 1969; Fletcher, 1957), so the question of direct and unequivocal evidence of it is bound to involve reinterpretation of findings from differently-oriented studies. In view of this it may be appropriate, for the final part of this section, to revert to some recent discoveries relating to the transmission of behavioural novelties among other primates.

The influence, upon the effectiveness of spread of behavioural innovations, of the rank in the social-dominance hierarchy of the initiating individual, postulated in the previous section, is confirmed in startling clarity by various findings of Kawamura and his colleagues at the Japanese Monkey Centre (see Kawamura 1959). Each colony of the species studied (the Japanese Macaque) develops its own characteristic 'food culture', with its own special selection of foods from the total range of species, its own peculiar methods of preparation of material, and so on. Against the constancy of the colony-specific food-getting patterns, any innovation stands out. The initiation and spread of several new feeding or food-getting patterns has actually been observed. A particular immature female, for example, in 1953 started carrying sweet-potatoes to the nearest water-source and washing off the soil and grit adhering to them. The new habit spread to the other juveniles, and from them to the group of mothers. Thereafter it was transmitted by the mothers to each new generation. The then reigning adult males, however, did not acquire the new habit, mainly, it would seem, because they did not participate in caring for the juveniles in the colony in question, hence would have had little opportunity for the relatively intimate contact by means of which the 'socially upward' transmission might have been effected. (No doubt the males of later generations as they became dominant would retain the habit they had acquired in youth.) In another colony, by contrast, an 'upward' transmission occurred from juveniles to adult males. The new habit involved the eating of candy (provided by human visitors), and transmission to the adult males was facilitated by the fact that in this colony the males did participate in caring for the juveniles.

The candy-eating innovation was slow to spread. Only 51 per cent of the group had accepted it after one and a half years. In marked contrast to this, a wheat-eating innovation spread right through a different group in only four hours! The innovator in this case was one of the adult males, from whom the new habit spread rapidly to the dominant male or 'overlord' of the colony, then to the dominant female, her family, the other females and the rest of the group. In this case the innovation could be said to have spread *with* a 'mimetic gradient' within the society, whereas in the potato-washing and candy-eating innovations the transmission had to be *against* this gradient.

Mead (1956) shows, within a relatively simple human society (still immeasurably more complex than any infra-human social group), both the inherent conservatism and the individual contribution to mechanisms of social change. While it is not altogether clear whether the effective innovator-leader was effective because of prior high status, or whether he achieved high status because of his innovative and other powers, it does seem clear that he possessed and used mechanisms of social dominance that were, in themselves, instinctive, besides having the creative intelligence to conceive the new policies.

Consideration of these findings brings out very clearly how a hierarchical social structure, plus the tendency of the individuals to form relatively fixed habits, sets up an inherent conservatism within a society. Exploration and innovation are functions of youth. With increasing age, the individual's variability of behaviour tends to decrease, while habitual activities take over a larger and larger part of his total day-to-day repertoire. Since social rank tends, in general and up to a point, to correlate with age, it follows that the most influential members of a social hierarchy tend to be those who are also most fixed and habitual in their behaviour. Two important theoretical consequences stem from this, which will be investigated in the next section.

3. DESIDERATA FOR SOCIETAL CREATIVITY

The combination of societal and individual characteristics tending towards conservatism, adduced above in relation to monkey societies, can be observed also in the vastly larger and more complex human societies. The loss of 'creativity', diminution in the ability of a society to respond to challenges, which for Toynbee (1947) is the basic cause of the 'breakdown' of a civilization, may thus be ascribed, in part, to the actualization of these tendencies. It has been suggested that the 'vitality' of a nation may be gauged by the age of its leaders. The English-speaking nations were led through the period of rapid political and social change round the turn of the eighteenth and nineteenth centuries – the century from 1750 to 1850 saw an enormous variety of changes in different spheres of activity – largely by men in their thirties and forties. Present leaders, on the other hand, are with few exceptions men in their fifties to early seventies. John F. Kennedy, for example, is one of the notable exceptions in the present era; and in terms of the thesis in question it is possible to argue that the fact of his election may be an indication of sustained or renewed 'power to respond' in the United States. (Whether his assassination might be taken as an argument in the opposite direction is, presumably, equally open to speculation.)

There is a considerable volume of evidence to support the view that in quiet times the age of leaders, whether of nations or of institutions, does tend to increase. Not only that, but the individual leaders tend more to orthodoxy, even to a stereotyped conservatism. In times of crisis and challenge, on the other hand, leaders tend to be younger and/or more heterodox, more individualistic. Whether this is enough to constitute a form of natural selection, in psychosocial if not in genetic terms, is difficult to say, but there would appear to be a clear analogy, in terms of intermittency of demand for adaptation, between the processes operating in human societies and those operating to effect natural selection in the societies of infra-human animals. (Natural selection is reduced during an 'up' swing of the population graph, but operates with extra severity – relative to equilibrium conditions – during the 'down' swing. In fact, it is the differences in selection pressures which cause the population changes. The results, on the other hand, are increased individual variation during the 'up' phases, and reduced variation through the population during the 'down'. For a present-day human example, see the beginning of Chapter X.)

The question of the 'youth and/or heterodoxy' of leaders during times of crisis or challenge is susceptible to further analysis, and, if posed in terms of the personality characteristics (notably those of the factors of evolutionary intelligence) present in the leaders, may lead to the elucidation of several specific desiderata.

Under present conditions of ever-increasing complexity of social organization it seems unlikely that leaders of such extreme youth as at some periods in the past could operate effectively. Leaders in their twenties may well be adequate, may indeed be ideal, for leading a relatively simple society through a relatively simple and straightforward challenge such as external aggression. Pitt rallied Britain, and through Britain the rest of Europe, against the aggression of France under Napoleon. He did more, of course, than merely resist the threat of Napoleon. The fiscal and constitutional changes made under his aegis were certainly of great significance but they were not exclusively his, neither were they all unequivocally beneficial in the long run. He seems to have been more an opportunist than is usually admitted. In fact his startling power seems to have derived from the extent to which he could call upon the abstract intellectual resources of the preceding century and apply them to the immediate political exigencies of his own time and circumstances. He was, perhaps, the paradigm of the interpreter-implementor of other men's creativity, and to a considerable extent, then, a creator in his own right. His gift was to see that an intelligently-conceived policy, consistently pursued but intelligently modified when necessary, is the best strategy for the leaders of a democracy; and to have the intelligence to develop a reasonable policy and the fortitude to carry it out, when necessary in the face of strong opposition.

But it seems most unlikely that national leaders can emerge now at the age of twenty-eight. This is perhaps less a matter of there being more to learn than it is a matter of change in the mechanisms of actual (as opposed to nominal) selection. A highly intelligent individual might still be able to formulate an adaptive response-policy by this age – the difficulty would be, to publicize his ideas and convince a sufficiently influential proportion of the Establishment to support them. This might generally be expected to take at least twenty years. The time might be shortened for an individual born into an influential pressure-group, or who inherited great wealth or a respected name, but even if this were the case, success would still depend on the possession of unusual personal qualities (as it did with John F. Kennedy, who satisfies the other 'contextual' criteria).

Creatively conceiving a new response-strategy is one thing; gaining its acceptance, even if only by a pressure-group, is something else again. For the latter, consistency is the pertinent characteristic. A policy has to be consistently held and advocated, perhaps over a long period of time, before it can have any hope of gaining support. The consistent advocacy of a policy demands in

H

turn a consistency or firmness of character, and the self-confidence to persist in the face of indifference and attack; both of which will inevitably be experienced, successively, and often before any appreciable measure of support has been obtained.

The creative formulation of a response-strategy, on the other hand, sets up other demands on the individual. Since the challenge is as yet unsurmounted, the problem as yet unsolved, it follows that the extant accepted thinking on the subject must be either inadequate or mistaken, otherwise there would be no need for a *new* response-strategy. If current views have to be known to, understood by, but rejected by the potential leader, it follows that he must have high intelligence in the ordinary 'I.Q.' sense – and current views must be understood very thoroughly, otherwise they could not be contraverted when the new strategy has to be argued through to acceptance – it also follows that he must have a strong 'negating' capacity, the power to withhold assent to the persuasiveness of commonly-held belief. Since an appreciable part of this 'persuasiveness' would appear to involve the group-cohesive social instincts, here again we are faced with the instinct-resisting function of the P-factor of evolutionary intelligence. The tendency of the individual to conform to the social group and the group's established beliefs must be countered. All sorts of acquired or environmentally-induced characteristics of personality may contribute to this – but beneath them all must be some intrinsic basic capacity for holding off the responses and the ideas to which the individual himself must previously have adhered. This 'basic capacity' must be regarded as just another way of describing the Power of Abstention or P-factor.

Apart from its function in the creative formulation of new response-strategies, the P-factor must be necessary to the would-be leader in another way, at the stage where his ideas, already made known, have to be advocated and sustained under the attacks of both conservatives and competing innovators. It is desirable, from the point of view of gaining ultimate acceptance and implementation of the new response-strategy, that certain instinctive tendencies activated in the course of controversy should be given only controlled, or even little or no, expression. The fierce irrationality of public assault on the new idea must not be allowed to elicit either of the extremes, of retreat, or retaliation in kind. The essentially anti-rational devices used by the neophobic defenders of any established 'faith' are well-enough known. Appeals to emotion, logical sleight-of-hand, *argumentum ad hominem* – the last, the attack on the innovator rather than his innovation, is perhaps the most obvious danger (though not necessarily in the long run the greatest). The innovator may be scared off, may abjure his advocacy; at the other extreme he may be provoked into retaliation by the agonistic activities of his opponents. Retreat is obviously fatal to the new response-strategy. But violent polemics, denunciation of conservatives and competitors, may be just as bad. Extensive resistance may be provoked. A public altercation can on occasion mean good publicity. The innovator's counter-attack can, however, if he hits out too freely, lead to the permanent alienation of large sections of the community. So it pays the innovator to be selective in his instinctive responses in agonistic situations. Which again entails P-factor control . . .

The need for such control is enhanced by the desirability that the advocate of a new response-policy should not be totally devoid of aggression. Too much sweet reasonableness can be a handicap. People tend to be suspicious of any extreme departure from usual behaviour-patterns. They expect an attack to provoke a counter-attack. Hence the desirability of reasonably strong instinctive forces within the innovator, and of his giving a certain amount of expression to them. But not too much. The requirements here would seem to parallel the requirements for effectiveness in teaching discussed in Chapter X, with relatively high levels both of instinctive drive and of P-factor being desirable.

Thus the features desirable in the leader of a democratic society include high general intelligence and high levels of both instinctive drive and P-factor control. It must be emphasized that these are necessary but not sufficient conditions for good leadership. An individual might possess the right attributes but wish to use them for other purposes, e.g. simply to domineer over his fellows, rather than lead them constructively. The direction of motivation, the basic attitudes and values of the individual, are of decisive importance for the moral or the existential quality of the leadership. The outline of the intelligence-instinct complex given above relates rather to the *efficiency* of leadership in relation to the need to meet novel challenges with novel responses. It is a truism that what is new is not necessarily good: an innovation is not necessarily an improvement. But, in a democratic society the role of would-be leaders is expected to comprise only the proffering and advocacy of new response-policies – the decision as to which policy to select and implement is supposed to be taken by the society as a whole (or at least by substantial minority groups within it). The decision as to the *direction* of the society's response, and its moral/existential status, depends not on the leaders but on the society as a whole. This is a function, therefore, of the general level of effective intelligence within the society, and of the social mechanisms that have been set up for improving or degrading the general intelligence and for facilitating or hindering its effective use. Legal and political institutions are important here; but more important, probably, is the nature of the education system and the manner in which it actually operates. Some aspects of this will be discussed in Chapter X. It will be appropriate to conclude the present chapter with a discussion of the desiderata for inter-institutional relationships, and the concomitant implications for individuals, within societies.

4. POLICIES FOR SOCIETAL RESPONSIVENESS AND
   SURVIVAL

The key to the future, it has been said, is an understanding of the past. This chapter has been an attempt to understand the past and present in terms of interactions, within societies, between mechanisms of instinct and intelligence. A real and detailed continuity has been indicated between human and animal social interactions. Such continuity is retrospective. It relates the present with the past. But if from this we want to seek a 'key to the future', what form can it take? How are we to know what we are looking for?

The same basic feature of adaptive variability would seem, *prima facie*, to be an appropriate general guide for the future as it has been, as we have demonstrated, in the past. The objection may be made, that it is too general to be useful; and it might be argued that its use as a guideline for the future would commit our species to mere orthogenesis. Both of these objections can be countered in one argument. The essential premiss is that 'adaptive variability' *is* general. In a sense it does not form a guideline or policy, but rather a criterion upon which guidelines and policies can be evaluated. A policy which seems likely to achieve a specific objective and to retain or enhance the possibilities for subsequent adaptive variation is a good policy; whereas one which achieves a specific objective at the cost of reduction or loss of scope for later variation is a bad one. There can be no sure-fire recipe for making these evaluations. The only thing we can do is to exercise intelligence, individually and collectively. Here the doubting, negating, critical function of the P-factor can be most important, in leading to self-criticism and to a continuing reappraisal of social and political institutions. For it is through our institutions as emanations of our underlying attitudes that intelligence must work and through which it may be stifled, diverted, or stultified.

The first object of critical and intelligent scrutiny must therefore be our institutions.

By exposing through discussion of institutions the underlying attitudes, themselves not directly available to therapeutic reappraisal by reason of their being largely unconscious, it is possible to effect adaptive changes in the attitudes themselves (cf. Stenhouse 1968, 1972). This mechanism of indirect adaptation is exactly parallel to the mechanisms of scientific progress by the successive explication, testing, and modification of 'ideals of natural order' (Toulmin 1961) or of 'paradigms' (Kuhn 1962). The general requisite is that institutions should facilitate rather than hinder adaptive variability. It is not just the adaptive variability of individual behaviour which is in question – for it might be objected, with justice, that relatively few institutions attempt to restrict the variability of individual behaviour except in a narrow range of specific situations – but the range of variability of behaviour between individuals and even between institutions themselves. The appropriate model for understanding the needs for future human evolution is, paradoxically obviously, evolutionary theory itself (cf. Pringle 1951).

The central concepts of evolutionary theory, those of individual variation and of competitition, can be applied to institutions just as to individuals. It must be pointed out that natural selection does apply to human institutions and individuals, whether we like it or not and whether we are even aware of it or not. There are absolutely no grounds for assuming, as some seem implicitly to do, that human evolution ceased with the advent of civilized man. This is an anthropocentrism as ludicrous as that of the Ptolemaic astronomers – who are often ridiculed so patronizingly by the very individuals who themselves exemplify the same fallacy! That it is difficult to detect any genetic changes that may have occurred within recorded history proves nothing (but see Darlington 1969). In the first place they may well have taken place without being detected; and in the second, the few thousand years of historical time is so short, in the evolutionary scale, that few changes in the species *Homo sapiens* as a whole could be expected anyway. On the other hand, natural selection can reasonably be expected to operate, in this psychosocial phase of human evolution (Huxley 1964), not so directly upon ordinary phenotypic characteristics of individuals but rather through the epi-phenotypic characteristics of culture and institutions. Only at second remove will the individual phenotypes be affected (and the characteristics affected are likely to be psychological and temperamental rather than physical); while genotypes will be modified only at third remove.

Of the action of natural selection upon human cultures and institutions we have abundant evidence – this is the subject-matter of history. . . .

It has been alleged that speculation about 'human survival' is futile, since extinction is unlikely: remote possibilities are compared to 'pie in the sky'. But even if it were admitted that the extinction of the human species is so remote and/or so much outside the control of the man in the street as to be not worth discussing – and this, in the nuclear age, is far from being the case – there still remain questions of possible extinction and survival which are not at all remote and not obviously utterly intractable. For many nations, for example, the question of their own political, cultural, and/or economic survival is all too uneasily non-remote. Distinguishable though not always separable from questions of national survival there are questions, important to those directly involved, of the survival of specific institutions, formal or informal, within national societies. The realization that extinction of these is possible, that their progress and survival are not inevitable, that behind the demise of a particular institution may come the extinction of a national society or even, as a real possibility nowadays through one cause of another, the extinction of the human species itself, cannot but have a salutary effect on complacency. We can come to see the necessity of lifting ourselves, however painfully, out of the trenches of habit and up into the trackless plain over which, by intelligence, we have to navigate and lay down new roads. Basic to this is the need to set up institutions which, like Socrates' gadfly, will provide the irritation and worry to keep us intelligently awake and prevent our sinking back into the rigid sleep of habit and instinct.

Which brings us to the desirability of building competitive

relationships into our institutions, so that intelligence may be stimulated and so that selection mechanisms may be provided to enhance its effectiveness.

If this is done, on the evolutionary model suggested earlier, we shall in fact be creating not an analogy of evolution, but a homology within it. We shall be giving recognition, implicit or explicit, to a fact of history: namely, that institutions *are* in competition with each other. The objection is sure to be raised, that much now-abandoned nineteenth-century social, economic, and political theory was justified (or rationalized) by reference to the Darwinian theories. The 'struggle for survival' was invoked to bolster the respectability of *laissez-faire* and 'the survival of the fittest'. This simplistic attempt to apply evolutionary theory within human affairs has very rightly been discredited. But just because a premature and misconceived attempt to integrate scientific with socio-political theories has been proved wrong, it does not follow that the several bodies of theory are and must remain for ever discrete and isolated (see MacRae 1958). With the enormous development of the social sciences in the twentieth century the demand must eventually be made, and met, to show their continuity with the 'basic' and in particular the biological sciences. Unless this can be done, and discrepancies explained, the claim of the social sciences to be justified in calling themselves sciences must be held to be doubtful. The dilemma posed in the Introduction cannot ultimately be ignored.

The means towards effecting a new *rapprochement* have in fact been provided, also in the twentieth century, by numerous advances in the biological sciences. Our understanding of the mechanisms of evolution has been amplified out of all recognition. While Darwin's own theories have remained unshaken in the sense that little of substance has had to be negated, they have been particularized and extended so enormously that facile and misleading extrapolations from them, while not totally prevented, are rendered openly implausible to everyone with a reasonable acquaintance with modern biology (see Eiseley 1958; Barnett 1858c).

To illustrate the changes that have taken place, consider the interpretations which can be given to the notion of 'competition' in an evolutionary context. In the nineteenth century the 'struggle for survival' could be, and was, interpreted as involving actual physical combat. Nowadays, thanks to the findings of half a century of research in ecology, ethology, population genetics and other fields, we know that not only is the true state of affairs immensely more complicated, but furthermore, that the 'physical fighting' picture of the struggle for survival is factually incorrect. Even intraspecific fighting in most animals under natural conditions is rare – except for a limited amount of fighting over food during times of shortage (see Gibb 1954), and for fighting associated with reproduction. The latter is not necessarily insignificant *as* fighting just because it occurs normally only for a short period of the year. It is limited in time, however, and thus in 'quantity'; and even more important, its effects are mitigated by a variety of devices and mechanisms which may be summed up under the headings of territoriality and ritualization. As Lorenz (1966) has emphasized, under natural conditions few fights are to the death. Even substantial physical damage to combatants is relatively rare. The loser retreats, or else placates the victor by the use of an appeasement display involving ritualization. Displays are used also in the 'fighting' through which territories are set up; and once territories have been established, the amount even of display is reduced to a minimum. In fact, if the agonistic behaviour of competing infra-human animals is to be used as justification or taken as a model for anything at the human level, it would have to be for bluff, for 'talking big' and putting on a show of force, rather than for the direct use of force in physical combat or coercion between individuals.

Similarly, in the concept of 'survival of the fittest', the nineteenth-century notion that this justified the use of force and unscrupulous belligerency, can now be seen to be quite untenable. That 'fitness' should be equated with large size and physical strength in the individual is now seen as a grotesque distortion. 'Fitness' in infra-human animals might just as well be equated with fecundity; though in fact, in the present *human* situation fecundity is coming to be regarded as just as much of a menace as is aggressiveness. Either could lead to extinction. What is really required for the human species is what has been found adaptive in most other species: a compromise mixture of a variety of traits (see Dobzhansky 1956), including many that are the opposite of those prized by the nineteenth-century simplistic social theorists. Cowardice rather than courage, in many though probably not all situations; small size for going round and between obstructions, rather than large size for bulldozing through them; amicable co-operativeness rather than bullying social dominance – these are the pictures that emerge from the twentieth-century extension of the Darwin-Wallace theories (cf. Leach 1967). And because mistaken and misapplied abstractions from the early theories proved inadequate as a basis for socio-politico-economic models – they were inadequate in biology too, as has been indicated – there is no reason to assume that a more sophisticated reintegration will be unsuccessful. In fact, there are many indications that a return to biological models and concepts has already started in many fields (see Kuhn 1962; Toulmin and Goodfield 1965).

Medawar, in his influential *The Future of Man* (1960), critizes the argument from biological to cultural (i.e. human psychosocial, Huxley 1964) evolution. Although he stresses (p. 97) that '*both* are biological', he suggests that 'by calling attention to the similarities, which are not profound, we may forget the *differences* between our two styles of heredity and evolution; and the differences between them are indeed profound' (p. 98). His argument is based upon Lederberg's (1958) distinction between 'elective' and 'instructive' relationships between the environment and the behaviour of an organism. An 'elective' stimulus (e.g. the releaser of an instinctive sequence) merely selects from a range of already-prepared responses on an all-or-nothing basis. The response cannot be adapted in detail. An 'instructive' stimulus, on the other hand, causes a detailed adaptation of the response to the situation in which it is elicited. Medawar points out that biological evolution (Darwinian) is 'elective', in that natural selection either accepts or rejects the individual variations 'offered up' to it, without being able to influence the nature of the mutations and resortings which determine individual variation. He goes on to argue that cultural (=psychosocial) evolution, by contrast, is Lamarckian and 'instructive': acquired cultural features are passed on; and they have been acquired as the result of 'learning', hence are to be regarded as responses to 'instructive' stimuli. Medawar says that 'the higher parts of the brain respond to instructive stimuli: we *learn*'.

It may be questioned, in passing, whether an unequivocal definition of 'learning' would necessitate that the stimuli for it should always be classifiable as 'instructive' (but see Chapter XI for further discussion of 'learning'). The main objection to be made at present however, is against the two apparent assumptions:

(*a*) That the two categories or types of evolution, Darwinian-elective and Lamarckian-instructive, are mutually exclusive; and

(b) That cultural or psychosocial evolution is to be regarded as exclusively of the Lamarckian-instructive type.

It seems to be not in the least impossible that a particular cultural response could arise by a stimulus or challenge being taken 'instructively', and that it could be passed on by 'Lamarckian' transmission for a certain period. After a while it could well prove to be dysgenic, however, and could be abandoned. In such a case a Darwinian mechanism would have supplanted and over-ridden the Lamarckian ones, and an 'elective decision' would have been made *against* the particular cultural feature in question. The cultural institution of slavery, for example, while for millenia a virtually universal feature of human societies, has in the last two centuries become dysgenic – and has been recognized as such – in most 'advanced' societies. It has been eradicated from these societies, despite the fact that it did on occasion have its good features, e.g. the formation of sometimes large associations of unrelated individuals for mutual benefit. Thus although some owners made of their slaves an 'instructive' stimulus for a benevolent paternalistic response, the institution as a whole has been 'electively' rejected. Again, modern forms of democratic government, although largely formed of 'instructed' responses in the first place, have been adopted 'electively', the disadvantages along with the advantages, in many parts of the world – and it is possible they may be abandoned in the same way. Yet again, hereditary autocracy has been decisively rejected as a form of government, after several millenia of 'experiment'.

Many other examples could be cited from human evolution, to show that while Lamarckian-instructive mechanisms are indeed important, they operate only within a wider framework of Darwinian-elective mechanisms. In short, cultural or psychosocial evolution has not replaced biological evolution, it has merely been added to it. And although our most immediate concerns are with the Lamarckian-instructive level, we cannot on that account afford to ignore the more fundamental and slow-moving processes of Darwinian-elective evolution. This is perhaps easier for us to appreciate now, with the population explosion and attendant pollution and other problems becoming ever-more-urgent realities, than it was ten years ago for Medawar. We must try to be 'instructed' – but we may find ourselves 'elected', one way or the other. Appreciation of this may generate the motivation to try harder for an 'instructed' response.

(It may be pointed out that in addition to questioning the adequacy of 'learning' as a conceptual implement for building an explanation of Lamarckian-instructive mechanisms, this chapter has already pointed out that the transmission of the acquired characteristics is to a considerable extent effected by instinctive mechanisms – which further blurs Medawar's over-sharp distinction.)

To take an evolutionary model as a paradigm for a societal/institutional blueprint is, therefore, not to be discredited by the fact that evolutionary models have proved inadequate in the past. In particular the notion of competition is hugely modified by the acceptance of the idea that the struggle need not involve outright conflict *and need not be 'to the death'*. Competition between institutions (and between the individuals within institutions) need not result in one supplanting and destroying the other. The 'vanquished' can 'run away and live' – but not to 'fight again another day' – by altering its function so as to reduce the overlap of function, and hence reduce the competition. This is analogous with the evolutionary process of niche differentiation (see Lack 1954). Alternatively or as a supplement to this, the less efficient institution or individual can simply fly appeasement signals and allow outward dominance to be vested in the 'victor'. The 'subordinate' can still perform the very important function, however, of providing competition to keep the 'dominant' up to scratch. If it is recognized that the peaceful co-existence of a variety of partly-competing institutions provides a reservoir of 'function availability', then each institution can achieve justification by performing well a number of functions, rather than by being the sole performer of one or a few. The situation envisaged is analogous to the built-in redundancy of function which appears to be such an important principle of brain organization (see Ashby 1960). The brain copes with the multiplicity of demands upon it, makes provision against injury or malfunctioning of some of its own components, and provides a reservoir of 'open-endedness' in the form of unutilized potentialities for action, all by having more 'circuits' available than are strictly necessary for the day-to-day demands of ordinary life.

Another analogue of a human society is the ecosystem. As Elton (1958) has shown, the stability and adaptability of an ecosystem is a function of its complexity. In a 'rich' ecosystem consisting of a large number of species, there are many contestants available to fill any niche that should happen to become

236 THE EVOLUTION OF INTELLIGENCE

vacant. Conversely, any invading organism is likely to strike very severe competition for at least some of its life requirements. Thus the balance of the system is unlikely to be much affected by either extinctions or invasions. Furthermore, even gross changes in the physical environment are more likely to be met by organisms that can adapt to them, in a 'rich' than in a 'poor' ecosystem. In human societies, by analogy, the questioning and innovating activities of high-P individuals are less likely to be thwarted if each of them has available a diversity of institutions and associations through which to work.

Interchangeability of function within a society, or total versatility, is possible only when excess energy, time, etc. are available over and above what is necessary for routine maintenance. This depends, therefore, on the prior achievement of efficiency at the maintenance functions – but over-emphasis on maintenance can reach the extreme of conservatism, where maintenance becomes the only consideration and there is antagonism towards innovation. Both the 'maintenance' efficiency of an institution and its capacity to alter and to innovate depend heavily indeed upon the attitudes and skills of the individuals within it. Attitudes are much more important than skills, since a skill is relatively easily acquired by anyone who has a pro-attitude towards learning it. The inculcation of a particular skill, on the other hand, may tend to draw with it an over-valuing of *that* particular skill and a concomitant dis-valuing of other skills and, more importantly, dis-valuation of generalized critical and creative thinking. Thus a great deal depends on the actual functioning of the educational institutions within a society. They can promote attitudes in favour of general critico-creative thinking, of diversification and variation within a society, and of innovation and non-conformity. This will tend towards adaptability, versatility, and strength within the society. They can on the other hand emphasize specific skills, technical expertise, to the detriment of long-range general thinking. This will tend towards rigidity – and though a quick return may be obtained in the form of immediate technical and technological efficiency, this is likely to be paid for in the long run by inability to cope with altered conditions and new challenges.

Thus the need for adaptability in institutions points to a need for overlapping of functions and intentional 'redundancy', and this in turn to the demand that each individual institution have a variety of different functions. Narrow specialization in institutions has the same limitations as it has in individuals. Narrowness, in a context of change, tends towards increased probability of frustration, thence to pessimism, then either to negativism and apathy or else to desperate extremism and violence. Institutions and individuals having a diversity of competencies, on the other hand, can easily shift the emphasis of their endeavours and thus attain the satisfaction of progress even if not that of final and total achievement. But for an institution to 'shift its emphasis' it is necessary that individuals within it resist and break the institutional 'habit' of perseveration – and to be able to do this they must have retained the effectiveness of their P-factor 'habit-breaking' capacity.

In brief, then, the foregoing argument suggests that human adaptation (hence survival) at the present time and in the foreseeable future depends upon the retention and enhancement of individual and collective adaptability. This has always been the prime hominid characteristic. The analysis of its behavioural prerequisites (Chapter III) revealed the importance of a factor, hitherto neglected in theoretical investigations of this area, which must have been necessary to enable our ancestors to break free from the previously-standard instinctive behaviour patterns. This factor, the P-factor, can be detected by various means in human beings at the present time (Chapters IV, V, VII and VIII). Although it still has a significant function in the withholding or control of instinctive responses and proclivities, possibly its chief importance, now and certainly in the future, is in breaking the dominance of acquired (i.e. 'learned') patterns of action, belief, and attitude. Only by loosening the force of habit can we be versatile and adaptable. Through our intelligence we have made civilizations, and a culture in which science and technology are causing an accelerated rate of change in our circumstances. We depend upon our intelligence, and especially on the P-factor within it, to enable us to increase our response-rate to match the rate of change we have generated in our man-made and man-altered environment.

What effect is exerted upon this aspect of intelligence, at the present time, by our educational endeavours both formal and informal?

X

Biological Intelligence and Education

1. INTRODUCTION: THE FOUR FACTORS AND TRADITIONAL EDUCATION

Perhaps the most effective way to set the stage for a brief examination of the relationship between evolutionary intelligence and education is to present a thumb-nail sketch of the history of Western education in terms of the four factors.

Formal education has always leaned heavily upon the C-factor: children are always expected to remember what they have been taught. Often, formal education has involved virtually nothing but the memorizing of information. In proportion as the information is without significance to the child, the A-factor is neglected. This is not to say that memorizing has always been, or need always be, a bad thing. If what is memorized will prove beneficial in later life, it is better to learn than not. And it must be remembered that some types of information which are meaningful to adults, and which it is desirable for adults to possess, are unlikely to be meaningful to children due to limitations in their experience. Yet childhood may be the best, perhaps virtually the only, time to acquire such information. Thus there may be an inherent conflict between two assumptions of modern education, viz. that the child should be given what he will need as an adult, and that everything given should be meaningful to him at the time.

That education should be meaningful at the time is, in fact, a demand that education should involve the A-factor of intelligence. But this involvement can come about, it should be noted, in two ways. Under the influence of Dewey and others, it has been accepted that the child should be able to see relevancies, similarities, etc. between the subject-matter of formal education and his various other activities and interests. Curricula are thus designed to ensure coherence between formal and informal education. Similarities and relevancies can be sought not only between the major divisions of formal and informal education, however, but within formal education itself – and it is important that they should be found there, especially when a diversity of subject-areas is being presented and when many of them are not related in an obvious and direct way to the child's (or, more urgently nowadays, the young adult's) other interests. Thus A-factor functioning can be encouraged within formal education, provided that educators are prepared to foster it.

Sensorimotor development involving D-factor was emphasized in the Hellenistic ideal of 'mens sana in corpore sano'. Its value is implicitly acknowledged also in the inclusion of sports and physical education, art, music, handcrafts, woodwork and other 'trade' experiences, within the curricula of general education. Extensive interaction with major natural phenomena, as in sailing, mountaineering, camping, etc., is clearly important, too, even though this may not come within the range of formal education.

It is the P-factor, the Power of Abstention, which appears to have been largely overlooked in the practice, as it has in the theory, of education. (Its neglect in education and psychology is an interesting example of the need for conceptual analysis in the empirical and practical sciences. In the absence of an adequate investigation of the logic of 'innovation', 'creativity', and similar concepts, the observables of behaviour which point to the P-factor have been misinterpreted. They have been *observed*, in a minimal sense – witness the passages quoted, for example in Chapters V, VI and VII – but their significance has not fully been appreciated) (see also Stenhouse 1971.)

How *does* the P-factor manifest itself in formal education? How do traditional and current practices impinge upon it? And what changes, if any, should be made in order to optimize P-factor utilization by individual, society, and the human species in general?

The initiating function of P-factor doubt in creativity has been mentioned in Chapter V. Traditional education has had little room for creativity. It has not necessarily been totally inimical to the P-factor, however. The emotions have had to be controlled. Drives towards physical activity, towards 'play', towards spontaneous exploration, and towards social dominance contests with teaching personnel, have had to be controlled. This has often led, no doubt, to a variety of bad effects, but among the bad effects it would seem that there must have been one good one, namely 'selection pressures' upon individual development in favour of enhanced P-factor control.

Also good has been the fact that these 'pressures' have been intermittent. While educational goals were circumscribed, pupils could adopt strategies which would achieve the goals expected of them – the passing of the examinations, adherence to the laid-down rules of conduct – yet at the same time leave themselves free, in the substantial sectors of their lives where adult authority did not penetrate, to develop all sorts of personal interests and idiosyncrasies. Thus the individual of high instinctual/emotional motivation combined with high P-factor could afford to con-form to adult standards in the 'official' part of his life, because he could 'let off steam' in the 'unofficial' part. Much that occurred was undoubtedly bad – and this sort of regime has in-built tendencies towards hypocrisy – but many individuals did manage to develop strong independence in thought and action, and did as adults manage to express their creativity in substantial ways.

Their creation is in fact our whole contemporary world. It must not be forgotten, when we seek ways of improving our educational policies, that the colossal cultural explosion of science and technology in which we live, is almost entirely the emanation of what we now call, condescendingly, 'traditional education'.

We must now attempt a brief examination of the reciprocal interactions likely to be occurring, largely through our systems of formal education, between the contemporary world and human intelligence.

2. EXPANSION AND INCREASED VARIABILITY IN EDUCATION

The most notable feature in contemporary education, as with human societies and the homosphere generally, is its rate of expansion. This is due to a number of causal factors, and 'expansion' can be interpreted in a number of different ways. The purely quantitative expansion is part of the general population explosion. There are many more people to be educated, and the rate of increase is high in most parts of the world. In addition to the sheer increase in number of the people to be educated, however, there are increasing demands for everyone to receive more education, for school-leaving ages to be raised, for more young people to go on to tertiary education, for more opportunities for continuing formal or semi-formal education throughout adult life. This is part of a cultural explosion or revolution which has itself resulted from the still-continuing Scientific and Technological Revolutions; and in this cultural revolution education is intimately embedded. New branches of knowledge are being proliferated, new methods of transmission of knowledge are racing to keep pace with the explosion in subject-matter and in demand. This is part of the qualitative expansion in education, and is directly linked with the quantitative aspects. There are other sorts of qualitative development, however, which can be regarded as in some sense at least reactions *against* the quantitative developments; and it is these which are from the evolutionary point of view the most significant.

To explain this last point: consider the education explosion literally as a form of population explosion. As such, it is subject to evolutionary mechanisms and must be assumed to have similar characteristics to any other biological explosion. One of these is

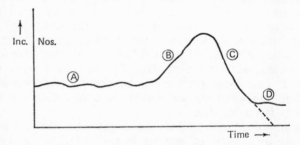

Figure 15. Generalized graph of changes in a natural population under 'natural selection'.

A. Stable population. Minor fluctuations due to variation in selection pressures.
B. Phase of rapid increase – due to lessening of selection (= mortality) pressures. Increased individual variation.
C. Population 'crash' – phase of massive decrease, due to extra-heavy selection pressures (compared to those at A).
D. Population may stabilize, or disappear. Characteristics of individuals at D may be different from what they were at A, due to:
   (i) Extreme selection pressures at C acting upon:
   (ii) Highly variable population at 'peak', due to:
   (iii) Relaxation of selection pressures at B.

that during the phase of expansion of population, on the steeply-rising part of the graph, the mortality factors through which Natural Selection operates have relaxed their force – this is in fact what must happen to allow the population to build up. As a consequence of this, many individuals who in 'normal' circumstances would have been eliminated in the 'struggle for survival' are not eliminated; and these 'less fit' individuals are found in greater numbers than at other times. In purely descriptive terms, the amount of variability within the population is greatly increased. When the population stabilizes, i.e. when it ceases to increase and either remains steady or declines, the mortality factors must have gone back to 'normal' and the death rate comes up to equal or exceed the birth rate. (We can neglect immigration and emigration for the moment; but see p. 269). In a stable or a declining population a larger proportion of individuals is eliminated than was eliminated during the phase of increase. Natural Selection goes back to its previous severity; and in general it is the 'less fit' individuals, who survived during the phase of increase simply because things were easy for them, who are cut down. It must be remembered, of course, that Natural Selection may not be operating on quite the same basis after the population explosion as it was before: the characteristics which make an individual 'more fit' or 'less fit' may have altered in the meantime. But irrespective of these details, what is important is that there must come an end to every population explosion, and when the end comes the competition gets tough and many fail to survive.

In terms of education then, we might well interpret present conditions as a phase of increase in which individual variation is rampant (new subjects, new teaching techniques, larger proportions of populations being educated, etc.) and in which the stringency of natural selection has been much relaxed. We can predict, however, that there must come a time when natural selection will reimpose a 'weeding out' process; and at that time many of our 'innovations' whether deliberate or accidental will be scrapped (along with the people who have been involved with them). One problem is this: can we predict, now, which of our innovations will prove to have survival value, which will not?

The question of the survival value of particular innovations is a peculiarly urgent and personal one for those professionally engaged in education. It also impinges upon everyone else, though perhaps less directly, in that the education of our own children must be affected. Through them, the survival of our forms of culture and ultimately of the human species itself is in question. Is it impossible to gauge in advance which features are likely to survive? Must we simply await the event, hopeful but passive? Or can we on the basis of a present evaluation predict at least what sorts of approach are likely to prove survival-worthy?

On a basis of evolutionary biology and in terms of the principles (if not the details) of human evolution we may be able to work out such an estimate.

The salient feature of past human evolution has been adaptive variability of behaviour. Let us take as an axiom for the present inquiry into contemporary educational phenomena: that whatever seems likely to preserve or enhance adaptive variability (cf. the concept of 'effective intelligence', Stenhouse 1966) is to be taken *prima facie* as having positive survival value; and that whatever seems likely to reduce adaptive variability is to be taken as having negative survival value, i.e. as being harmful in the long run irrespective of whatever merits it may have in the short run.

Since the adaptiveness of a variation can often be perceived only long after the event, the 'axiom of adaptive variability' must be translated, at the level of practicality, into an *axiom of mere variability*: variability in itself must be preserved and increased, irrespective of its apparent short-run adaptiveness positive or negative. It should be reduced in particular instances only upon the strongest proof of definite contra-adaptiveness; and mechanisms to review previous variability-reducing decisions should be set up and kept functional. (The parallel between my 'axiom of variability' proposed for human societal policy, and the 'law of requisite variety' in cybernetics (Ashby 1964) will be obvious.)

3. SIGNIFICANCE OF CREATIVITY RESEARCH

On the basis of evaluation suggested above at least one contemporary educational phenomenon can be recognized as having potentialities for survival and progress. I refer to the widespread interest in creativity – whether called by this name or not.

Creativity increases diversity and variability, almost by definition, even though the adaptiveness of the creative outcome is not guaranteed. (Adaptiveness is a function of the survival of what is created.) And increased diversity comes not only in what is created, but also in the people who create: encouraging creativity entails encouraging people to develop and enhance individuality, differentness. The 'creativity movement' may give rise to an increased range of variation in social mechanisms, too, in that it has encouraged the exploration of new teaching methods, assessment techniques and so on. Thus diversification at various levels – behavioural, individual, and societal – may result. Whether these tendencies will continue, and what use will be made of them, are questions still to be answered.

Besides the hopeful tendencies, there are others of more sinister import. The P-factor of intelligence is mainly in question here, since some trends appear likely to involve contra-selection, possibly severe, of this rather than the other factors.

A number of issues relating to creativity and evolutionary intelligence were discussed in Chapter V, and it seems appropriate to develop at this stage some of the detailed educational implications of the arguments there put forward.

Firstly, in supporting the creativity researchers (e.g. Getzels and Jackson 1962) against some of the criticisms of their work, namely its being done on 'non-representative' samples, atypical minority groups, and hence as being non-egalitarian and possible 'undemocratic', I have already indicated that evolutionary theory supports the view that the *future* characteristics of any group of evolving organisms are to be found not among the average or typical characteristics of the organisms as they are at present but rather among the characteristics of atypical minority groups. (Some characteristics, eventually destined to become typical, may not as yet have appeared at all – this depends, obviously, on how far ahead one wants to make one's predictions.) Thus we might postulate that if *Homo sapiens* is going to have a future, the future *Homines sapientes* may have some of the characteristics at present confined to a 'high creative' group. It must be kept in mind, of course, that the future characteristics of the species need not come from only one of the present minority groups. Just as the birds incorporated in a new type of organization, indeed in each one of the 'new type' individuals, features (such as hollow bones, tetraradiate pelvis, separation of pulmonary and systemic blood circulation and so on (see Colbert 1955)) which were previously found, separately and severally, scattered through various assemblages of saurischian and ornithischian dinosaurs and other archosaurian groups, so the future characteristics of our descendents are likely to be derived not from any one existing sub-group but from a number of the sub-groupings which might be distinguishable (but not separated or separable) at the present time. Thus there is no question of the future of man being in the hands of an 'elite', of any particular 'dominant minority' (this phrase in a non-Toynbeean sense) or a 'master race'. Especially, there is no question of selective *breeding* to produce a master race – the message from evolutionary and genetic theory is that the intermingling of minorities (in the plural) is more significant than the preferential survival of any one minority – and this takes a long time.

Thus it is not suggested that we should try to turn everyone into a 'creative'. It cannot and should not be the intention to spread the characteristics of one sub-group – namely the creative minority – through the whole human population. The practical difficulties and the time required are only part of the problem. More important is the fact that we do not yet know enough about the creative minority – we cannot yet diagnose individual creativity in advance (see Hudson 1966 and above Chapter V – and it is more than doubtful if we would be wise to aim at a whole population of 'creatives' even if such a goal could be achieved. Even in an age of automation there are many jobs to be performed by humans which require, for example, the toleration of monotony, of the dullness of routine. For this, the low threshhold of rejection concomitant with a relatively high P-factor is a poor qualification. And the sort of job in question is not just the mechanical repetitiveness of a factory assembly-line: even such high-grade and important functions as that of the medical practitioner, dentist, accountant, and lawyer, are probably best performed by individuals whose endowment of P-factor is not especially high relative to the other factors.

Another major consideration is that if substantial creativity is dependent upon high P-factor, and if the latter is genetically restricted to a smallish minority of the population, an educational policy which holds out creative achievement as the universal goal and the criterion of personal worth is going to condemn a majority of persons to frustration and a sense of failure.

One question which may be asked, then, is whether our systems of formal education could or should be altered so that they would foster creativity in the majority of children and young people. It is necessary to be cautious at this point, and to make as clear as possible just what seems to be involved. Earlier discussions (especially Chapter V and references given therein) indicate that the causal factors which may result in real-life creativity usually interact in ways which are intricately complex and subtle. Notably, perhaps, there seems to be a good deal of evidence pointing to a need for tension, for an opposition between the creative individual and at least some parts of his social context, if creative potential is to be actualized in the production of substantial innovation or discovery. It has been suggested, for example (Chapters V, VII and VIII), that the operation of the P-factor of evolutionary intelligence, the factor of postponement and especially of doubt and provisional rejection, plays the initiating role in the cycle of creativity, and that this is so both within the immediate mechanisms of creativeness and also within the period of growth and development of the creative individual. But since *both* a high level of affectional motivation *and* a high level of P-factor appear to be necessary to real-life creativity, and since the motivation seems likely to be actuated at least in part from social conflict, it seems to follow that the more-or-less complete elimination of conflict between potentially-creative individuals and society would actually remove the generating mechanisms of creativity (In fact, a high endowment of P-factor 'contrariness' would probably ensure that in a context specifically designed for creativity, the potential high-creative would take pains to be bland, stereotyped, and totally unproductive!)

It would be a mistake, therefore, to think that major creative potential could be much increased as a result of changes within a system of formal education. In this sense, the amount of creativity in a social group, a nation, or the human species as a whole is not in the short run susceptible to cultivation and propagation. But a great deal could be done, it appears, along two other lines:

(a) To reduce the inhibitory factors which tend to thwart or distort whatever creative potential *is* present; and
(b) To enhance the effectiveness of the social mechanisms for accepting the fruits of creativity.

These two possible functions are not unrelated, and the educational policies conducive to achieving the one are also likely, as will be seen below, to be helpful with regard to the other. The same mechanisms of 'effective intelligence' must operate in both – but since they will operate at different levels and with different emphases in the two cases it will be convenient to consider them separately.

4. REDUCTION IN THE DISTORTION AND THWARTING OF CREATIVE INTELLIGENCE

One of the great merits of the 'child-centred' approach in education has been its recognition and acceptance of individual differences among children. The acceptance has not, perhaps, been universal in practice but there has, nevertheless, been a considerable change in emphasis especially with regard to, say, the first six to eight years of schooling. Children up to the age of about eleven to thirteen in many 'Western' countries are encouraged to develop their individual potentialities – within limits – and the aggregate of emotional trauma resulting from pressures to make everyone fit a single pattern must be a fraction of what it was even twenty years ago. Many changes in classroom practice have contributed to this: 'streaming', promotion on achievement and stage of development rather than chronological age, emphasis on interest groups, widening the curriculum and gearing subject-matter to community interests. . . . The potentially creative child has no doubt been enabled, as a result of all this, to survive *as* a potential creative – at least until he moves into the secondary level of education where pressures towards conformity, towards intellectual and social orthodoxy, become more severe. These pressures are generated largely though not entirely by the orientation of secondary-level schooling either towards the obtaining of formal marks of qualification or competence, such as school-leaving certificates (or whatever they may be called), or towards entry to the tertiary level of education. There is a great deal of polemic against 'the examination-orientated curriculum', 'exam dominance', and so on. Much of the criticism is misplaced, not because there are no faults to be criticized – there are many, and bad ones – but because the basic 'faults' are not the mere fact of examinations coming at the end of secondary schooling but that:

(a) The mechanisms of assessment, i.e. of examining and grading, are often defective in detail and mistaken in basic assumptions; and
(b) If there is to be any evaluative assessment at all, the criteria used, or believed to be used, are bound to generate a certain amount of pressure towards conformity with these criteria.

With regard to (a), it would be inappropriate to attempt a discussion of details. The main point is, that it is the *bad* exam,

248 THE EVOLUTION OF INTELLIGENCE

not the exam *per se*, which is the evil to be ameliorated. (Discussion of some of the issues involved is offered elsewhere, see Stenhouse 1969 a and b; Hoffman 1962). The only further point to be made here is that any assessment procedure, even a 'good' one, involving a formal verdict of attainment, must inevitably reinforce (b) – so that total elimination of conformity-pressures would demand the complete abandonment of a verdict on the educational achievement of the individual. This would appear a revolutionary notion, utterly impractical and undesirable, and I do not propose to argue for it. What I shall argue for is that the approach of 'evolutionary engineering' be adopted: that reflexive 'selection and correction mechanisms' be set up within the existing systems, to alter and improve them.

Getzels and Jackson (1962) and others have remarked that 'high-test creative' ('diverger') children tend to be less popular with class teachers than do their classmates whether of 'high I.Q.' or just average or low ability (see also Hudson 1968). The suggestion is implicit in Chapter V and may be made explicit now, that the relative unpopularity of high-potential creatives is likely to be due mainly to the high endowment of 'P-factor doubt' postulated as occurring in such children (see Stenhouse 1971). There is nothing to be gained by failure to recognize that 'divergers' and high-P children – and adults too – are often highly irritating people to deal with. The scepticism and irreverence generated by high levels of P-factor must tend to actuate opposition and distrust among those in contact with a potential high-creative. If these people are required to furnish reports on the creative individual there would seem to be a strong probability of their being adverse, at least in the sense that the creative individual would be given less than his due, would be 'written down'; and it must be remembered that such 'writing down' can be accomplished in subtle yet effective ways which may be very difficult to penetrate. The crucial feature of the situation is that many of the characteristics of a high-creative *do* make him difficult, do make him problematical to fit in as a component in a group or an organization; and any report on his personal characteristics can very easily stress these ones, and miss the other valuable ones, without telling an untruth and without appearing to distort the picture in any material way. For in fact it should be notorious, but at the present time is not, that true creativity is very difficult indeed to detect in advance of its actual manifestations: for the very reason that the earliest manifestations are often negativistic and superficially contra-creative (see Hudson 1966: 121).

While at the upper end of the scale, major creative talent will make its own way (more or less), there is a much larger section of the population in the middle and lower reaches of ability where the encouragement of creative talent is desirable not directly for societal adaptation but rather for ameliorating the quality of life of the individual. In the vast societies of the present and future, the individual is apt to lose any sense of personal worth and achievement (Riseman 1953). Increases in scale, automation and technication of vocational activities lead to difficulties in personal identification with vocational outcomes. These increases therefore demand – and also make possible – a transfer of achievement-satisfaction from the vocational to the extra-vocational activities of the individual. Hobbies, sports, music and the arts – and perhaps most important of all for many young people, continuing education – must increasingly offer the avenues for achievement that once were provided by a person's primary life-work. But a regime in which income-producing activities are divorced from those producing personal satisfaction is bound to generate its own stresses. This has not yet achieved adequate recognition. Policies for education must recognize the need that individuals should be able to minimize and/or cope with such stresses. They must be educated to do so. The problem is: how?

It must be reiterated that the total removal of factors apparently inimical to individuality and creativity is not in question: this is neither possible or desirable. What is possible, and also eminently desirable, is that there should be mechanisms to safeguard *all* individuals, of whatever talent or lack of talent, from being consigned to educational environments which are unsuitable to them. And it cannot be too much emphasized, that in making the treatment accorded the creative individual a test case, it is not for a moment being suggested that non-creative individuals are of lesser significance or value. It is a good systemic strategy to concentrate upon the creatives. They are recognized as being valuable. It has already been indicated that they are not readily identifiable. Therefore, if a society wishes to strengthen itself by encouraging primary creativity, it must encourage *all* individual variation, since the primary creatives cannot be isolated for special treatment. As indicated earlier, there are good grounds for believing that many potential high-creatives are likely to be among the minority discriminated against by current techniques of assessment. A trend away from so-called 'objective' criteria, apparent for example in the procedures by which some universities are now selecting their undergraduate intake, and towards increased dependence on personal reports given by teachers, headmasters, and others, can be seen to have disturbing implications of social and academic conformism when considered in the light of the evidence that teachers and others are likely to react adversely to 'divergers', 'high creatives', 'high P-factor' individuals (but see Hudson 1968: 76–82). There is little the individual can do, in the immediate sense anyway, about those who dislike him. Usually dislike has no one specifiable cause. An attempt to find a cause, by means of frank person-to-person discussion, usually results in finding a rationalization rather than a real cause; and the 'frank and free discussion' beloved of the counselling psychologist frequently has no result other than an exacerbation of difficulties. In any case, school pupils and university undergraduates are not usually in a position to impose 'frank and free discussion' upon those who might otherwise give them adverse reports; so the safeguarding of high-P 'gadfly' individuals must be accomplished, if at all, by other means.

It would appear that the attainment of 100 per cent fairness, even if only in principle, must be dependent upon an 'objective-type' instrument. (The very use of the term 'objective', relative to the type of test in question, is in fact a mark of belief in its '100 per cent fairness' – in the literal sense, the answer-sheets (scripts) of a traditional essay-type examination are of course not a whit less 'objectively real' than are completed 'objective'-type test sheets.) One problem which arises if a special 'creativity' test is incorporated in a large-scale assessment programme (as I.Q. tests were incorporated in the British 'Eleven-plus'), is that candidates may be coached to improve their performance at it. Another problem, intrinsically more difficult is not impossible to overcome, is that the very awareness, 'I am doing a creativity test' is itself likely to inhibit or otherwise distort performance. The effect of P-factor in the 'Test-situation Triviality' principle (see Chapter V) is a special case of this.

The aim of 100 per cent fairness is not only a worthy social and educational ideal in its own right. It is also, from the societal point of view, good business. By helping to optimize the accommodation of individual variation within systems of formal education, the fairness ideal helps to preserve an optimum spectrum of individual capability which forms the reservoir of contribution to social well-being and advance. And in proportion as the ideal of fairness of opportunity is actualized in social practice, a society will win the active support and loyalty of its individual members.

Thus the fruits of the creativity-testing movement would best be utilized, not in an attempt to build up an elite group of creatives or potential creatives to be given special treatment in special schools or classes, but rather in enabling educational groupings to be kept flexible. The storms of abuse which have from time to time been directed against various measures designed to separate children into 'streams' to receive differentiated treatment intended to be appropriate to the particular 'streams' or subgroups, has to a considerable extent been mistaken. 'Streaming' in itself is desirable. It is not desirable, certainly, to give the impression that the children in one 'stream' are morally, socially or intellectually inferior to those in any other; and it is absolutely necessary that the 'streaming' arrangements be prevented from becoming rigid. *This* is the real fault: rigidity. If groupings are flexible, if in fact there is a steady interchange of members between streams as individual interests change and effective abilities fluctuate, the objections to streaming on both ethical and efficiency grounds must fall away. The criterion of the need to move an individual from his present group to a new one must be a discrepancy usually between some part of individual potentiality and the average potentiality (and/or actuality) of the group. The actual transfer of an individual from one group to another may not always be possible; and the decisions as to which group would be most appropriate, and how and when best to effect the transfer, can seldom be easy. Nevertheless it is essentially a 'discrepancy mechanism' which must be used to detect the anomaly in need of specific investigation. Anderson (1960) has shown that individuals with high I.Q. but low academic attainment scores on entry are unlikely to do well in university studies, and Harper (unpublished M.A. thesis, Massey University, NZ.) found the highest failure rate and lowest average mark in a university biology class in a group comprising individuals in the top 20 per cent for both intelligence and divergent thinking scores. The wastage represented by the diversion or rejection from intellectual work of these apparently-gifted people is likely to be significant: numbers may be relatively small, but even a few talented and thoroughly disgruntled individuals may have a widely corrosive effect. Apart from the general social disruption to which they may contribute, many of these 'high-ability drop-outs' are likely to have a strongly antagonistic attitude to education, especially perhaps to the institutions of higher learning among which their own personal 'failures' have become manifest. It seems likely that many of the harmful influences on tertiary education – the forces of anti-intellectualism which have been growing disproportionately fast in recent years (see Barzun 1961) – have been due less to the influence of low-ability sectors in the population (as seems to have been assumed by many commentators) but rather to the influence of embittered minorities of disenfranchised intellectuals.

Returning to the problem of preserving potential creativity from the pressures tending to reduce or destroy it, the conclusion of this section of the argument must be, that an educational system must avoid rigid barriers, rigid selection mechanisms which are likely to close the doors of opportunity in the face of the potential creative and keep them closed. On the one hand it is necessary to improve selection mechanisms (for entry to particular 'streams' in secondary education and to tertiary institutions, also to employment opportunities); and on the other, maximum provision must be made for people to change tracks (in accordance with what they feel to be the needs of their own development) even after settling into the tracks they themselves had chosen. Changing tracks would not need to be accomplished entirely without penalty – young people need some degree of protection from their own passing whims and fancies – but it is important that penalties should not be too great.

In this way the 'axiom of variability' can be given practical expression in the social policies of education, with obvious practical benefits both to individuals and to society.

5. MECHANISMS FOR EVALUATION AND ACCEPTANCE OF INNOVATION

The paradox of innovation is that all societies need it to survive, yet they all tend to penalize those creative individuals who produce it. The paradox is to be resolved – though not always, in practice, completely – by the fact that while innovations and discoveries are usually resisted initially, the rewards for the innovator whose creation or discovery is recognized and accepted are often very great indeed. Much creative activity, probably almost all creativity at the most fundamental levels, is certainly disinterested. The creative genius is driven by something which he feels is greater than and often in some way external to himself (see Beveridge 1950). To some extent, therefore, creativity and discovery and innovation will always occur independently of social policies and attitudes. But while this is true, it is also true that much is lost by social inertia in the realm of ideas. The tragedy of the rejection of the techniques proposed by Semmelweiss for the control of puerperal fever, for example – Semmelweiss died insane as a result of the rejection and persecution to which he was subjected (see Beveridge 1950) – may be mentioned, as a striking instance from recent times and within the biological sciences, of wastage of the fruits of creative endeavour. Instances could be multiplied indefinitely: the real-life creative who achieves recognition usually does so as much because of his hardihood and stamina in withstanding abuse and neglect as because of his discoveries or innovations!

Education is commonly held to have two main functions with regard to those products of creativity which collectively form a 'culture'.

(a) That of transmitting the culture from one generation to the next – a maintenance function.
(b) That of providing cultural 'raw material' for further creative innovation – a function of *advance*.

These two functions are certainly performed. Education tends in practice to concentrate rather heavily upon the transmission of already accepted knowledge and skills, while innovation to an appreciable extent occurs in spite of, rather than because of, the formal institutions of education. The need of the present time, urgent because of the great rapidity of change of all sorts, is for adequate provision to be made for a function of *assessment and acceptance* in relation to the products of creative innovation. It has been pointed out (Stenhouse 1969 a) that the ability to deal fairly with new ideas, as with personal criticism, is likely to be a function of the individual's total personality rather than of his qualifications in formal education. This is not to say that the formal education he has received is irrelevant, far from it. But the aspects of formal education relevant to 'critical receptivity' are not necessarily the ones which at present lead to the bestowal of marks of recognition. It is to the elucidation of the relationship between education and this 'critical receptivity' that we must now turn.

The function of *advance* must, it would seem, be subdivided, so that we recognize as its components:

(i) Individual creativity,
(ii) Critical receptivity.

These are two distinguishable functions within any society, a deficiency in either of which must retard its advance and reduce its viability. (Note that 'advance' is being used in a completely general sense at this point. It is emphatically *not* to be understood as limited to technological advance, neither must it be thought that the practical criteria for assessing 'advance' must include 'dominance over competing societies'. While it may be true that the society which fails to advance will eventually come to be dominated by, and ultimately supplanted by, its competitors, it does not follow that dominance is anything more than a negative criterion. Specifically, the type of 'advance' most needed at the present time would appear to be the abandonment of dominance of one society by another.)

The emphasis, in the phrase 'critical receptivity', must be on 'receptivity'. It is clear that societies generally reject many innovations which would, had they been accepted, have been beneficial. Logically, one must receive a new idea before it can be criticized, so that 'receptivity' has logical as well as temporal precedence over 'criticism'. Criticism must not be forgotten, of course, and in some fields, especially perhaps in the social sciences at the present time, there is need for more and better criticism. As Passmore (1967) has emphasized, criticism must not be understood as 'picking faults', as a mere 'capacity to think up objections': it is itself a form of creative activity. True or proper criticism, as distinct from mere disagreement or unthinking rejection, involves coming to understand what is criticized. Thus 'criticism' as well as 'receptivity' entails that new ideas must not only be 'entertained' but also *understood*, and the crucial point is, that to get people even to entertain let alone attempt to understand a new idea is a major and unusual event in most fields at the present time. If this can be appreciated the magnitude of the educational problem becomes apparent.

Besides the transmission of knowledge and skills, and giving creative individuals the wherewithall to create, then, the third function of education must be to inculcate not only an understanding of the need for critical receptivity of new ideas but also an actual willingness and ability to receive and then criticize them in practice. What is involved in this third function?

6. EDUCATION FOR CRITICAL RECEPTIVITY

The inculcation of understanding of the need for new and sometimes disturbing innovations to be received and evaluated by a society would appear to present few problems in principle, although major changes in emphasis and approach at several educational levels would be necessitated. The need for receptivity can be demonstrated in relation to any of a number of subject fields. The comparative study of the histories of various nations can bring it out very strongly – even the parochial history of a single nation probably furnishes sufficient material as the causes of its vicissitudes over a long period of time are explored. The message can come through from comparative studies in human or cultural geography, as soon as these get past the barely descriptive stage. The history of science and technology, or of literature or of art, indeed of virtually any field of endeavour, can certainly be used to illustrate the need both for individual creativity and for social acceptance of its fruits. And if the historical perspective be enlarged to include archaelogy, and from that to human evolution, the story becomes even plainer in outline, even though its details must of necessity be neglected. In a different dimension, comparative psychology – if it gets beyond the limitation of considering only man and the white rat – must attempt to explain human achievement, must indicate the nature of human culture and of psychosocial evolution, and again must come to the mechanisms of individual creation and social acceptance. Similarly, any teaching in biology or zoology which fails to give an adequate treatment of human biology must stand condemned as biology – and an adequate account of human biology must include discussion of the cultural and behavioral mechanisms of creativity and innovation.

Thus, enhanced understanding of the need for and the nature of individual creativity could be fostered even in school children, at the secondary and perhaps even the primary level. The very topic of this book – human intelligence and its relationships with the totality of phenomena – could actually be the focus of an overall integration between all the major fields of knowledge. It is in the tertiary level of the educational system that the most important changes must take place, however; and here again, human behaviour especially in its aspects of intelligence and creativity could provide the focus for integrated studies, with an enhanced understanding of the vital issues in this area as the most important outcome.

An appreciation of the abstract and general need for intelligence, in both its critical and creative exercise, is the first desideratum. It is not sufficient, however, if the educational achievement stops at this level. What is needed, beyond this, and perhaps even in the absence of the abstract appreciation, is the actual willingness in practice to tolerate and indeed welcome both the innovations themselves *and also the innovators*. This last requirement is what hurts. We are all prepared to pay lip-service to the need for individuality, intelligence, thinking-for-oneself, and – especially nowadays – creativity. But when it comes to the point of tolerating the scepticism and 'disrespectfulness' often found in diverger-type or high-P children, or the turbulence and 'arrogance' often found in adult 'real-life' proven creatives, the spikes of hostility stab quickly out through our thin garments of broad-mindedness. We can always concoct some agile rationalization to explain away our antagonisms. And there is this much truth in Plato's old allegory of the cave, that the creative person is often too much blinded by the sun of discovery to be able to see clearly the traps that are being digged for his feet. He is often easy enough to fool; and even when, as not infrequently happens, he sees and even predicts the machinations of his fellows, as like as not he finds the evolutionary regression to the instinctive level of behaviour, the level at which intrigues and manipulations are conducted, so uncongenial that he disdains to slide back to it, even for self-defence. Socrates said at his trial (*Apology*, Section 35): 'There seems to be something wrong in asking a favour of a judge, and thus procuring an acquittal, instead of informing and convincing him.' That is, Socrates deplores playing upon the judge's instinctive dominance mechanisms instead of appealing to his intelligence. Socrates himself refused to compromise, refused to give his detractors the satisfaction of having him come down to their level – they hated him for it and murdered him.

Toleration and support for the person as well as for his innovation, is the need. The lives of the two great Lawrences of this century, D. H. and T. E., illustrate clearly the dependence of the creative individual upon the affective support and social mediation of those percipient enough to see his potentiality. D. H. leant heavily upon his Frieda, and earlier, presumably, upon his mother (see Murry 1954). T. E. owed a great deal to a number of people, notably perhaps D. G. Hogarth, Ronald Storrs, and General Allenby, but also John Buchan and Mrs Bernard Shaw among others (see Payne 1962). The 'difficultness' of the high-P creative has already been discussed, see Chapter V. It would be futile, indeed positively harmful, to minimize the problem of dealing with such people. One inference that can be made with confidence is that a purely intellectual and abstract appreciation of the need for creativity in general is not in itself going to ensure that anyone will be more competent than before at actually 'getting along with' creative persons. Knowledge alone does not make us more tolerant. And tolerance itself is not enough. Personal acceptance is necessary. This does not mean that we have to *like* every creative individual we meet, any more than the injunction 'Love thy neighbour' entails sexual promiscuity in the suburbs. The combination of 'acceptance' with even a fair amount of dislike is not, however, at all uncommon. Most people prefer to work for a 'boss' who is efficient (in the broad sense, not the 'busywork' sense), even when they could never be friendly on a personal basis; and on the other side, it is no slur on a friend to decide that he would be a poor risk as one's supervisor at work. We accept this sort of relationship in our professional and business associations largely because we perceive and acknowledge the functional necessity of efficiency at the planning, organizing, and administrative levels. The need for efficiency is obvious and immediate – most of all, perhaps, in its absence. The need for creative innovation, by contrast, can never be so immediately obvious. It can be appreciated, in practice, only retrospectively. Therefore to achieve increased acceptance of creatives must become a specific objective of education.

Better acceptance of proven and more especially of potential creatives is a particularly urgent problem in relation to fields of activity carried on within institutions. Education itself is a case in point, scientific research another. Whyte (1963: 190) points out that in the United States in mid-1950s: 'Of the $4 billion currently being spent on research and development by government, industry, and the universities . . . less than 4 % is for creative research. The overwhelming majority of people engaged in research . . . must now work as supervised team players, and only a tiny fraction are in a position to do independent work. Of the 600,000 people engaged in scientific research it has been

I

estimated that probably no more than 5,000 are free to pick their own problems.'

He goes on to discuss various factors which inhibit and distort the independence and creativity of scientists: the need to obtain money for research, which gives rise to the need to 'keep in with' the project leaders, administrators or 'organization men' who control the purse strings, the need to build up research projects often far beyond the optimal size simply in order to qualify for a grant, and so on. Without going into details, the problem in general is that the demands of administration cut across the peculiarly personal and idiosyncratic needs of individual innovation. As Whyte points out, the result in a great many cases is that a person who started as a potentially-creative scientist ends up as an administrator or as an organization- rather than discovery-orientated scientist (see also Hoyle 1963.) Further, the 'leaders' of research are likely to be appointed more on their 'public relations' abilities than on their capacity to conduct research or to organize others into conducting it.

The last point is important.

The capacity to lead and direct research is not guaranteed by personal competence at conducting research, still less by efficiency at 'public relations' and navigating the corridors of power. While proven competence in first-order research is a necessary condition for true leadership – it puts the leader in a position where instinctive social-dominance mechanisms can reinforce his intellectual stimulus and control – it is not a sufficient condition. Many an outstanding individualist has turned out mediocre or worse as the leader of a team; conversely, there are others who achieve their greatest productivity in the effect they have on colleagues and subordinates. Here, as elsewhere, the policy of appointing an individual to perform one function, on the basis of his performance of other (and often quite different) functions, is not likely to be adaptive in the long run. What is needed is again flexibility: mechanisms to allow individuals to develop and demonstrate their talents in a variety of fields. And the evolutionary 'trial and success' model (see Pringle 1951: Ashby 1964) is again relevant. Given time and sufficient diversity of opportunity, the individual will gradually narrow his field of endeavour to the field(s) in which he can make his most satisfying contribution (see the excellent discussions on this area by Sanders, Siepert, Kaplan, and especially Shepard, in Hill 1964.)

7. CREATIVITY, 'FACT'-LEARNING, AND REASONING

Intelligence in the creative sense is one thing, intelligence in the ratiocinative or 'reasoning' sense another. The latter is likely to be heavily underestimated in the next decade, creativity having become the focus of interest; but just as the power of 'reasoning' was at one time held to be the supreme mark of human nature and excellence, and its actualization, development and refinement the chief end of the educative process, so we can again turn to 'efficiency in reasoning' to redress the over-emphasis upon 'fact-learning' at one extreme and upon creativity, in the 'mere' sense, at the other. Formal education can do little in a positive sense to foster real-life creativity at the fundamental level. The causes behind this have already been indicated. Formal education is at the present time falling ever more heavily into the other extreme, of factualism, 'learning that' rather than 'learning how'. University graduates tend, increasingly, to be 'knowledgeable' rather than 'capable'. This is the real point behind the demand, from employers of graduates in industry and technological fields, for more specifically technological training to be given by the universities. The training in basic sciences, and also the social sciences to a considerable extent, is felt to be inadequate in some way. And so indeed it is. Its inadequacies are not to be cured, however, by giving heavy doses of instruction or 'drill' in specifically technological activities. What is required by industry – and what is in fact often generated in the course of active experience of practical problems and their solution, though it can also be generated to an enhanced level by 'pure' education of the right sort – is really *effective intelligence*, the ability to 'think through' a novel problem-situation and come up with a practicable policy. The ability to deal with what is recognized from the first as a problem-situation does not depend to the extent that *de novo* creativity depends upon the sceptical pre-conception-removing function of the P-factor. The P-factor does come in to the extent that over-precipitate action must be avoided and the claims of alternative strategies must be held off while they are subjected to critical appraisal. But effective intelligence in solving already-recognized problems does not entail an endowment of P-factor at the erosive and subversive intensity of creative genius. 'Problem-solving' in this relatively straightforward sense is amenable to improvement through education – and one most important sort of education is, of course, practical experience. 'Experience' is ambiguous however. In what might be called a quantitative sense, a minimal sense of 'bare description', experience accumulates simply with the passage of time. The drawback of this 'quantitative experience' is that it need not involve improvement of judgement or greater efficiency of performance: we all know individuals who lay great stress on their 'thirty years' experience, man and boy', but who appear to have learned less in thirty years than some people do in three. So we turn to 'experience' in a qualitative sense, in which an interaction of quantitative experience with evaluative intelligence produces enhanced understanding and competence. This 'qualitative experience' is not a product only of the passage of time; so the relative 'experience' of several individuals cannot be assessed by the mere enumeration of their years of service. It can be assessed only in terms of competence on the job, ability to vary techniques and strategies with changing conditions – in short, only in terms of effective intelligence. And increase in effective intelligence, while it does demand *some* real-life 'on the job' experience, is not thereafter tied quantitatively to such experience. A good deal of subsequent 'experience' can be vicarious. This is what people generally mean when they talk of 'theoretical' experience. Theory can be vicarious experience only when it builds on a certain amount of prior *real* experience; and it can be significant, i.e. it can produce 'qualitative experience', only if the individual is continually exercising his powers of critical evaluation upon the stream of quantitative experience and/or its surrogate in the form of reviewed and restructured remembrances and imaginings.

In short, real life experience of coping with refractory practicalities can force people towards *learning* the value of the P-factor 'pause and consider'. The individual who has to encounter and cope with a large number of novel problem-situations at a practical level is under pressure to acquire the habit of looking – and planning – before he acts. He is under pressure, in fact, to actualize whatever P-factor potentialities he may possess.

(The dilemma faced by the potential creative innovator in any field is illustrated very clearly in the criticisms made of the anthropologist Lévi-Strauss by Edmund Leach (1970). There can be no doubt that Lévi-Strauss is an important figure in theoretical anthropology – otherwise, Leach would scarcely be bothered to write a book about him – yet his first-order direct field experience was, on Leach's showing, very slight. This would account for the various faults of detail that Leach is able to demonstrate in his work. Yet it is arguable, I suggest, that a more extensive and sustained first-order experience would, by allowing greater domination by the 'factual' with its built-in orthodoxy of terminology and categorization, have led merely to reduction in or perhaps the complete disappearance of theoretical adventurousness: Lévi-Strauss would have become merely another of the thousands of 'normal science' anthropologists (Kuhn 1962). Clearly a balance has to be achieved between the quantity of the individual's experience and his level of P-factor endowment, if he is eventually to effect a theoretical re-orientation and/or restructuring. Too much experience leads to diminution in creativity; too little means that the new formulations may be faulty. It is only the supremely powerful high-P creative who can immerse himself in the detailed real-life minutiae of his subject and still come up, in the end, with a new and substantial restructuring. Koestler (1964) in *The Sleepwalkers* shows very convincingly how the scientific revolutionaries Copernicus, Kepler, and Galileo came to their great restructuring of astronomy and cosmology slowly, only half-aware of what they were doing ('like sleepwalkers'), and only reluctantly being forced into realization that the 'facts' of the time, embodying as they did the presuppositions of the previously accepted paradigm, were not in accord with the actual observations which they and others were making. Here was no facile negating of accepted theory; here, rather, was a protracted struggle over accepting its negation: but the negating was, nevertheless, seen and eventually accepted as real. Newton perhaps exemplifies, by contrast, a bold high-P willingness to negate orthodox presuppositions – but then, the way had to a great extent been prepared for him.)

Returning to the question of technological training, the education of graduates in, say, the pure sciences, should be criticized not so much on the grounds of its being non-technological, as on the ground that it has been insufficiently orientated towards effective intelligence. 'A learned blockhead is a greater blockhead than an ignorant one,' as Benjamin Franklin remarked; and education, especially perhaps in science, increasingly produces learnedness rather than the critical flexibility of intelligence. The specific loss appears to be in the inhibition of the P-factor, this negativistic-seeming sceptical capacity for doubting the authorities and not accepting the standard techniques. Within pure science it is probable that the atrophy of the P-factor can pass unnoticed

262 THE EVOLUTION OF INTELLIGENCE

in the short run, in most graduates. There is plenty of new 'raw material' for the old research techniques to be practised upon; and in teaching, the recipients of dogma are usually in no position to dispute or even to recognize it as dogma. Those among the taught who do react against dogma and authoritarianism can easily be written off as 'poor material', 'maladjusted', and so on; and unfortunately for them as individuals, but also for the system and the society in general, there is usually plenty of 'objective' evidence, e.g. in the form of poor exam results, to support their dismissal. Few question the rightness of the accepted criteria, the accepted techniques of assessment, until the time when their inadequacy has become painfully obvious. By that time, however, so much harm has probably been done that it is excessively difficult or even impossible to rectify things. As pointed out elsewhere (Stenhouse 1966), the reactions of biological systems are often slow, but they have a strong inherent 'momentum' so that once a deleterious change has started it may be impossible to avert disaster – we usually do not read the subtle 'danger signals' in time. This 'momentum', where human institutions are concerned, is normally to be equated with the tendency to form self-fulfilling 'closed' behavioural systems rather than self-correcting 'open' ones. A self-fulfilling system is self-perpetuating over the short run. In the long run, since no system can be isolated indefinitely from the other systems collectively forming the largest system of all, the biosphere, any system which has closed itself to corrective influences is likely to prove incapable of the large-scale self-correction, amounting often to self-reorganization on a major scale, which would be necessary for survival. Since the majority of systems are continually changing, a non-changing system is likely to lose its capacity for adapting: an unmoving object is likely to be a dead one! Thus while it is very easy for an educational system to become self-fulfilling, in that its mis-assessment of individual persons is often concealed by their being denied opportunity of demonstrating anything contrary to the assessment, the non-adaptive nature of such a system will be demonstrated macroscopically in the failure of its products to cope with the challenges arising outside the system.

A system of tertiary education need not, then, set before itself the task of producing armies of creative geniuses. But it must try to ensure – and for its own long-term survival it must succeed in ensuring – that:

(a) Potential high-creatives are not consistently rejected; and
(b) That *all* the personnel forming the output of the system are in fact *educated* rather than trained, indoctrinated, or 'programmed' (see discussions in Peters 1967).

Interpreting these requirements in terms of the four factors of biological intelligence, the first requirement (a) is that the system must be able to tolerate the disruptiveness and disturbance of high-P individuals. In particular, the system must incorporate mechanisms for self-correction: all individuals, but especially those with the awkward burden of high P-factor endowment, must be given continuing opportunity to prove wrong any negative assessments that might have been made against them. The dangers of the contrary policy have already been pointed out – and are proving themselves in practice, in the form of massive dissatisfaction and protest movements in many institutions of higher learning at the present time. But in some respects requirement (b) is of wider scope. The basic difference between education and indoctrination, training, etc. is that education tends to be open-ended and self-correcting, whereas training and especially indoctrination tend to be closed, because they are aimed at specific performance. Indoctrination and similar processes are indeed by their very nature aimed at excluding from consideration everything outside what is dogmatically held to be 'relevant'. Contrary to this, in education the basic intention must always be the extension of the C-factor to cover anything whatever; the free employment of the A-factor to discover similarities and differences, relevancies and significancies, in any and every part of the range of raw experience; and the exercise of P-factor 'doubt', suspension of belief and withholding of response, in any part whatever of the individual's experience and activities. The operation of P-factor negativism is a necessary prerequisite to the 'evolutionary' model of intellectual functioning (see Pringle 1951): it is only if an item of belief or behaviour can be 'negated' that its contribution to the 'adaptiveness' of one's system of beliefs or of one's behaviour can be assessed. And only so can the 'adaptiveness' of new formulations be tested (see Chapters III and V). The D-factor is involved, of course, in all transactions between individual and environment whether social or 'natural' – and through consideration of the D-factor contribution we can come to a reconciliation between 'training' and 'education', and to a resolution of the paradox of the demand for less technicism in university graduates employed in industry and technology.

Perhaps the first point to be made is a correction or amplification of the statement, above, that training results in a specific and 'closed' performance. It is true that when we speak of 'training' we usually have in mind the inculcation of specific and determinate skills. This need not exclude, however, the possibility that the trained person may go on to develop some new skill, some improvement on the procedures which were taught to him. The important word here is 'improvement'. Variations in the performance of a trained person are looked upon with disfavour, they are viewed as definite faults, except when they result in clearcut improvement. This is because of the connection between training and practicality. In activities which one learns through training the pragmatic test of practical workability is never far away. This contact with primitive reality is both the strength and the limitation of any performance developed through training. 'What works' is immediately apparent, and cannot be gainsaid by the dogmas of orthodoxy. But 'what doesn't work' at this level is not necessarily to be accepted as decisively negative. Innovations which work can be, indeed must be, accepted immediately – but the non-workability of an innovation may prove nothing more than that the changes made were not radical enough. The anti-theoretical 'practical' man is likely to discard prematurely any innovation which seems to him, in a very immediate context, not to work or not to be useful. The classical instance might be the 'useless' experiments of Galvani. But on the positive side a truly pragmatic approach involving direct experience of the refractory complexities of raw nature can generate a real and valuable – if usually non-explicit – recognition of the limitations as well as the strengths of particular theoretical formulations. This first-order experience, mediated by the D-factor, impinges therefore upon the interacting complex of C- and A-factors to provide corrective feedback. The corrective mechanisms must be actuated, as usual, by the detection of discrepancies between what was expected to happen and what actually did happen, between the theoretical construct and the reality (cf. Chapter VII). In real-life situations, the individual is under some pressure to look for discrepancies before they become serious. In short, there are pressures of 'natural selection' tending to make the individual self-critical even when he is not inclined to be! It is of benefit to him if he is able to simulate the action of the P-factor, and suspend belief or final acceptance of his hypotheses until they have been given a maximum of testing. (It is doubtful whether such 'simulation' could occur if the P-factor were, in fact, low or absent. Probably the 'simulation' involves rather a removal of the suppressors of P-factor activity.) Thus in real-life situations there is still this form of 'natural selection' in favour of the P-factor of intelligence.

The defect in the education received by many graduates, of both schools and universities and other tertiary-level institutions, is therefore that they have been exposed neither to sustained and rigorous argument at a theoretical or philosophical level (such as leads to the development of a 'personal position', see Stenhouse 1968) nor to the corrective influences of free interplay with an open environment. Instead they have all too often been subjected to the mutually reinforcing effects of a limited set of theoretical positions and a selection of laboratory, observational, or experimental 'experiences' chosen precisely because they accord with the theories. Since there cannot, under such a regime, be any substantial conflict between the several theories, or between the theories and the circumscribed versions of 'reality' specially chosen to illustrate them, it is no wonder that the persons 'educated' in this way are inept at dealing with larger, untidy, and generally awkward sweeps of reality when in their subsequent professional lives they encounter them. The remedies are of course obvious: provide practical experience of unselected and unstructured complexes; and provide educative experiences in sorting out real and substantial conflicts at the theoretical level. This latter requirement must now be explored in greater detail.

If education for the perception and resolution of methodological conflicts and for the evaluation of proposed changes is to become a major objective, the conceptual frame of reference of each subject taught must be as wide as possible. Science teaching, for example, especially on the biological side, must concentrate much more on general methodology, much less on technicalities at the practical level. On the humanities or arts side, subjects such as history should desirably involve a look at the widest perspectives, the longest time-scales. Detailed critical work on specific smaller issues can continue much as at present – but here again, the explicit discussion of methodology must also be incorporated.

8. NEED FOR NEW PHILOSOPHY IN TERTIARY EDUCATION

If a shift of emphasis to include methodological appreciation along with technical competence and factual knowledge is desirable on the grounds that it would promote
receptivity to and better evaluation of new ideas, it would appear
logical to extend this line of development to its widest scope. The
exploration of questions of widest scope is one of the traditional
functions of philosophy. Are we then to include Philosophy
within the curriculum of the sciences and the liberal arts, perhaps
as an obligatory component within all university degrees? Should
professional scientists be required to take courses in philosophy
not as 'culture' or for 'broadening the mind', but as an integral
and essential part of their *professional* training?

One might point out that in one way or another precisely this
sort of requirement has been built into an increasing number of
degree prescriptions over the past ten to twenty years. Science
degrees with built-in 'generalist' requirements fall into one of
two patterns. In many American liberal arts colleges and in some
of the recently established universities in Britain and Commonwealth countries, 'generalist' studies which may include pure
philosophy form the bulk of the first one or two years of undergraduate work. These years are followed by a narrowing and
intensification of interest in selected 'specialist' fields, culminating
in a certain amount of nominally independent research on a
chosen topic at the postgraduate level. The alternative pattern –
in the degrees which encourage or require 'generalist' studies at
all – has the 'generalist' element coming in late, often only at the
postgraduate level. (The 'reading foreign languages' requirement
so rigorously insisted upon in many universities for their science
graduates cannot be regarded as having liberal or generalist
influence at all. It is nothing more than a technical adjunct, like
learning to use a calculating machine or computer or even a
typewriter.)

Each of the above patterns has both good and bad features.
There is much to be said for giving a strong dose of general and
critical training in the latter stages of a course, for the students
can then appreciate the need for it, also they have more background to enable them to cope with it. On the other hand, there
is evidence indicating that many at this stage have got the 'set' of
their minds fixed so that they cannot easily cope – in which case

either the 'generalist' courses tend to become a rather empty formality, or else otherwise-competent scientists (or technicists) are held back, in some respects needlessly, from their chosen careers.

'Generalist' and diversified education could be expected to have the following effects in terms of the four factors of intelligence: A greater range of subjects studied leads to greater 'quantity' and variety of items in the C-factor 'information storage system'. Their greater variety provides greater need and opportunity for A-factor re-sorting and re-classifying; and this can be directly encouraged by formal requirement for cross-disciplinary comparisons. The changes in approach and mental set associated with switching repeatedly from one subject to another impose demands on the P-factor, to withhold the patterns appropriate to subject A every time the field of study is changed to subject B. Thus an intellectual habit of P-factor utilization may be fostered. And D-factor stimulation can follow from increased variety in types of educational activity: laboratory and field work in the sciences, discussion groups, writing, speaking formally, reading parts in (and perhaps acting out) dramatized incidents, and so on.

The key factor is the actual nature of the courses intended to inject 'generalist' considerations. If these involve no more than learning up a certain amount of potted 'history of science' or 'Ancient Greek Thought', if they are no more than name- and 'ism'-dropping accompanied by waffle in a cultured accent, they are indeed of no value, or rather they are of negative value, for they perpetuate the superficialities which cause the 'Two Cultures' gap (Snow 1957). But they need be neither waffle nor stodgy recitals of names-and-dates. In philosophy the twentieth-century Revolution (Ayer *et al.* 1956) has been towards teaching the subject as an *activity*, 'knowing how' rather than a 'knowing that'. Much teaching, admittedly, is still of the old sort, in which competence is measured in knowledge about the arguments of the past rather than in being able actually to conduct arguments and investigations in the present. If in 'generalist' courses the approach is made through the examination and analysis of urgent contemporary problems, problems which are real to the students themselves and in which they can be shown hitherto-unsuspected complexities and implications, their interest can be aroused and they can be led into the acquisition of 'generalist' skills of thought and attitude. The lead-in must be through students' already-existing interests and commitments: it follows from this that the persons teaching 'generalist' courses must be able to catch the students *within* their 'home range' of interests and activities. This can be done in either of two ways. On the one hand there is the possibility of starting within the areas of subject interest which are already active. Thus science students can be led to generalist philosophical considerations through work in the philosophy of science, students of education can be educated through work in educational philosophy, historians can broaden their horizons through study of the philosophy of history. All of these 'philosophies' can and should consist largely of work in pure philosophy (Stenhouse 1968) although they start in so-called 'applied' philosophy; and students can be attracted from these courses into further intensive and rigorous philosophical investigations, in 'pure' philosophy courses designated as such.

Initiation into philosophical 'generalism' and critico-creative thinking can also occur in an alternative way via students' extracurricular interests. There are plenty of these. Problems of sexual morality, of individual-*v*-society, of whether we can ever be *really* sure of our knowledge or of our own personal worth – these and other questions are the virtually universal preoccupation of young adults. Certainly they receive plenty of publicity at the present time. The teacher looking for a lead-in to philosophy has an abundance of interests upon which to play – but . . . he himself must have the personal qualities to be able to do the leading effectively. Without them all efforts are useless. Insincerity is fatal. The teacher must have a real interest in his students and the problems confronting them, a put-on simulation of interest is stripped away sooner or later, with fatal results to the enterprise of teaching. He must have real competence at arguing, too, on any issue, without appealing to the authority of his position, and without becoming over-vigorous to the extent of being aggressive.

Good intentions in teaching are not enough. Neither is expertise in the technicalities of teaching. While it is perfectly true that an intelligent and highly motivated student *can* learn from the bad teaching of a bad person, it is also true that bad teaching has cumulative indeed a multiplicative effect on students. There are two main types of effect. On the one hand, teaching which is bad in the technical sense will tend to cause an immediate drop in recruitment to the subject in question. On the other hand, teaching which is technically effective but wrong in its basic presuppositions may keep up recruitment for a while; but this merely results in the wider dissemination of the false presuppositions, and causes a worse 'crash' of recruitment when the deficiencies begin to be perceived. We are seeing this phenomenon on a worldwide scale at present, the deficiencies in science education having caused the massive 'drift from science'. The 'drift' might indeed be called a cataract! It can also be interpreted as a form of 'emigration' from an uncongenial environment (see p. 242).

9. INSTINCT AND INTELLIGENCE AMONG TEACHERS AND TAUGHT

But while the intrinsic evils of bad teaching must be recognized – and while it needs to be recognized that bad teaching is likely to be widespread at the present time, for reasons indicated earlier in this chapter (pp. 240–3 – it is also of fundamental importance that the prerequisites for good teaching should be understood. The most basic of all such prerequisites is that the teacher should be a *good person* (cf. Stenhouse 1969a). This does not mean that he needs to be 'good' in an orthodox, conventional sense. But there are 'goodnesses' which he must have. He must, for example, be sincere, as was indicated above. He must have a deep understanding of his subject. (Too much weight is allowed to the view that particulate factual or technical ignorance destroys the teacher-pupil relationship. On the contrary, one of the important functions of the teacher should be to demonstrate that mere details *are mere*, i.e. that provided one knows and understands a subject properly it is no fault to forget some of the details. And in fact highly effective teaching can result from the simulation of ignorance of details or of terminology.) Above all things the good teacher must be prepared to hazard *himself* in the teaching relationship: he must lead and demonstrate, rather than exhort and direct from the sidelines. He must speak with authority (cf. Barzun 1961), yet he must not use his position for ego-aggrandizement and 'one-upmanship' (Potter 1948). Without developing this line of argument further (but see Stenhouse 1968, 1969a) it should be obvious that the qualities required for good teaching, for effective education especially (but not exclusively) at the tertiary level, are not qualities which can adequately be assessed by most current procedures focused as they are on narrowly academic achievement. The need for a better understanding and assessment of the personal as against the narrowly academic or technical qualities of the teacher gains added force from the consideration that 'wastage of creativity' must be reduced not only by improving receptivity to the *products* of creativity but also by increasing receptivity and tolerance of the *creative individuals* themselves. And as I have been at pains to emphasize, creative individuals are likely to be difficult, spiky creatures, largely because of their P-factor contrariness. So those who are to teach them, teach them without harming them, have all the more need of those personal qualities which generate strength and security, and which allow the teacher to engage in critical discussion – and to be himself criticized – without arousing the social-dominance mechanisms of self-assertion and aggressiveness. This suggests that a relatively high P-factor endowment is desirable in teachers, especially tertiary-level teachers where a large age-and-status gap does not give them the protection it affords to teachers of younger children. I have suggested elsewhere (Stenhouse 1965a, 1969b) mechanisms which would account for various currently-observable educational phenomena; and Hudson (1968) cites experimental findings showing the likelihood of general relationships between teachers and pupils in terms of the converger/diverger continuum. Briefly it seems likely that diverger pupils learn best from, and are least likely to be intellectually harmed by, diverger teachers. If we assume that a substantial contribution to the diverger syndrome is made by the P-factor, a conceptually precise mechanism for teacher/pupil interactions may be postulated. In the pupil, high P-factor makes for a questioning or doubting attitude which shades over into irreverence, indifference, or outright rebelliousness. The 'independence in the pupil' to which such lip-service is paid in educational discourse is in practice usually taken as insubordination: it is often construed as subversion of the teacher's authority, and penalized accordingly. But if P-factor 'suspension of belief' and 'withholding of the standard response' has this tendency in the pupil, it can have a compensating effect when present in the teacher. The high-P teacher has the capacity to withhold his agonistic reactions to the 'threat' of the high-P pupil. He has the power not to respond, to ignore the challenge. Thus the P-factor can be built into a dynamic self-compensating and self-correcting system within the process of education. High-P teachers can preserve the effective P-factor endowment of their pupils. Even more, by their non-reactance (relative or absolute) to the instinctual-level challenges of their pupils they can reduce the total 'amount' of instinctive transaction within the teaching situation and thus allow more scope to intelligence-based transactions. It terms of presently-orthodox human psychology, a reduction in the affective tensions allows more freedom for cognitive activity. But this is an oversimplification. It seems likely that the complete reduction (i.e. elimination) of affective tensions would result in the cessation of all activity. What is needed is, in ethological terms, not the removal or elimination of all instinctive drives, but their control. Intellectual activity demands a certain level of motivation from the emotions or instincts – it is the mode of expression of this motivation which must be controlled. If we postulate that at least some of the motivation of both parties within a teaching situation is provided by the 'social dominance' complex (itself made up of tendencies to attack, to flee from, and to make friendly approaches to, the other members of the social group), we can see that this motivation could be discharged in the form of overt dominance displays of one sort or another. The form of these displays in humans is largely learned, expressed in culturally-derived patterns of behaviour. Among academics from the British Commonwealth, for example, the Oxbridge mannerisms of understatement and exaggerated courtesy often form part of the dominance display; while among those from the American orbit and elsewhere the display is often closer to basic patterns, involving relatively emphatic utterances in a confident tone, and using 'learned' phraseology and the current 'in' words from the jargon of the sub-culture. Instinctive components are usually present in a pure form, also for example uprightness and relaxation of posture in the dominant individual, and with frequency of blinking and of eye-movements increasing with decreasing dominance-rank. It seems likely that these 'pure' signals are often more decisive in rank disputes than are the learned components – but however this may be, it is certain that if a dominance-challenge from a pupil is taken up by a teacher, and if it elicits an instinctual counter-reaction, what follows will be either a series of dominance contests which will tend to exclude the educationally desirable cognitive developments (i.e. the 'learning tasks'), or else the complete instinctual-emotional subjugation of one party by the other, with again a reduction in cognitive potentiality. If, on the other hand, the dominance-challenge issued by the pupil is met by the withholding of dominance displays, and their substitution by an intelligence-actuated response, i.e. a rational response, then the pupil is put under pressure to establish his position, relative to the teacher, in terms of intelligence rather than instinct. And it is not only social position relative to the teacher which is in question – the relationship between the individual pupil and his peers is affected by his relationship with the teacher; and vice versa. This brings in the possibility that instinctual transactions among the group of pupils may generate pressures on the individual pupil to openly challenge, or attempt by surreptitious means to subvert, the authority of the teacher. That the teacher should be able to withstand, and if need be to overcome, this assault at the instinctual level, would appear to be essential – and this in turn would require the teacher to have strong instinctual motivation but not, usually, to allow it to be used for instinct-level activities. This requirement for 'force held in check' is in fact a requirement for high P-factor withholding power, along with high instinctual motivation (in this case presumably for social dominance). The general picture that emerges is that the ability of a teacher to lead his pupils into intelligence- or intellectually-orientated rather than instinctual-orientated activity must be roughly proportional, other things being equal, to his endowment of P-factor. A person with a relatively low level of instinctual motivation can be a good teacher of amenable classes; but turbulent and refractory classes will require the stronger leading-power of an individual whose levels of instinctual dominance motivation *and* P-factor control are both high (see Hudson 1968: 70–71). Within an educational system minimal levels of general intelligence of teachers might be expected to rise with the increasing age of the pupils – certainly a rise in general intellectual background and competence appears desirable – but the maximum of instinctual-dominance-balanced-by-P-factor might be expected to be necessary in the upper-middle part of the range, where among children of middle and late teens the burgeoning instinctive drives actuate the sharpest challenges to 'authority' in the form of society's dominance systems.

10. IMPROVEMENTS IN SYSTEMIC MECHANISMS

Several implications have been mentioned regarding improvement in mechanisms within systems of education. Notable requirements are: the need for flexibility of assortment of the recipients of formal education; relating to this, the need for considerable improvement in examination and other assessment procedures; and as a special case of this last, the need to improve mechanisms for the selection, training and retention of teachers. This point has been little elaborated in the arguments so far, and cannot fully be dealt with here. However, some relevant implications may be mentioned, expecially since they lead on to some significant general considerations.

Selection, training, and retention of teachers – this field has not been accorded a particularly balanced discussion in recent times. The training or education of teachers has been over-emphasized. This is presumably because a large number of other teachers and administrators have a personal vested interest in teacher training. The emphasis is an *over*-emphasis not because teacher education is unimportant but because selection and retention are in fact very much more important than has been allowed to appear. It may be objected that most teacher-training institutions do in fact operate selection schemes before and during training; and that many educational systems incorporate mechanisms intended to facilitate retention of teaching personnel. This is of course perfectly true, but true only in a context of limited scope.

The emphasis on training presupposes a 'silk purses out of sows' ears' principle: that within broad limits anyone, given the prescribed training, can be turned into an adequate teacher. Selection is generally made on only two bases: absence of criminal convictions; and academic competence as assessed mainly by examination. It is difficult to see that selection of teachers could have been much more stringent than it has been in recent years, in view of the vast 'education explosion' mentioned at the beginning of this chapter. As the population and then the educational explosions come increasingly under control, however, selection pressures will inevitably become more severe, and the question arises: in what ways, precisely *how*, will they and should they become more severe? So far, almost the only type of change implemented or even proposed has been to raise the hurdle of academic entrance standards a little higher. There are grounds, as indicated earlier, for believing that current academic assessment procedures (notably examinations supplemented by personal reports) tend to discriminate against high-P, and diverger and non-conformist individuals. And there are grounds for believing (see earlier argument, also Hudson 1968) that these are the types of individuals who can perform most effectively the most vital function of education, of transferring the orientation of their pupils' reactions from the instinctive to the intelligence systems. There is the further probability that pupil creativity is best conserved and fostered by high-P teachers. If, therefore, increased severity of selection is to mean merely a raising of the 'pass mark' in selection procedures of the present type, all the indications suggest that the proportion of high-P and diverger and nonconformist individuals will be further reduced. This would seem likely to push educational systems into a regressive cycle of increasing conformism and rigidity, hence of decreasing power of adaptation. To avoid this, it will be necessary to emend and improve assessment procedures, to make them more favourable to the categories of person already indicated. That this can be done, and on a large scale, seems not impossible.

Briefly, the solution appears to lie in 'multi-modal' examining, especially in the use of appropriately structured objective tests as a standardizing component within a complex of other measuring devices operating upon both examinee *and examiner*. High-P individuals are likely to have relatively low tolerance of 'exam-situation triviality', and this will tend to depress both I.Q. test scores and scores in orthodox essay-type academic achievement tests. They are likely, on the other hand, to enjoy risk-taking and gambling on their own abilities (see Chapter V, Hudson 1966) and if of good basic ability will tend to do better on well-structured objective tests than they do on more tedious essay-type ones. Thus a characteristic profile of test results may be expected to be diagnostic of many (though probably not all) high-P individuals. This would enable them to be identified, and any discrepancy between overall assessment – especially in the form of personal reports, from teachers and others – and real-life performance, to be investigated.

Thus selection-procedures could be, and needs to be, made more complex, more sophisticated, and more subject to subsequent check-up and correction. Which brings us to the second major issue, that of *retention* of teachers within an education system.

The retention of teachers within an education system cannot be allowed to mean the mere retention of a list of names on the payroll, of bodies gesticulating before classes. Qualitative considerations must never be allowed to be supplanted by the merely quantitative. The teacher must be retained *as a teacher*. The qualities conducive to his proper functioning must be unimpaired by his professional employment: if possible they should be enhanced. He must not be treated as an expendable resource, a sort of raw material to be used up. He must be treated as a biological resource – which can be self-renewing and even self-improving, but which if treated wrongly can become self-destructive and even destructive of the social system. The key factor here is self-respect. And for a teacher to retain his self-respect he must be free to exercise professional responsibility. This is especially important for high-P individuals, who tend, as pointed out earlier, to sit loose to the esteem of their fellows, and even to antagonize them (especially those superior in substantive rank but inferior in ability to themselves).

The retention of a teacher in this qualitative sense must necessitate, then, that he achieve recognition reasonably commensurate with his deserts. This entails that the system must incorporate self-corrective mechanisms of assessment, preferably continuous in operation, so that educational effectiveness may be encouraged both immediately and on a long-term basis. Virtue is to some extent its own reward, in teaching as in most other fields, but the overt marks of achievement are not nearly so obvious, or so quick to appear, as they are in most other industries, and it is therefore necessary that extrinsic mechanisms of assessment be set up. 'Good teaching' cannot be quantified. It does not reduce to 'teaches so many students' or 'achieves such-and-such a pass rate' or 'is popular' or 'is unpopular' or 'holds so many degrees'. Neither can the 'product' of teaching be used directly or immediately as a measure of success. There is in fact only one way in which the professional performance of teachers can be assessed, and that is for them to do it themselves. Multiple and mutual evaluative judgements can be incorporated into a self-correcting system, utilizing the reciprocal or reflexive logical property of the evaluative judgement: that a judgement which turns out wrong penalizes the judge and not the person judged. 'Memory' can be built into the system using information storage and retrieval techniques, and computers can be used to correlate the very large quantities of information involved and to perform the 'spadework' of assessment. (An outline of a 'multiple and mutual' plan for teacher evaluation is given, in editorially distorted form, in Stenhouse (1968a), and an amplified account is in preparation.)

Systems of evaluation as outlined above can be used at any level within an educational organization. At the level of tertiary or higher education there is of course already in operation a mechanism providing motivational feedback to the individual and corrective commentary on the evaluative judgements made within the system. This mechanism is that of international publication in the learned journals. The reactions to a particular individual by those in his 'home' institution and even his home country may be largely determined on an instinctual basis. For example, an individual in whom the social-dominance displays of self-confidence, etc. are notably absent may find that his intellectual products are not even properly examined, are not considered worthy of careful scrutiny, by those in personal contact with him (or by others whose judgement has been 'contaminated' by the instinctual-reactors). But while under present circumstances the teacher in a school can generally do little about such dis-valuing, the university teacher can establish his own reputation, and at the same time (whether he will or no) use the reflexivity of evaluative judgements to impugn those who have depreciated him, by achieving publication in centres well removed from personal contact. At a distance, the intellectual quality of the work produced is the only criterion that can possibly be used, and the effects of instinct are at a minimum.

An appreciation of the fact that human behaviour still includes substantial instinctual components, and that these can and do cause distortion of functions supposedly based upon intelligence, can thus lead to greatly amplified understanding of operations within educational systems and other human institutions. In particular, as has been illustrated above, the requirements for retention of effectiveness among teachers need to be appreciated. The current over-emphasis upon training as against 'selection' and 'retention' (adherence to the 'silk purse out of sow's ear' principle) can be seen as the result of simplistic assumptions inherent in the learning theory approach. As Lorenz (1965) points out, motivation for learning is derived from the instincts – it follows that an activity which cannot be linked to an instinctive drive cannot be learned. And that instincts can negate learning. The various provisos and qualifications necessary in relation to this point are discussed in Chapter XI. For the moment the bare statement must stand on its own. The theoretical deficiency is practically exemplified in this way, that while a person can learn to be a good teacher, the fact that he has *at some time* learned to be a good teacher does not in any way ensure that he will continue to be good. Continuance or retention of 'goodness' depends on an interaction between factors of intelligence and of instinct which has been sketched, in barest outline, in the foregoing. Changes in instinctual orientation, for example, into an agonistic attitude towards the Establishment or towards his own students, can vitiate a teacher's worth – and this can be permanent, even though his purely technical skills in teaching may be unimpaired.

Table I

Factor-group variations and some possible concomitants

	cf. Guilford 1967	Some possible phenotypic outcomes*
1. c a p d		
2. C a p d	M	'Know-all' (in extreme form, idiot-savant)
3. c A p d	E, R	'Developer'-type scientist
4. c a P d		EITHER: Drop-out. Sceptic or cynic. OR: very contented 'acceptive' person.
5. c a p D	B	Artistic performer, sportsman (locally renowned).
6. C A p d	C, N	Conventional intellectual. 'High I.Q..' (Getzels & Jackson 1962)
7. C a P d	D	'Disaffected intellectual'
8. C a p D		'Rhodes Scholar'
9. c A P d		Slow real-life creative (intellectual, abstract)
10. c A p D		Nationally or internationally renowned artistic performer, sportsman
11. c a P D		Sporting champion, world level Artistic performer – innovative (Jazz)
12. C A P d	S, D	Fast real-life creative (intellectual)
13. C A p D		'Rhodes Scholar' – high ability, both physical and intellectual, in conventional lines only. No 'P' disturbance.
14. C a P D		Conflict due to 'a' handicap: reduces coherence and integration, e.g. in artistic production. 'Frustrated creative'
15. c A P D		Slow creative – ('concrete' rather than abstract). Sporting champion – 'all time great' level
16. C A P D		'Allrounder' (Hudson 1966, 1968)

* Some of the extreme possibilities of achievement for each category have been indicated. These are very tentative and generalized sketches, and the phenotypic outcomes are not to be regarded as exclusive or exhaustive.

'Creative' implies 'real life' rather than 'test situation' creativity; and creativity is to be understood as more substantial than mere innovation.

11. A SIMPLE FOUR-FACTOR MODEL FOR INDIVIDUAL DIFFERENCES IN INTELLIGENCE

The interaction of four-factor intelligence with education has been approached, so far, from the educational side. Implicitly, we have looked at existing educational policies, and have asked the question:

What modifications are needed in order to meet the general evolutionary requirement for adaptive variability, or 'requisite variety' (Ashby 1964)?

This approach is essentially a conservative one. It is not at the extreme of conservatism embodied in the dictum 'When not necessary to change, necessary not to change'; but it does assume that the overall strategy should involve modification of existing policies, practices, and institutions, rather than abandoning them and starting again from scratch. It is thus evolutionary rather than revolutionary. But it is desirable to look also at what is revolutionary within the evolutionary continuum; and in particular, to examine the possible relationships between four-factor intelligence and education from the 'evolutionary intelligence' and 'diversifying' side of the fence.

If we stipulate that education is *for* intelligence, i.e. that we educate in order to enhance our intelligence, individual or collective, what are the needs of four-factor intelligence as against other views of intelligence?

Much educational practice leans to the tacit assumption that 'intelligence' is unitary. On this view differences are merely quantitative, and are adequately accommodated by varying the amount of formal education. The qualitative differences which are allowed – arts or science, technology, music, languages, physical education, and so on – are largely justified in terms of interest rather than intellectual predilection (though Hudson, for example, has shown that arts/science choice is largely determined by basic individual differences in intellectual functioning, see Hudson 1966, 1968). No doubt the use of I.Q. ratings as the only measure of intelligence has been a major influence towards the acceptance of 'unitary' preconceptions, but even the 'creativity' movement has not altered things in this respect. The expression of creativity is accepted, within limits, as being idiosyncratic, hence as demanding a *laissez faire* regime in certain areas such as painting and imaginative writing, but outside these areas, the tacit assumption is still made, that a measurement on a single scale is an adequate characterization of the individual.

Against the 'single scale' views, whether of I.Q. or 'creativity' or of divergent thinking, may be set a multi-factorial view of human intelligence such as that elaborated by Guilford (1967). A multifactorial theory appears in principle to offer more prospect of doing justice to human diversity than does a unitary theory. Apart, however, from questions of the biological and adaptive significance of the factors Guilford proposes, the sheer number of categories offered by his theory appears to be an embarrassment if one attempts to relate them to feasible educational policies. With 120 possible combinations of various types of intellectual activities to be considered, the drawing up of broad alternatives of educational policy must remain problematical. Guilford's 'structure of intellect' theory, useful though it must be at the individual level, does not within itself offer criteria for the aggregation of the multifarious intellectual activities into functional groupings. Some activities may be of more frequent occurrence than others, some are more important than others for particular modes of adaptation, but they cannot readily be organized into a lesser number of functionally-significant constellations.

The present four-factor theory of evolutionary intelligence, on the other hand, not only offers a more manageable basis of grouping individual types (as will be indicated below), but also a basis for the functional grouping of Guilford's intellectual activities. The evolutionary-adaptive foundation of my own theory means that it has a built-in orientation towards the functional relationship between individual and environment (including social environment); and the significance of such relationship is already assessed in the minimum number of general terms. To put the issue obversely, the Guilford 'structure of intellect' theory can be seen as offering detailed particularizations of some of the more generalized functions discussed so far with regard to evolutionary intelligence; and the research findings upon which the Guilford theory is based could therefore be regarded as evidence supportive of both Guilford's theory and, at second remove, my own.

In order to indicate the nature of the relationship, it will be necessary to set up a simple model of the range of individual variation which can be accounted for in terms of four-factor evolutionary intelligence.

If we simplify the range of variation of each factor by assuming it to be present either at high level (denoted by capital letters appropriate to the several factors) or at low level (denoted by small letters), a range of 'collective' possibilities of sixteen is provided. An individual can be low on all factors ('capd'), high on only one ('capD' or 'cApd', etc.), high on more than one ('CApd', 'CApD', etc.), or high on all ('CAPD'). The full range of possible factor-group variations, on these assumptions is presented in Table 1 along with some 'commonsense' identifications of particular groups and a few apparent correspondences with the theories of Guilford (1967).

It is clear that, while there are substantial similarities between the present schema and that of Guilford, the categories of the present theory to a considerable extent cut across the categorizations of other workers. The various functional expressions of the P-factor, in particular, appear to have been overlooked in earlier works (see Chapter V); so |that Hudson's 'diverger' category, for example, probably includes both individuals of high P-factor endowment some of whom will become 'real life' creatives and others of low P who will remain 'mere' divergers. (The latter are the 'interesting' people not infrequently encountered, who have just sufficient P to 'loosen up' though not enough to negate the conventional structurings of information, and sufficient C and A to bring out unusual relationships and unexpected information.)

Little consideration is needed to show that most of Guilford's categories of intellectual operations would probably occur, at some time or another, in most of the persons corresponding to the several four-factor groupings. The different four-factor types would be characterized not so much by the presence or absence of particular Guilford categories, but by their relative frequencies of occurrence. It is in this sense that the column of Guilford similarities in Table I is to be interpreted: the nominated Guilford category is one characteristic of, rather than peculiar to, the corresponding four-factor grouping.

With regard to strategical (rather than tactical) considerations in education, it is suggested that the crucial issue in the first instance is the presence or absence of a high level of P-factor; and secondly the presence or absence of high A-factor.

Individuals relatively low on P are in no need of major strategical restructuring of education. This is not to say that educational practices should everywhere stay as they are. Considerable improvements are desirable in most systems of education, even for the average low-P individuals. But most of the needed improvements are already in operation, somewhere or other; and what is required now is merely to spread universally those practices which have already been found successful in the places where they have been tried. Education in a wide variety of fields and postponement of 'specialization' is desirable, not with the objective of turning out polymath-jacks-of-all-trades but rather, to give the individual a wide range of actual experiences within which he can match up, as effectively as possible, his own abilities and interests with the realities of the various subject-disciplines. In this light, extension of curricula to include mathematics, science, social studies, languages, arts including music, crafts and trades, and outdoor activities like sports, farming, sailing, mountaineering, etc. can be seen to be of particular value. Further extensions, some of which are very urgently needed, include the study of political, economic and legal systems and of their mutual relationships with one another and with human ecology generally.

Individuals of the several high-P groups demand special attention if their potential contributions are to be actualized and if their potential destructiveness is to be turned into productive channels. It will be argued that it is unnecessary, however, to work out different educational/social strategies for each of the eight factor-groupings containing high P. A crucial difference is made, it is suggested, by the presence or absence of high A-factor in association with high P. Before going on to investigate the likely differences in detail, the suggestion may be put forward categorically, that individuals with high endowment of *both* P and A are adequately accommodated within existing systems *provided* that there are abundant mechanisms for independent achievement and corrective re-assessment of the sort indicated earlier in this chapter. Thus it is only the individuals of high P but low A whose needs require further discussion at this point.

The 'caPd' individual has relatively low general intelligence but high P-factor. Depending on the mode of expression of the P-factor, this individual may at one extreme become an anti-social drop-out from society, possibly a wastrel or criminal; or at the other extreme, the epitome of the contented worker, well adjusted to a relatively humble social role. Which extreme he tends towards will depend upon his social, familial, and educational circumstances. The high level of P-factor confers on the one hand a tendency to break the accepted conventions of knowledge, behaviour, etc. and thus to set up the conditions which, for a person of high ability, may lead into the cycle of creativity – but 'caPd' person, by definition, is not of particularly high ability, hence is unlikely to achieve success in creative innovation. Therefore there is a distinct likelihood of alienation from society, and the genesis of an asocial or antisocial pattern of behaviour which may be very harmful in one way or another. (Those who break the law are not the only, perhaps not even the most dangerous, 'criminals' in a society.)

A high level of P-factor also confers ability to control instinctive reactions and hence to remain relatively unaffected by many of the petty annoyances of everyday life. Notably, perhaps, it can enable a low-ability individual to be unworried in the acceptance of a socially subordinate social position and role. It seems likely that actualization of this latter possibility would be dependent on the presence within the society of more senior individuals that the 'caPd' individual could take as models to be imitated. This in turn would appear to be dependent upon a certain stability of values and roles within a society. Thus the 'caPd' individuals would tend to act as an agency of 'momentum' in relation to social change: they would tend to be a conservative force in a stable society, to be 'multipliers of change' in a changing but still coherent one, and to be disruptive in a fragmenting one.

Turning now to the individuals of type 'CaPd', high C-factor would enable them to achieve considerable success in an education system orientated towards 'knowledge' and 'learnedness'. Against this, the high P would tend towards scepticism, cynicism, and flippancy. Again, the other extreme is that of calm acceptance; and again, which way the individual turns is likely to be determined by various factors of his upbringing, education, and social environment.

The 'caPD' individual is likely, providing his peripheral physical equipment is up to the standard of his central sensori-motor D endowment, to have a more straightforward life strategy. The high D is likely to channel him towards immediate sensori-motor intercourse with the environment, of one sort or another, say through the performing arts or sport for example, while the high P is perhaps more likely to contribute positively rather than negatively, by allowing the postponement of immediate satisfaction in the interests of better long-term performance. Since C- and A-factors are both low, there would be reduced temptation to intellectual aspirations; while high D would be conducive to success in active physical and/or sensory activities.

The 'CaPD' is perhaps the most difficult combination for its possessor. Given by P-factor the impetus towards creativity, given good powers of memory and effectiveness in sensorimotor interactions, the adaptiveness of his creations and even of his everyday responses is apt to be vitiated by inappropriateness and irrelevancies stemming from relatively poor A-factor functioning. He creates, but his creations are faulty. With poor A-factor, it is difficult for him even to perceive the flaws in his creation. Individuals of this type are indeed fortunate if they can be associated with friendly and supportive critics, who can help to prevent 'closure' in the creative processes, so that continued reworking can slowly compensate for the want of quickness in the perception of relevancies and irrelevancies. Failing this, the 'CaPD' individual is likely to be increasingly frustrated if he persists in the attempt at substantial innovation. Other possibilities are open to him, of course. His positive qualities could make him an excellent schoolteacher, for example: knowledgeable (C), forbearing and controlled (P), good at sports or music perhaps (D), though sometimes a little obtuse and impercipient (a). They could at the other extreme make him, through frustration, an embittered and influential subversive.

A general similarity in the problems faced by all 'aP' individuals will have become apparent even from the foregoing cursory sketches. They all face difficulties intrinsic to the peculiar balance of their intellectual endowment, which are additional to and likely to exacerbate the difficulties of upbringing, education, and circumstances which are the common concern of everyone. But the peculiar and the general difficulties, however, appear susceptible to amelioration by appropriate influences from upbringing and environment. As Hudson (1968: 83) remarks: 'We are bound to envisage the intellectual growth of the individual . . . as the product not only of his genetic endowment and hormonal secretions, but of a continual traffic with his context – with parents and teachers, examinations and curricula, prejudices and myths. And even when his frame of mind is firmly established, it seems that an individual's intellectual performance is partly conditioned by the audience for which, and the setting in which, it is produced.'

With regard to the special difficulties of low-A high-P ('aP') individuals, then, it would appear that adaptive educational and social strategy would need to incorporate:

(a) The inculcation of understanding within society generally but especially among educators, of the nature of the problem.

This in itself would lead to:

(b) Enhanced insight into their own nature, potentialities, difficulties, and possible strategies, on the part of the 'aP' individuals themselves. Guilford (1967:476) suggests that 'It might make a significant general contribution to intellectual development to inform all children, as soon as they are ready, concerning their kinds of intellectual resources.' He also states that 'There is a little evidence (Forehand and Libby 1962) that instruction concerning the nature of the abilities in question can be even more important [for the development of these abilities] than drill exercises aimed at those abilities' (p. 476).

It seems likely, then, that 'aP' individuals are in greater need than most of instruction regarding human abilities in general including their own. It would be possible, no doubt, to devise tests for the identification of these individuals, so that they could be given special instruction. While the devising of tests is desirable for the further study of the various constellations of ability, here as elsewhere I do not advocate the isolation of a particular group for special treatment. The preferable strategy is to improve each whole system of education, especially in its internal flexibility, so that every individual has a good chance of selecting, from the 'offering' provided, whatever he feels is conducive to his further development and wellbeing.

12. GUIDANCE BY INSTINCTIVE IMITATION

There is one important proviso to this policy. Clearly, inexperienced young people, especially those already having difficulty in adapting, are unlikely always to make wise choices regarding their own needs. They must be guided by their elders. Parents and teachers are especially important. From the strategical point of view there is little that can be done, apart from general slow improvement in education, to improve the guidance available from parents. Therefore the most immediate and speedy advance must come through the teachers. The provision of specialist 'guidance counsellors' is a short-term expedient of inherently limited value, indeed it has positive dangers. Limitations and dangers arise from a form of 'sampling error' in the 'guidance' available to each individual child or young adult; and this applies to guidance from parents as well as specialist 'counsellors'. Firstly, since a great deal of 'guidance' is implicit and tacit, learnt by example and imitation rather than by explicit precept and exhortation and discussion, it follows that the adults who have most contact with the child will in general have most influence. Secondly, since a great deal of human behaviour is adaptively sub-optimal, it follows that an appreciable proportion of the 'guidance' available to a particular child from the adults with whom he comes in contact is likely to be deleterious rather than beneficial. This is dangerous in proportion as the 'guidance' is intensive and sustained. Little can be done about deleterious parental influences – except in extreme cases and except by the general dilution of parental influence in school and other social groupings – but it seems unwise as a general policy to subject children to a second set of *intensive* 'guidance' through a small number of specialist counsellors. Better that the children be subjected to a diversity of 'guidances', in the form of the twenty to forty schoolteachers with whom they have contact, anyway, in the course of their schooling. If the quality of *this* 'guidance' can be improved, by the mechanisms of selection and encouragement as well as training discussed earlier in this chapter, both immediate and longer-term prospects of adaptiveness are likely to be much improved.

In simple terms, the 'normal' child is best to receive guidance from properly functioning schoolteachers; and the badly maladjusted child needs the psychiatrist not the guidance counsellor.

The importance of this 'learning by imitation' (often discussed in terms of 'role learning' (see Biddle and Thomas 1966)) is that, of the models offered by adults exemplify intelligence, adaptive variability rather than stereotypy, then the *instinctive* activity of imitation can lead directly to an increase in the *intelligence* of the behaviour of the child.

Since this has been a long chapter, a short abstract of its main points may be useful:

1. Creative innovation is necessary for continuing adaptation (= survival) of societies and the human species in general; and P-factor functioning is essential for creativity.
2. Substantial real-life creativity cannot be the goal of every individual; but:
3. P-factor attributes are valuable in other functions besides creativity.
4. High-P individuals are both valuable and vulnerable; therefore:
5. It is desirable to safeguard high-P individuals; but since they are not all readily identifiable:

6. Safeguards for high-P individuals must operate, in fact, as safeguards for all individuals, of individuality in general.

7. Improved societal 'systems engineering' is required in order to:

(a) Preserve individualism and diversity;

(b) Provide alternative activity-channels for everyone; and

(c) Provide corrective feedback on:
   (i) Verdicts on individuals
   (ii) The mechanisms of the system.

8. As part of 7. above, education needs to be improved:

(a) To enhance 'effective intelligence' partly through:

(b) Improvement in general understanding of the nature of man, intelligence and creativity, human society, the homosphere in relation to the biosphere in relation to the universe. . . .

9. Education should be improved through improvement in the quality of teaching resulting from 7. and 8. above.

10. The net result of achieving all the above should be:

(a) Enhancement of human adaptability hence of prospects of survival;

(b) Improvement in the likelihood of satisfaction for the individual.

XI

Instinct, Intelligence, and Learning Theory

1. PREVIOUS APPROACHES

Earlier attempts at providing an account of the evolution of intelligence have assumed 'intelligence' either to be a unitary capacity relatively opaque to analysis and peculiar to the human species, or to be a behavioural capacity intimately and necessarily connected with 'learning ability'. The first assumption (or anything approximating to it) is apt to lead nowhere, almost by definition, as regards an evolutionary account. The theoretical difficulties of macro-evolution are generated in an extreme form. A major, one might well say a massive, change in behavioural mechanisms has to be postulated, on this sort of view; and while this might have seemed not too implausible when dealing in a purely descriptive way with the behaviour of living species especially at the time when little was known in detail of the behaviour of the other primates, a 'unitary' or 'monolithic' view of intelligence has to be abandoned as soon as the realities of an evolutionary transition from instinct to intelligence are properly appreciated. As soon as one begins to think in terms of the adaptiveness, the survival value, of particular behaviour patterns or behavioural capacities, the difficulties inherent in the 'unitary' view of intelligence are realized to be virtually insuperable. At one extreme the notion of a *sudden* change to a well-developed intelligence would, perhaps, overcome some of the problems generated by the necessity of behaviour always having to remain adaptive. But while this gets over the problems of adaptiveness within the transition period between instinct-controlled and intelligence-controlled behaviour systems, it does so only by dodging or ignoring them; and it leaves entirely out of account the problem of the causation of the change. Why *should* there be a sudden change from an instinct to an intelligence system? The supposition that it could be due to random genetic changes would appear really to be little more than an assumption of 'miraculous intervention', since the number and extent of the changes necessary would appear to have an improbability of a vast order of magnitude. It is of course impossible to assert that miraculous intervention did not take place; but to assume or assert that it did is automatically to remove one's speculations from the realm of scientific enquiry. And the assumption of 'miraculous intervention' (the phrase can be used simply to denote the 'sudden jump to unitary intelligence', without theological overtones) is not forced upon one, anyway, since alternative explanations can be worked out.

To consider now the second assumptive strategy adopted in previous attempts to explain the evolution of intelligence.

If intelligent behaviour is first defined as 'adaptively variable behaviour', and if the attempt is then made to explain this behaviour in terms of one of the orthodox learning theories, confusion is apt to be generated, in the first place, as to which type of learning theory is to be employed (see Hilgard 1958, on the variety of theories extant). The *ad hoc* nature of current learning theory in relationship to the evolutionary frame of reference becomes apparent. Different theories are employed to explain different 'learnings' by different animals at different levels in the phylogeny and in different habitats and ways of life. In the 'evolutionary tree' analogy, it is as though we would describe the distal portions of each branch, separately, but could not describe the general growth-pattern of the tree as a whole. The difficulty cannot be overcome by looking for common features or common patterns among the theories of learning. What is required is a theory of a different order of generality; and this cannot be synthesized from a set of lower-order theories having heterogeneous presuppositions (some of them perhaps mutually contradictory).

In most theories of learning it is assumed that an activity once learned is fixed and unvarying. This is the direct implication of the 'learning curves' which figure so largely in learning-theory arguments. The invariability or 'permanence' of what is learned is indeed often built into the very definitions of learning that have been proposed (see Thorpe 1963 for review and references therein). If invariability is part of the operational criteria which enable us to say that an individual animal has learned a particular behaviour, it is clear that adaptive variability in that particular behaviour is impossible except on the assumption that the animal can *unlearn* the version of the behaviour it learned first, and then learn a new version, and so on for each subsequent variation.

On the line of argument which would equate 'intelligence' with 'flexibility of learning and unlearning successive adaptive responses to changing situations', then, we might expect to find the possibility of quite significant advances in the understanding of intelligent behaviour at the level of detail, but a deficiency in explanatory power at a more general level – including, significantly, the power to explain why particular detailed changes *are* significant!

The titles of two papers by Bitterman reveal common presuppositions with regard to relationships between learning and intelligence. 'Toward a Comparative Psychology of Learning' (1960) is followed by 'The Evolution of Intelligence' (1965). The basic assumptions and approach of the two papers are the same. Since the 'learning curves' of several species taken from various vertebrate classes 'do not differ significantly', Bitterman argues (1965: 92), this has previously been taken to imply 'intellectual continuity throughout the evolutionary hierarchy of animals', and that 'differences in intelligence from species to species are differences only of degree'. He says that 'learning was thought to involve qualitatively similar processes throughout the evolutionary hierarchy'. Bitterman is at pains to dissociate himself from this 'continuity' view of learning-cum-intelligence which he ascribes to Thorndike. He states that his own researches 'were inspired by the conviction that the traditional [i.e. Thorndike's] theory called for more critical scrutiny than it had received'. Advance has certainly been made in the sophistication of experimentation, and one can agree in a general way with Bitterman's conclusion, that his 'studies of habit reversal and probability learning in the lower animals suggest that brain structures evolved by higher animals do not serve merely to replicate old functions and modes of intellectual adjustment but to mediate new ones (a contradiction of the Thorndike hypothesis)' (p. 100). But it must be noted that the 'critical scrutiny' does not go so far as to question the equating of intelligence with learning. Neither does it question the assumption that 'evolutionary continuity' is to be equated with 'qualitative uniformity' – for this is the assumption being foisted, implicitly, upon Thorndike. This is presumably due partly to methodological preconceptions. The last sentence of the 1965 article is revelatory: 'Clearly bringing

K

the study [of behaviour] into the laboratory was the first real step toward replacing guesses with facts about the evolution of intelligence and its relation to the evolution of the brain.' It would appear that results emanating from laboratory investigations are factual while those of field studies are not! It would clearly be unreasonable to expect evolutionary and ecological issues to receive adequate consideration within a context in which that sort of statement could be made. But there is little point in becoming embroiled in detailed criticism: the general position is clear enough. A devotion to 'facts' and laboratory work reinforces the simplistic reductionist tendency to equate 'intelligence' with 'learning'. Not only is there philosophical naivety with regard to 'fact', there is also a biological unsoundness in the assumption that an animal's behaviour can be studied in the laboratory before and without reference to the investigation of its natural behaviour in the field. These two faults reinforce each other to doubly confound the work. In a laboratory study, for example, how is it known what to count as a 'behavioural fact' unless the normal behaviour in the wild is known and understood? As Hinde (1966:6) shows, it is sometimes necessary to know not only the natural behaviour of the species in question but the behaviour of related species as well, before certain behavioural 'facts' can even be perceived:

'*Even in the initial stages of selective description*', he says, 'knowledge of the behaviour of one species may be helpful in the study of a related one. The present writer would certainly never have noticed the occasional pivoting movements made by Green-finches (*Chloris chloris*) if he had not already been familiar with the highly ritualized and thus conspicuous pivoting display of the related Goldfinch (*Carduelis carduelis*).' (Emphasis added.)

Thus Bitterman allows his investigation to be shaped by the expediencies of laboratory experiment. It may be expedient to allow 'the investigation of intelligent behaviour' to be reinterpreted as 'the measurement of learning and unlearning', because the latter appears straightforward, and *easy* in technique even though all the important methodological questions have been begged. It is of course perfectly true that comparative investigations of 'learning ability' are worthwhile in their own right. But it looks as though Bitterman at least, and probably other investigators as well, want to draw conclusions *about intelligence* from work on 'learning'; and this raises the very important question of the relative significance of the behavioural categories involved. Is it feasible, is it legitimate to treat 'intelligence' merely as 'learning ability'? Are they, in reality, synonymous? Bitterman and others, it seems, implicitly assume that they are – but if so, why bring 'intelligence' into the discussion at all? The fact that 'intelligence' does so often creep into the discussions – and not only does it creep in, sometimes it is blazoned in the very title of a paper – suggests that it is implicitly accepted by the investigators themselves as being the more fundamental category. (This in turn suggests that it is coming to be widely if unconsciously assumed that the traditional dichotomy of instinct and learning is less significant than other possible schemata, such as a dichotomy which puts 'intelligently variable' over against 'fixed and invariable' as basic categorizations of behaviour.)

If intelligence really is a more fundamental category than learning, then an investigation which purports to illuminate the former by means of the latter is under an obligation to demonstrate the relevance of its findings to the more fundamental issue. This Bitterman scarcely even attempts to do. His remarks on the relationship between intelligence and learning do not touch the principle issues: as suggested above, they merely beg the questions. *Why* are 'habit reversal learning' and 'probability learning' relevant to the evolution of intelligence? They are relevant, but Bitterman does not demonstrate why they should be. One of the functions of this chapter might be taken to be the demonstration of such relevance. Looking at the argument in this way, however its dependence on general evolutionary theory must become apparent. We are quite unable to appreciate the relationship of 'habit reversal learning' to intelligence except by making use of such concepts as those of 'adaptation,' 'selective advantage,' and so on; and it is necessary that we keep very much in mind the operational context of these concepts. They have no meaning in complete abstraction. Thus to understand 'adaptation' it is necessary to have recognized and understood a considerable body of knowledge of detailed relationships between different animals and their physical and biotic environments through time. These detailed relationships must immediately involve each animal's natural behaviour by means of which it makes its adjustment to the environment; so that unless we can specify how 'habit reversal learning,' for example, could fit into an animal's normal way of life, it is impossible that we should do more than guess at the appropriate sensory and motor modalities, etc., which should be used in experimental investigations of that animal. And in more general terms it is necessary to know whether 'habit reversal learning' is likely to be of adaptive value to a particular animal, *and in what contexts*, before it is worth beginning a laboratory study on that topic. Even more generally it is necessary to delineate clearly the conceptual relationships between 'habit reversal learning' and 'intelligence' before conclusions about the evolution of intelligence can be derived from comparative studies on learning. The function of P-factor in allowing for 'habit reversal learning', and conversely, the results of experiments on habit reversal learning as evidence of P-factor operation, will be discussed in a later section of this chapter (p. 305). Before that is attempted, some general clarification of relationships between instinct, 'learning' and intelligence must be undertaken.

2. DEFECTS OF THE 'LEARNING THEORY' APPROACH

Kuo (1967:140) remarks that 'most, if not all, of the experiments on learning have been based on the oversimplified S-R formula. In the case of operant conditioning, many advocates prefer to carry out their experiments without reference to *pre-experimental history* [of the individual subject], *extra-experimental conditions* (environmental context or setting), and, of course, what happens under the skin of the subject. . . . Their oversimplified conceptualization has led the learning experimentalists into a cul-de-sac from which their only escape is by *postulating* motivation, learning set, sign Gestalt, insight, or various schemata of cerebral organization, and now, more fashionably, information processing DNA or RNA' (p. 141; my emphasis added). What Kuo is talking about when he says that learning' psychologists are forced merely to *postulate* motivation, learning set, etc. is the fact that false (or premature) abstraction leaves no alternative but to accept these features as 'given', as being irreducible and not susceptible to analysis or explanation. But there is no need for abstraction to be false or misleading. If the behavioural capacities of the experimental animal are related to its normal ecological setting, it becomes a straightforward matter to give an account of behavioural tendencies (motivation, learning set, etc.) in terms of the functional requirements imposed by a particular way of life in a particular habitat. Presumably it is to considerations of this sort that Kuo refers when he talks of 'extra-experimental conditions'.

These functional requirements imposed by the environment, and the precise means used by a given animal to adapt to them, can in turn be explained only in terms of the phyletic history of the animal's ancestors and the evolutionary processes to which they have been subjected. It is true that an explanation of behavioural capacities in evolutionary-adaptive terms must necessarily involve more inference and even outright speculation than does explanation in terms of directly observable mechanisms. Nevertheless it is easy on the one hand to over-stress the degree of uncertainty which is generated by increased dependence on inference as against observation; and on the other hand to neglect the fact that a comprehensive account must *necessarily* include the phyletic/evolutionary side of the story. One's explanations may have greater superficial neatness if the untidy evolutionary considerations are simply omitted. The superficial neatness is bought, however, at the price of an increased risk of experimental results being spurious or beside the point – and at the price of necessary incompleteness.

We have so far been considering the possibility that 'learning processes' as understood in a reasonably restricted sense (at something approaching the level of 'operational' definitions) can give an adequate account of behaviour. It becomes obvious that they cannot. Too many factors have to be accepted simply as 'given', as not requiring or not being susceptible to explanation. 'Learning processes' are therefore capable of explaining only limited and circumscribed changes in behaviour. It might be thought that this difficulty could be overcome by widening the concept of 'learning'. The trouble is, that as the concept is widened, its operational utility becomes ambiguous and suspect – and even then the evolutionary side of the story remains unexplained. The limiting condition is found where 'learning' is given the widest possible meaning, to include *all* factors, ultimate as well as proximate (see Lack 1954); and at this stage, 'the widest possible meaning' becomes 'no meaning at all'. Kuo (1967) argues this point very well:

The fact is that if we treat the animal as an ever-changing organism and if, whenever we study behaviour, we take all . . . determining factors into account (instead of just isolating certain elements, such as the effect of reinforcement, maze learning, sensory discrimination, imprinting, maturation, critical period, etc. and leaving out an enormous number of other influencing factors), the term 'learning' or 'learning process' becomes meaningless. We will then be dealing with the re-organization of behavioural gradient patterns at a particular stage of the epigenetic process, seen in dynamic relation to the changes of the environmental context. Such a concept of patterned reorganization of behavioural gradients is far broader, far more comprehensive, and far more inclusive than any definition of learning. Therefore, the concept of learning will be scientifically superfluous, as all the phenomena of the classical experiments of learning (including imprinting) will be readily absorbed and integrated into the patterns of behavioural gradient, and will not reappear as independent entities (pp. 141–2).

As Lorenz (1965:81) remarks, in criticism of assumptions which had at one time been accepted by himself as well as by 'learning theorists': 'I came to realize rather late in life that "learning" was a concept illegitimately used by us as a dump for unanalysed residue and that . . . none of us had ever bothered to ask *why* learning produced adaptation of behaviour' (My emphasis added.)

This last question of why particular learnings are able to produce adaptation of behaviour is an absolutely basic one, and will be taken up below – for the moment the similarity may be pointed out between the present situation with regard to learning and the situation at the height of the 'reflexology' movement. Before 'reflexology' collapsed as a general theory, the concept of the reflex had been stretched to include virtually all types of behaviour. 'Reflex' had come, in fact, to be synonymous with 'behaviour'. It had thus necessarily lost its power to explain behaviour and to serve as a basis of classification (cf. Chapter II). At the present time the concept of learning has been similarly overstretched. It can regain its utility in a similar way to the reflex concept, by reverting to a more limited meaning and to a circumscribed range of application. The question is, precisely what meaning? And what range of application?

It is not intended to offer, in this context, a systematic argument directed toward answering these questions. Some points may indicate positive possibilities for a sounder and more restricted concept of learning; my main intention, however, is from the learning theorist's viewpoint the negativistic one of chopping off various areas where 'learning' notions are inappropriate and misleading. For example, the use of 'learning' in relation to problem-solving in humans causes logical anomalies (Green 1966). If we talk of 'learning to solve problems', we must motivate the human subjects to our experiments: 'they must *try* to learn, whatever that means', as Green remarks (p. 7). The absurdity of ignoring the factor of motivation he demonstrates in the following anecdote: 'I once had a subject in a finger maze learning experiment who, unfortunately, had read and believed the theories of incremental learning. He passively ran his finger up the maze, trial after trial confidently expecting his finger finally to learn the maze. After his finger had had four times as many trials as the other subjects had needed, I persuaded him to take a more active role in the proceedings!' (*ibid.*)

This illustrates the point made by Kuo, above, that a 'learning theory' which simply postulates as 'given' or takes as axiomatic such matters as motivation, learning 'set', and so on, is likely to be wildly astray in at least some of its explanations. Even the classifications of behaviour based, implicitly, on theory of this sort, are likely to be bogus. Ryle (1967) exposes one of the logical anomalies generated by current 'teaching and learning' notions in relation to higher-order human activities. 'Teaching fails,' he says (p. 105). 'That is, either the teacher is a failure or the pupil is a failure, if the pupil does not sooner or later become able and apt to arrive at his own solutions to problems. But how, in logic, can anyone be taught to do untaught things?' This is the logical problem at the heart of notions like deutero-learning (see Barnett 1967), 'learning to learn', 'learning to think for onself', so on. Ryle instances the pupil who 'learns', by himself, to read a word he had never met before. 'There is the semblance of a conceptual puzzle here,' he says (p. 106). 'For we seem to be describing him [the pupil] as at a certain stage being able to teach himself something new, which *ipso facto* was not yet in his repertoire to teach. Here his teacher was as ignorant as the pupil, for they were the same boy. So how can one learn something from the other?'

The logical problem is dissolved, it appears, if two provisos are made. First, that *what* is taught is not only a particular and specific operation, but also a pattern of procedure, a *modus operandum*. Second, that the pupil brings to the 'teaching-learning situation' the motivation to acquire new skills, to increase his behavioural repertoire. On the latter point, Ryle dissects out very clearly the relationship between compulsive and voluntary components in teaching-learning. Some parts are within the power of the teacher to compel; others are not. 'How can the teacher make or force his pupil to do things he is not made or forced to do . . . How can the teacher be the initiator of the pupil's initiatives . . . He cannot. I cannot compel the horse to drink thirstily though I can take him to the water. I cannot coerce Tommy into doing spontaneous things. Either he is not coerced, or they are not spontaneous' (p. 112).

The practical problems of teaching do not concern us here. Suffice it that there are various activities into which the teacher can entice or if need be coerce his pupils, which once they have started usually generate the motivation for further 'exploration' and 'learning'; but that in some circumstances teaching may be impossible. As Ryle says, 'a total absence of eagerness or even willingness spells total unteacheability'.

The problem of motivation for 'learning' may be represented in the extreme case as a vicious infinite regress. If Tommy has no desire to learn a particular activity we want to teach him, we may exhort him to *want* to learn. He may then reply, quite truly, that he is trying to want to learn, (or, that he wants to try to learn), but that the second-order motivation does not generate the required first-order motivation. On the other hand, it may be that even the second-order motivation is absent; and an exhortation to try to generate or acquire it is clearly committing us to the slippery edge of the precipice . . .

This argument needs greater elaboration than it can receive here, but even from this thumbnail sketch it is apparent, I suggest, that in the human teaching-learning situation it is impossible to produce the requisite motivation by moralistic exhortation. The would-be learner cannot 'switch on' motivation for specific learning even if he wants to! (No wonder Green (1966) is puzzled over what it could mean to 'try to learn'.) The appeal to the pupil's conscious mind, the 'appeal to reason', must in this context fall on deaf ears.

But is the case hopeless? Is it in fact impossible to remedy a deficiency in, or more especially a total absence of, motivation towards learning'?

I shall argue that it is not; and in particular that if the problem is formulated in terms of instinctive mechanisms and the four-factor theory of evolutionary intelligence, a fruitful strategy will be suggested by the very clarification of the problem which results from its being conceived in these terms. It will be found that a number of practical measures deducible from the theory correspond with what has, in fact, been found pragmatically useful, in some instances for many years past. In this context, the present general theory of intelligence can thus be regarded as a general theory of 'learning' and education which unifies several previous disparate areas of explanation.

In order to present a theory of the interaction of instinct and evolutionary intelligence – which will serve as a basis for a general theory of human behaviour as well as for education – it will be necessary to enter the latest phases of the learning-v.- instinct controversy.

3. LEARNING AS AN INSTINCTIVE ACTIVITY

The assumption of an exclusive antithesis between instinctive (equated with 'innate') and 'learned' behaviour results in a pernicious begging of all questions. It is simply not true that, as Hebb (1953) contended, the one can be defined only as the negation of the other. It has long been recognized, in fact, that a great many instinctive behaviour patterns, which in their general outlines can be regarded as innate, yet depend for their detailed form and orientation upon specific learnings (see Thorpe 1963 for discussion and further references). What has not so far been generally recognized and accepted is what Lorenz (1965) has pointed out: that each 'learning' must be dependent upon an innate or instinctive 'capacity or propensity to learn'.

The key issue is that of motivation or disposition.

In instinctive patterns which incorporate learned components it is clear that if the final 'innate plus learned' configuration of behaviour is to be adaptive the specific learnings within it must have been guided and directed so that they do in fact contribute (in the majority of cases anyway) to the final adaptiveness. Except on the assumption that the activities leading to the learnings are initially entirely random – which would raise virtually insuperable problems for the achievement of adaptiveness under natural conditions (see Chapter III), and for which there is no empirical evidence whatever – it must be conceded that guidance toward the appropriate (i.e. adaptive) learnings must be incorporated in the non-learned parts of the total pattern. That is to say, guidance towards the appropriate learnings must be, in some sense of the word, 'innate'. It need not be argued that this 'guidance' should be determinative and rigid. It will not be an argument against its existence, that maladaptive learnings do on occasion occur (see remarks on new habits in birds, Chapter III).

The famous imprinting of newly-hatched goslings to human surrogates would be highly maladaptive, would in fact be lethal, if the humans in question were irresponsible in the discharge of their acquired 'parental' functions! In the nature of the case, evidence for the occurrence of such maladaptive learnings under natural conditions must be almost impossible to obtain, since the resulting relationships would be necessarily ephemeral: the goslings would die.

It is easy, on the other hand, to see why the question of the adaptiveness of learnings should have been neglected, or simply not perceived, by most of those who have hitherto been responsible for the elaboration of 'learning theory'. And 'adaptiveness' here must include, as has already been emphasized in this book, not only the final adaptiveness of what has been learned but also the rapidity with which the learning has taken place. Both final adaptiveness and appropriate speed of acquisition of learned patterns are difficult to assess under laboratory conditions. Lorenz (1965: 97–9) instances the mistaken conclusions of Riess (1954), criticized and corrected by Eibl-Eibesfeldt (1955, 1963), to show the importance of appreciation of the time factor in such experimentation. It is not impossible that useful experimentation should be carried out under laboratory conditions – but it cannot be carried out unless the behaviour and general biology of the species in question *in the natural state* is well known and understood. A practical corollary of this is that it is doubtful if really penetrating laboratory investigations can be carried out except by individuals who are themselves directly and extensively acquainted with their subject species' natural behaviour. The pioneers of ethology (and their 'F_1 generation' trained mainly at Oxford and Cambridge in the 1950s) have appreciated this very well, and their practice has resulted in the immense progress in behavioural understanding in the last two decades. (Unfortunately there are signs that later-generation ethologists fail to appreciate the need for slow holistic investigation. Particularism and technicism are creeping in, part of the evolutionary succession (see Chapter I) that leads to 'maturity' in a science (Kuhn 1962, e.g. p. 11). It is important to note that 'maturity' here is purely a descriptive term, with no tincture of approbation or prescription.)

The tacit assumption that speed of learning is irrelevant, hence that many adaptive activities could be left to 'undirected' learnings, can thus be seen as the outcome of the laboratory approach to the study of behaviour. A mistaken and premature emphasis on precision and quantification has dragooned behavioural scientists into the laboratory decades before the state of the subject allowed their findings there to be other than equivocal. The crux of the matter is that there is no natural selection in the laboratory. Learning experiments have therefore been conducted in an ecological and evolutionary vacuum. If an animal is given a 'learning task' in the laboratory, even one which may superficially appear to be related to its natural behaviour (food getting, nest building, and so on), there is little wonder that the many discrepancies from the true 'natural conditions' situation should disorientate its behaviour sufficiently so that its attempts at learning might well appear haphazard. The environmental cues which would normally direct its appetitive behaviour are largely absent. Yet since time and opportunity are effectively unlimited compared to natural conditions – where predators, intra- or inter-specific competitors, and a variety of environmental changes all combine to limit the amount of shilly-shallying a wild animal can get away with – the animal can eventually achieve completion of the 'learning task'. Even so, it is unlikely that the 'inbuilt guidance' of appetitive behaviour has been completely lacking; but it is unlikely to have been observed as such, especially in the atypical context of the laboratory, unless by an experimenter intimately familiar with the natural behaviour of the species in question. It must also be assumed that the experimental animals were in fact 'wild' individuals which could be expected to possess whatever genetic basis might be required for the non-learned 'guidance' behaviour in question. Lorenz (1965:100) points out the absurdities which result if this condition is not fulfilled.

Thus the debate between the ethologists and the 'learning theorists' can be resolved into a series of dilemmas directed at the extremists of the latter group:

(1) Either there is no 'innate' or endogenous behaviour at all, and all behaviour without exception is learned; or else some behaviour is learned and some is innate.

The first alternative is clearly so much at variance with the facts, with the very possibilities for survival, that it must be rejected. The second alternative must, then, be accepted. It is in close accord anyway, so far as it goes, with the views of most of the non-extremists on both sides of the controversy. (A third alternative at this level may be mentioned only to be dismissed, namely, that all behaviour is innate. Nobody, presumably, would entertain this proposition for a moment, yet it is if anything less unlikely than its contrary, that all behaviour is learned. There is no animal totally devoid of unlearned behavioural components. On the other hand, there are plenty of relatively 'lowly' animals, sponges for example, in whose behaviour learning appears to play such an insignificant part that it might well be said to be non-existent).

Granted, then, that some behaviour is learned and some innate, we can move on the the second dilemma:

(2) Within the total behavioural repertoire of a species or individual, the learned and the innate are functionally quite separate and unrelated and constitute different sections of the repertoire; or else they are interrelated within each of the adaptive behaviour patterns, and do not occur separately.

This dilemma incorporates a certain degree of over-simplification which is necessary in order that the issues involved may be clarified. Few people, presumably, would wish to take the first horn of the dilemma – as a categorical and general proposition it is simply not true, there are many patterns in which the interplay and functional interrelatedness of learning and instinct are well established (see Thorpe 1963). The second horn of the dilemma, on the other hand, is not as it stands wholly satisfactory. Extreme protagonists of 'learning theory' have been at pains to argue that even in behaviour which seems to occur without prior learning, the possibility of such learning having occurred, even in the egg or *in utero*, cannot be ruled out (see Lehrman 1953). Although this smacks of 'argument to limitation' insofar as it is made a *general* objection to the existence of innate patterns, it is true that in particular behaviour patterns the exclusion of specific learnings of specific components is often difficult and problematical. Lorenz himself pointed out (1965:96) the need for extreme rigour in investigation before assertions of the absence of learning in relation to specific elements can be made. There are on the other hand many behavioural elements (e.g. reflexes) which nobody would assert to have been learned, even though they can be and often are modified by learning.

The key issue with respect to the second alternative of dilemma (2) is, of course, the extent to which learned and instinctive/ innate are combined within the repertoire of a species. It is their relative importance which is in question. In order to short-circuit what might otherwise be interminable argumentation on details here, I propose to reformulate the position as follows:

In the majority of adaptive behaviour patterns upon which

animal species depend for the performance of their vital functions, there is a dynamic interrelationship between learned and innate components.

This leaves open that there might perhaps be a few patterns which are wholly innate or wholly learned, but assumes that the importance of such patterns for species' survival is relatively small. (Implicitly, I am neglecting the behaviour of 'lower' animals. In them, innate patterns are of great importance (Ewer 1957), but extremist 'learning theorists' are not likely to be familiar with them, hence are unlikely to bring them into the argument!)

If it is granted that the majority of adaptive behaviour patterns are comprised by some sort of blend of learned with innate, we can move on to the third dilemma:

(3) In the majority of behaviour patterns, adaptiveness is achieved by the interaction of innate/instinctive with learned elements – but either the innate/instinctive side is of more importance than the learning side, or *vice versa*.

The form of this dilemma is designed with intent to exclude the easy middle way of saying: 'Both are important, both are essential, it is impossible (or undesirable) to attempt to assign any priority between them.' Even though one had to return to this position after investigating the other possibilities, it would still be necessary to provisionally negate it in order to be able to do justice to them. It is not the mere generality of interaction between instinct and learning which is important, but the precise nature of the interaction. The argument for priority of instinctual guidance of learning has been outlined above in terms of the functional requirements of rapid achievement of adaptiveness. Several other arguments are briefly adduced below.

In terms of the general phylogeny of behaviour it is clear that the 'lower' forms from which the 'higher' have been derived must have been, as their living relatives of the same grade still are today, almost exclusively dependent upon inbuilt responses in the form of reflexes, tropisms, taxes, and instincts. These 'lower' animals are in general small, short-lived and rapidly reproducing. As Ewer (1957) has argued, and as has been mentioned already in Chapters II and III, they must be able to perform their vital functions and get them right first time. This means, in the extreme case, that there will be no learning because there is no opportunity for it. This entails that when in the course of evolution learning capacities are first developed, the principle of continuity of function demands that they be grafted onto and incorporated within patterns of behaviour which are still basically instinctive. Thus the initial temporal priority of instinct over learning carries over into a form of logical or functional priority, whereby learning is indeed a necessary part of particular sequences of behaviour but *what* is learned is determined by the appetitive (instinctive) sequences which lead up to it.

The same argument can be even more illuminating in converse. A behavioural system from which learning is entirely absent can adapt to changing circumstances only through the processes of evolution, i.e. natural selection acting upon individual phenotypic variations produced by random genetic changes. ('Changing circumstances' here indicates changes greater than those within the tolerances of the relevant innate releasing mechanisms – clearly, changes within these tolerances need no further adaptation from the animals in question.) If the releasers of adaptive activities are learned, however, rather than innately programmed, the way is open for each individual to adapt to circumstances as they happen to be when the releasers (and/or orientating stimuli) are first learned. Thus the minimum time-span for the adaptation of the species to environmental changes is reduced from whatever time is necessary for the evolutionary change (usually at least a number of generations) to that necessary for the production of a single generation (about one year in the majority of species). This undoubtedly accounts for the spread of learning capacity through all major groups of animals. A pervasive and powerful basis of selection pressure such as this would allow polyphyletic geneses of learning ability to result in convergence upon a similar *functional* result at the behavioural level even if the causal mechanisms at the anatomical-physiological level happened to be greatly diversified (as it seems probable they are). Thus the similarity in the 'learning curves' of different animals under laboratory test, to which so much importance is attached by learning theorists (see Bitterman 1960), could very well be extremely superficial in terms of proximate causal mechanisms. In terms of evolutionary mechanisms, of course, the similarity could be meaningful along the lines indicated above – but these are not the lines along which the learning theorists apparently want to work.

The replacement of innately programmed releasers by learned releasers, postulated above, is presumably what Lorenz has in mind when he writes: 'The obvious trend in the evolution of behaviour in the direction of greater plasticity and increasing influence of learning and insight has to be regarded at least as much as a consequence of reduction and disintegration of innate fixed patterns as of higher development of those functions which, in the individual's life, effect adaptive modification of behaviour' (1965:5).

For a general explanation of the evolutionary shift from innate programming to learning and *a fortiori* to intelligent behaviour it is highly misleading to talk of 'reduction and disintegration'. If there were a real disintegration the behaviour could no longer be adaptive, the individual would die and the species become extinct. The adaptiveness of the behavioural system must at all costs be preserved. That is why (as was argued in Chapter III) the instinct system must be preserved until the new system is sufficiently elaborated to supplement and finally, perhaps, supplant it. The fact that the instinct system must be kept in being is what generates the necessity for the P-factor of intelligence. The capacity to hold-off the still strong instinctive tendencies would be unnecessary if they had in fact disappeared. It is only within the last few years that the existence of strong instinctive mechanisms in humans has begun to be recognized. The enormous interest in Desmond Morris' two best sellers, *The Naked Ape* (1967) and *The Human Zoo* (1969) may herald a widespread abandonment of the 'learning theory' paradigm. Psychology and the other social sciences may be about to emerge from their pre-Darwinian assumptions, and may be about to accept and get to grips with the implications of the evolutionary continuity, *in instinctive mechanisms rather than learning*, between man and the other animals.

The priority of instinct over learning can therefore be argued on grounds of:

(1) *Evolutionary continuity.* If ancestral animals were dependent on instinct (or 'innate programming' as a more general term), similarly to modern forms of the same level of organization, then either the instinct-system must have retained its function in the overall patterning of behaviour while learning replaced or supplemented various details piecemeal, or else there must have been a sudden and enormous evolutionary jump from the old system to the new one. It is difficult to see how adaptiveness could have been maintained during the latter transition.

(2) *The adaptive requirements of ontogeny.* This may be argued as

follows. The human pre-eminence in learning, indeed an alleged dependence on it, is presented as the outcome of the extremely long period of juvenile dependency in the human. It is also claimed, by extremists of the 'learning' camp, that all or most of the other higher animals are dependent on learning. But they lack the juvenile dependency period of the human. The 'learning' extremists cannot therefore retain *both* universality of dependence upon learning among the higher animals *and* the relationship between learning and juvenile dependency. One or other must be dropped. The facts would appear to demand abandonment of the 'universal learning' tenet.

The basis of the above argument is, of course, the analysis of functional requirements during ontogeny which was presented earlier.

The functional requirements of adaptiveness during ontogeny come down to this, that learning can and indeed has been incorporated to an increasing extent in the behaviour of the higher animals, but only because specific learnings have been interpolated within a framework of instinctive or innately programmed behaviour which has preserved the adaptiveness of the total pattern.

Lorenz (1965) makes exactly the same point in a slightly different way when he talks about the 'innate schoolmarm'. 'None of us', he says (p. 81), 'had ever bothered to ask why learning produced adaptation of behaviour. All older ethologists, with the outstanding exception of Craig, had been confining their attention to the innate, while more or less neglecting all problems of learning, without realizing that in doing so they were neglecting one of the most important functions of the majority of phylogenetically adapted behaviour mechanisms: the function of teaching!'

He goes on to point out that the concept of the consummatory act, its mechanisms and functions, can be brought into significant relation with innately guided acts of learning. For this to be done, it is necessary that the concept of the consummatory *act* be altered. As was argued in Chapter III, if we think of a consummatory *situation* as being the end result of an instinctive sequence, with the termination of the sequence occurring as a result of sensory feedback rather than as a result of exhaustion of action-specific energy (see Chapter II), we are enabled to assume that any variant of the behavioural sequence which attains the consummatory situation will be at no disadvantage, in terms of probability of subsequent occurrence, in comparison with the original sequence. Thus learned variations can be incorporated. Since the consummatory situation tends to have reinforcing value, an animal will tend to repeat whatever behaviour it has found successful in the past. If the past behaviour has achieved adaptiveness (i.e. the consummatory situation) as a result of instinctive appetitive sequences leading to appropriate learnings, then these learnings will tend to remain as a permanent part of the repertoire. As Lorenz suggests, 'It is hardly an exaggeration to say that the special structure of a typical consummatory act is quite as much the result of the selection pressure exerted by its function of reinforcing appetitive behaviour as of the one exerted by its primary function' (p. 81). In short, the consummatory sequence, seen in reverse, as it were, is simply the link between the final consummatory situation (which is more or less fixed by the structure of the animal in relation to the function to be performed) and the 'crescendo of particularity' within the final phases of the appetitive sequence. The latter are equally necessary with the former, though of course only because they contribute to it. And if the consummatory performance has reward value, the appetitive phase can be variable and can incorporate a variety of learned components, even though the consummatory sequence itself has to be fixed.

4. ADAPTIVE VARIABILITY AND THE P-FACTOR

This brings us to the crucial issue of the relative utility of the concepts of learning and of intelligence. If it is granted that a learned component, once performed successfully, will be reinforced and probably retained in the repertoire – granted even that the instinctive components of the appetitive sequence perform their 'innate schoolmarm' function of guiding *what* is learned so that it is adaptive – the question remains: by what means can there come to be a particular act of learning, at some particular and appropriate place within the sequence, in the first place?

'Truly "new" learned movements . . . can evidently be developed only by very few animals' (Lorenz 1965: 72). The basic reason for this is that, before a new element can be inserted in a sequence, a gap must exist to accommodate it. One way in which this could occur would be if the genetic 'code' for the behaviour came to be deficient at the appropriate place. It is quite unnecessary to postulate a simple one-to-one relationship between genes on the one hand and behavioural elements on the other, and that a gene must be 'missing' before a new learned element of behaviour can be substituted for a previously-instinctive one. The real situation is likely to be both more complex and more subtle than this, with the 'deficiency' occurring perhaps as a pleiotropic effect of genetic interaction. Nevertheless the simple model does represent what is causally required: the absence of whatever genetic and/or epigenetic configuration previously determined the behavioural element now to be missed out. Thus the opportunity to learn something new is in this frame of reference directly dependent upon a prior genetic or developmental change. It is, in fact, an evolutionary change. The overall rate of change which is possible is therefore an evolutionary rate. Undoubtedly a great deal of behavioural change and development of learning, especially in the lower animals, has occurred as a result of mechanisms essentially of this sort. Its limitations must, however, be recognized, if it is proposed as a mechanism to account for all types of behavioural change and enhancement of learning capacity, especially for example, the changes and capacities exhibited in some mammalian groups. 'Gap dependent' innovation involving learning is unlikely to occur anywhere near as frequently as the genetic 'deficiencies' upon which it must be dependent. There are several reasons for this. One has been touched upon already: that even given the guiding function of the rest of the instinctive sequence (i.e. the unaffected parts of it around and especially preceding the 'gap'), the animal must still produce the new response to meet the two conditions of overall adaptiveness. The new response must be:

(a) Adaptive in itself, in detail; and
(b) It must have been produced quickly enough.

Another reason for the production of successful innovations falling short of the number of 'gaps' is that the genetic changes which cause the gap are likely often to cause disruption of other parts of the behavioural sequence in question. In such cases the 'schoolmarm' becomes inefficient; and even if an innovation is produced which *in itself* might have been advantageous, the adaptiveness of the total pattern may be lowered, perhaps to the point of becoming dysgenic, by changes in the other parts of the sequence.

In general, this mechanism for the genesis of innovation and the development of learning ability does involve in a literal sense, the 'disintegration' of instinctive behaviour to which Lorenz alluded.

If, on the other hand, the question of behavioural innovation is cast not in terms of learning theory but, as in the present investigation, in terms of the generation of intelligent behaviour out of instinctive, our conceptual armoury is extended to include the possibility of selective withholding *within* a non-disintegrated instinct-system. If, instead of 'gap dependent' and in fact 'gap subsequent' learning theory formulations, we set out prepared to accept the need for a *'gap creating'* mechanism, if we are led by our argument to expect one, it may be that we can formulate a theory which will surmount the limitations inherent in the 'learning theory' approach, and which will enable us actually to discover evidence of the existence of the type of mechanism required. The 'gap creating' mechanism (P-factor) not only allows the development of intelligent behaviour as we have defined it, stipulatively, in the present study, but also allows for much more rapid and extensive evolution of 'learning ability' in the narrow sense. This is because phenotypic (behavioural) innovation can occur much more frequently – even to the extent of occurring more than once, for a particular behaviour pattern, within the lifetime of a single individual, though such a high frequency would be rare – instead of having to wait upon the slow processes of genetic change as in the 'gap dependent' hypothesis. This in turn means that selection pressures will be enormously enhanced, even for 'learning' prerequisites such as the A- and C-factors. The first appearance of the P-factor must have acted as a 'multiplier' of the adaptive value of the other factors of intelligence and of learning ability. By stepping up the frequency of learning, even more of adaptive variability within the lifetime of the individual, the P-factor must have set up strong selection pressures in its own favour (see Chapter VI) and also in favour of the other factors.

Thus the P-factor can help to explain some of the particular puzzles of intelligence-cum-learning ability. The supreme infrahuman powers of learning ability or intelligence are found in the higher primates and in cetacea. Why in these particular groups? In discussing 'truly "new" learned movements', which as he points out 'can evidently be developed only be very few animals', Lorenz (1965: 72) states that 'Dissociated, or to be more exact, unassociated muscle contractions are at the beck and call of the pyramidal system and can, therefore, be directly and independently activated by our will. They are the raw material which motor learning strings, like beads on a thread, into a sequence and welds them into one skilled movement . . . Curiously enough, it seems that the beads which are strung into modifiable sequences by learning are, to a great extent, gained during phylogeny by unstringing phylogenetically adapted fixed sequences of coordination. The study of locomotor coordinations in primates makes this appear highly probable . . . As literally every single step and grip of the prehensile limbs must be spatially directed in its smallest detail, it is obvious that the length of fixed motor sequences must be reduced to a minimum and the mechanisms of spatial orientation expanded to a maximum. The "lack of rhythm", which is so conspicuous in exploratory movements of the great apes, is a consequence of this.' Now, some of the 'unstringing' is no doubt attributable to genetic changes. If this were the primary cause, or even worse the only cause of the unstringing, it seems certain that only chaos could result. Such 'unstringing' would be the opposite of adaptive. If, however, the genetic changes – such as were needed – could take place under the phenotypic cover of the operation of P-factor 'unstringing', the adaptiveness of behaviour could be maintained throughout and the genetic changes could follow 'at their leisure'.

Lorenz proposes a mechanism to account for cetacean skills, which is also susceptible to re-interpretation in terms of P-factor operation. I shall allow the reader to formulate the reinterpretation for himself. 'The possible explanation which I venture to suggest', says Lorenz (1965: 77–8),' is that the ancestors of the whales were, at the time they began to revert to aquatic life, carnivora with comparatively highly developed brains and were endowed with a fair portion of voluntary movement. When, consequently, a selection pressure was brought to bear on the development of motor patterns that were entirely new to a terrestrial animal, some of it, instead of creating new phylogenetically adapted fixed motor patterns, caused a higher development of voluntary movements which . . . could be turned to any purpose, including efficient swimming techniques. This would account for the fact that aquatic mammals, for example otters, sea lions, and whales, are able to produce such an amazing multiformity of newly created and elegant, skilled movements in their "play".'

The significance of Bitterman's (1960, 1965) 'habit reversal learning' can also be explained by reference to the P-factor of intelligence (see above, pp. 289–90). The learning of any habit *de novo*, if as argued earlier it is dependent upon an instinctive motivation, might entail the exercise of P-factor withholding of whatever 'normal' activity was usually associated with the particular instinct in question. It is in the *unlearning* of an already-established habit, however, that the most obvious exercise of the P-factor must be involved. In an animal totally devoid of P-factor, habit reversal learning would be impossible, since 'punishment' for the performance of the previously-established habit would merely generate a contra-motivation equal and opposite to the 'original' motivation. This would result, therefore, in motivational nullity, i.e. no motivation at all. In such a situation, learning a new habit would be impossible. What is required is that the 'punishment' for the performance of the first habit, should result in a withholding of that activity but *with motivation available for learning a different activity*. Two distinct mechanisms are required: one to withhold the 'old' habit, the other to learn the new one. The capacity for withholding the 'old' habit may be ascribed to the P-factor of intelligence.

Thus what is peculiar to 'habit reversal learning' as distinct from 'ordinary *de novo* learning' is the implicit recognition of, and indirect measurement of the withholding mechanism. And for the reason why this is significant, the reader is referred to Chapter III, above.

To sum up, then, it is possible to perform a 'voluntary' action only so long as one can *not* perform an involuntary one. This 'not performing' is the peculiar function of the P-factor of evolutionary intelligence.

As to the general question of 'learning' *v.* 'intelligence', the relationship to motivation constitutes the crucial difference. This was stated earlier, and can be explained succinctly now. Learning is in fact itself basically an instinctive activity. In its bare form it is at the mercy of instinctual motivation. It is only intelligence which, by the operation of P-factor inhibition of instinctive mechanisms, can cut across the 'tide flow' of motivation. Learning, in the 'higher' animals at least, is always contaminated by intelligence. The actual behaviour which we visualize when we talk of 'learning' is thus really a conflation of learning with intelligence – this is what has lent plausibility to several decades of Ptolemaic 'learning theory'. But inadequacies and anomalies have multiplied, even within the false abstraction of much that has gone under the rubric of 'learning theory'. With the concept of adaptiveness now breaking through the laboratory roof, behaviour theory is being forced to come to terms with the real world. Space, in the form of 'natural conditions' observations, and Time, in the form of evolutionary considerations, lead us inescapably to a reassessment of our origins in relation to them. One of the dark places in our origins has been the genesis of our intelligence. The present theoretical exploration has been directed towards the eventual illumination of this; and to the extent that the argument has been cogent, a modified theoretical framework for the understanding of learning, as well as of instinct and intelligence, should have been provided. The significance of such modification of theoretical framework thus comes into question. Does it matter what framework we use? Do not the 'facts speak for themselves', irrespective of context? Assessment of these and similar issues is forced upon us; and will be attempted in the next chapter.

XII

Science and Philosophy in the Understanding and Use of Intelligence

1. SCIENCE OR PHILOSOPHY?

The fact that reference is made, here and there throughout the book, to matters of fact and of direct empirical observation – and that my theoretical take-off point, outlined in Chapter II, has been the recently developed and even more recently accepted body of scientific theory known collectively as ethology – may have given the impression that I have been attempting simply to extend the range of applicability of ethological theory, as part of what Kuhn (1962) has called 'normal science'. In one sense I have, of course, been doing just that. The appeal to fact has been made – even though it has demanded, in most cases a reinterpretation of existing reports, the original terminology of these being tied to a different frame of theoretical reference. But this reinterpretation has been more than a matter of mere terminology, of changing one set of words for a different set. The use of one set of terms, one particular theoretical framework, rests on a substratum of unstated and often unconsciously held assumptions of a methodological and evaluative nature – we do have grounds for our methodological beliefs, they are not arrived at either by random chance or by arbitrary selection (cf. Stenhouse 1968). If a change is to be made in terminology or theoretical framework, then, it may be that a concomitant change will be necessitated in fundamental presuppositions.

When our field of action changes from the relationship between theory and observational fact, however, to the elucidation of unstated presuppositions and to the assessment of logical coherence between one set of presuppositions and another, it might seem that we have changed from doing science to doing philosophy. So the book might also be regarded as a contribution in philosophy. Not 'pure' philosophy, perhaps; and certainly not what would be regarded in many quarters as 'traditional' philosophy – but in being concerned with the inference as well as, and something apart from, observation, in attempting to reassess and alter criteria rather than merely apply them, and in working often at high levels of generality and abstraction, I feel that it accords with views of philosophy that are coming to be accepted at the present time. As Passmore (1969) remarks of Shwayder's (1965) approach, 'he does not explicitly deny the possibility of constructing a causal-explanatory science of behaviour. But his own approach is . . . "resolutely definitional". In the Aristotelian tradition, he sets out to clarify and classify rather than to explain.' Clarification and classification are necessary prerequisites for explanation. While the present book may be seen in terms of science as an attempt to meet Gregory's (1961) demand for 'functional block diagrams' ' to serve as models without which experimentation must be blind, it might equally well be regarded as an attempt, within philosophy, to follow Wittgenstein's advice (as reported by Passmore 1969: 8) to make 'not more experimental investigations or better theories but more careful conceptual investigations'. As conceptual relationships are clarified it would be expected and hoped, of course, that experimental work would be facilitated and would ultimately support the findings of the more abstract explorations. And while I would hope that the present work would contribute to both clarification and classification, perhaps the possibility of contribution along this line is not lessened by my having departed from the Aristotelian and Shwayderian fold to the extent of having attempted to be both 'definitional' *and* explanatory.

2. SCIENCE—'NORMAL' OR 'REVOLUTIONARY'

Reverting to the question of the scientific status of the present work, Kuhn (1962) argues for the distinction (mentioned above) within the general category of 'scientific activity' between what he calls 'normal science' and 'scientific revolution'. The former is what is commonly meant by references to 'science'; the latter must often be regarded as 'non-science', or even as antithetical to science, especially by practitioners of the pre-revolutionary 'normal' science – yet the 'revolutions' in science constitute its major advances. The Copernican and Darwinian revolutions may be cited as examples.

' "Normal science" means research firmly based upon one or more past scientific achievements, achievements that some particular scientific community acknowledges for a time as supplying the foundation for its further practice. Today such achievements are recounted, though seldom in their original form, by science textbooks' (Kuhn, 1962: 10).

The 'scientific achievement' in question here is the demonstration that a particular set of assumptions and approaches – collectively designated by Kuhn a 'paradigm' (pp. 33 ff.) but approximating in many ways to what Toulmin (1961) has called an 'ideal of natural order' – has the power to generate explanations and predictions of, and possibly ultimately a degree of control over, particular classes of natural phenomena. Many of the assumptions which constitute the paradigm are unstated and even unsuspected when it is first articulated; in fact one of the functions of 'normal science' is to explore and make explicit what was previously only implicit within the currently accepted paradigm. Other functions are to extend the range of phenomena to which the paradigm is applicable; and to improve the techniques by which its applicability is tested. As Kuhn remarks: 'There are seldom many areas in which a scientific theory . . . can be directly compared with nature' (p. 26). An essential part of the 'articulation of a paradigm', then, is the working out of a methodology and finally of specific techniques through which the coherence of paradigm with reality can ultimately be assessed.

It must be obvious even from the above summary account that 'normal science' is in an important sense ultimately self-negating. Except on the assumption – not so far actualized – that a particular paradigm furnishes the 'ultimate truth' about a particular segment of the universe, it must follow that as time goes on, the activities of normal science reveal a larger and larger proportion of the assumptions implicit in the original paradigm. The range of its application is also extended to greater and greater numbers of different classes of phenomena; and the precision and subtlety of assessment in its relationship with reality is extended closer and closer to the limits set by the available technology. Finally, the situation must be reached where no more implicit assumptions remain to be made explicit, where no more classes of phenomena remain to be tested for applicability of the paradigm, and where the technological limits of the testing processes have been reached. This is the limiting situation for 'normal science'. The paradigm has been worked out, exhausted. Nothing remains to be done. Science comes to a stop.

If science were limited to 'normal science' this would indeed have happened long ago. (Apparently it did, in the Middle Ages, when the Aristotelian and Ptolemaic paradigms were worked out, and when the technological and socio-political conditions in Europe prevented extensions in the range of phenomena available for assessment. It is significant that the Aristotelian paradigm of classification was revived to organize the glorious harvest made available as a result of the European voyages of exploration, trade, and settlement, themselves the result of technological and socioeconomic advances, in the sixteenth to eighteenth centuries. Linnaeus was the successor of Columbus and Magellan as well as of Aristotle!) But in fact science includes scientific revolutions as well as 'normal science', and the generation of new paradigms is just as important as the exploitation of old ones. The important question becomes, then: In what ways do 'scientific revolutions' occur?

A proviso must first be made about terminology. Although it is implicit in Kuhn's work that a 'revolution' in science can be small-scale and completely unspectacular, in a book directed towards readers from a wide diversity of backgrounds it is inevitable that most of the examples he discusses should be the better-known 'revolutions' e.g. the Copernican, Newtonian and Darwinian. His own background and interests direct him, too, toward the physical rather than the biological sciences. Since the theoretical bases of the physical sciences are more definitely structured than are those of the biological sciences at the present time, thus allowing any 'revolutions' within them to be more clearly perceived (and also to avoid so far as possible any suggestion of spectacular importance) I propose in the present context to use, instead of 'revolution', the term 'theoretical re-orientation'. This is deliberately intended to cut across the very important distinction established by Kuhn, between a theoretical change resulting from the explication of an existing paradigm and a change resulting from the adoption of a new paradigm. In short, I wish to leave open, for the present, the question of whether this present book is a further development of an existing paradigm, or whether it is in effect an outline of a new one. In the latter case I should hope to take comfort from Kuhn's plea that 'we must recognize how very limited in both scope and precision a paradigm can be at the time of its first appearance' (p. 23). For the moment I can claim simply to be attempting a certain amount of theoretical re-orientation.

The question of the nature of scientific revolutions – or in the present terminology theoretical re-orientations – can now be attempted.

3. ALTERNATIVE PARADIGMS IN BEHAVIOUR THEORY

Theoretical reorientation is demanded by anomalies of various sorts. These include: phenomena that do not fit the theories or paradigms currently available; inconsistencies of explanations or measurement; and conflicts between different schemata of description and/or explanation. The situation in the behavioural sciences at the present time might be characterized in any of these dimensions, or all of them. For example, with regard to human behaviour the major emphasis is still upon learning theory, no doubt as a result of the advances in understanding which accrued from learning-orientated experiments in the first half of this century. But various phenomena of human behaviour have proved recalcitrant within a 'learning' frame of reference. While 'learning theory' holds the stage for a great many academic psychologists and many theorists of educational psychology, psychiatrists and those clinical psychologists concerned with 'mental health' have found the phenomena with which they are concerned to be unsusceptible to 'learning theory' explanations. Also, in recent decades the phenomena of originality and creativity have obtruded into the 'learning theory' landscape and have refused to be either explained or explained away (see Chapter V). Whether one described this situation as 'failure of a paradigm to cope with certain phenomena' or as 'competition between paradigms to explain partially-overlapping classes of phenomena', must depend upon one's viewpoint. A devotee of 'learning theory' would presumably adopt the first formulation and emphasize that the 'failure' was only temporary (if indeed he even admitted the existence of phenomena which 'learning' could not explain), while an historian of science would undoubtedly take the second. The situation is also describable of course, at the level of detail, in terms of inconsistencies of explanation or measurement – this is probably the least illuminating way of characterizing it, though its 'low key' nature makes it useful as a non-committal lead-in to critical discussion.

The history of the behavioural sciences can show why there

should be several paradigms in part-competition at the present time. What might broadly be described as 'abnormal psychology', including psychiatry and much clinical human pyschology, has been concerned with the adaptiveness of human behaviour, even if only at second remove, in endeavouring to explain and correct what is maladaptive. This tradition goes back at least to Freud, Charcot, and other pioneers at the end of the nineteenth century. Its paradigm is that of homaeostasis, of the self-correcting system in equilibrium with the environment, i.e. an adaptive equilibrium.

In extreme contrast to this is the 'learning theory' tradition. This appears to have developed first in relation to the practicalities and expediencies of human education in the early phases of the 'education explosion' (see Chapter X). The 'intelligence testing' movement had its origins at the same time and in the same circumstances (see Chapter V): the need to achieve universal education which was demanded by the democratic ideals and the increasing technication of society, both emanating from nineteenth-century industrialization and liberalism. Learning theory, although gaining its first impetus as I suggest from the practical demands of human society, went off into intellectual orbit (or into intellectual 'outer space' in infinite regress, as some might assert) as the result of the facility with which its experimentations could be conducted upon laboratory animals. As Bitterman (1960) shows, it is easy to demonstrate a similarity between the 'learning curves' of a variety of organisms – pigeon, rat, monkey, man, and presumably most other animals – and one result of this has been a restriction in the activities of 'comparative psychologists', subject-wise, to man and the white rat. This is on the assumption that, since all learning curves are similar, there is no need for a wide range of species to be studied. Whether the similarity in learning curves is more than superficial and what its significance could be even if its reality were demonstrated, is perhaps open to some doubt. Two salient features of the 'learning theory' approach, already discussed in Chapter XI, are of great and undoubted significance, however: first the virtual absence of considerations of adaptiveness; and second, that the very criterion of successful learning is that the behaviour being learned becomes unvarying.

This stereotypy of behaviour resulting from learning is what is implied by the various forms of 'learning curve'. When learning is perfectly achieved, 'errors' are nil, a 'criterion performance' occurs on every trial. Some further implications of this will be discussed below. At this point a few preliminary remarks on the absence of considerations of adaptiveness will lay necessary groundwork for subsequent elaboration.

Although biologists who have been concerned with learning have consistently related it to adaptiveness (see Thorpe 1963: 55; Hinde 1966: 400), learning theorists from the psychological camp have generally neglected adaptiveness more or less completely. One psychologist who has displayed a better biological insight and a sounder methodological approach than most is N. L. Munn, who in working towards a definition of learning (1955: 200–1) considers adaptiveness and then, quite properly, excludes it from the *definition* of learning. 'We may define the process of learning as a more or less permanent incremental modification of behaviour which results from activity, special training, or observation. A given learned performance, whether motor or verbal, is often referred to as a *habit*,' (Munn's emphasis removed from the definition). We may note in passing the implications of fixity in the last sentence. It is of course necessary that the concept of 'adaptiveness' be excluded from the definition – otherwise it would be logically impossible for a maladaptive activity to be learned. But while it must be excluded from the definition of learning, it is of absolutely paramount importance to recognize that the very existence of capacities for learning, and the forms these capacities take in different species of animals, are intimately related not only to the notion of adaptation in general but also and in particular to the precise details of each species' way of life. By neglecting the adaptive aspects of behaviour and by restricting the investigation of learning to the laboratory, 'learning psychology' has cut itself off from a major and essential dimension, and has condemned itself to a massive and long-sustained false abstraction.

4. PROBLEMS OF CROSS-PARADIGMAL INTERCOURSE

It must be appreciated, however, that the devotees of 'learning theory' are unlikely to acknowledge this deficiency in relation to adaptiveness. Within their own frame of reference, on their own background of education and experience, the dimension of adaptiveness is generally invisible (see Chapter XI). Thus Kuhn's category, among the 'anomalies' which generate a need for theoretical re-orientation, of 'phenomena which do not fit the theories or paradigms currently available,' is of limited utility in the practical business of public argumentation towards changing the general orientation within the field in question. In arguing with a 'learning theorist', one cannot effectively appeal to phenomena of adaptation, because he is unlikely to recognize them. It is necessary, then, to use Kuhn's other two categories, and to demonstrate either 'inconsistencies of explanation or measurement' or 'conflicts between different schemata of description and/or explanation'.

The strength of 'learning theory' has been based on its precision and objectivity: the fact that observations can be quantified, and experiments replicated to obtain consistent results. Hypotheses can be tested, and an unequivocal verdict obtained in most instances. This depends upon the use of a clear-cut criterion of learning; and its testimony is the establishment of 'learning curves'.

We know when an animal has learned to solve a problem, even a very complicated one: it 'gets it right consistently'. This is so irrespective of whether we are talking about a rat or a cockroach learning to run through a maze, a fish or a pigeon learning to distinguish shapes and colours, a boy learning quadratic equations, a crow learning to count, or a monkey learning to solve Harlow's (1958) multiple-ambiguity problems. 'Getting it right consistently' is a criterion that applies to all sorts of problems and all sorts of situations, sensory modalities, etc. – with the one proviso, that we must in practice be able to equate the observed behaviour with the criterion. That is to say, we can apply the criterion in any situation and to any species of animal, provided we know what behaviour is to count as 'meeting the criterion' and what is not.

Is there any category of behaviour in which it would be difficult or perhaps even impossible to use this criterion?

What, for example, about deutero-learning, 'learning-to-learn'? 'A technique promising for the study of interspecies differences in higher forms,' says Hinde (1966: 423), 'is that of testing for the ability to "learn to learn" or "learning set".'

5. DEMONSTRATION OF INTRA-PARADIGMAL BREAKDOWN

It might seem, initially, that the question: 'Has this animal learned to learn?' could be given a straightforward though 'second-order' answer. If the answers to a sufficient number of first-order questions ('Has it learned new task A?', 'Has it learned new task B?', and so on) were in the affirmative, then the second-order question 'Has it learned to learn?' would, it might be argued, also be answered in the affirmative. But there are difficulties in this view.

In the first place, just what is involved in the so-called 'second-order learning'? What is its criterion? In particular, what is to count as a *sufficient* number of first-order learnings before we feel that the second-order criterion has been met?

In first-order learnings, the claim that the criterion has been met has predictive implications. If the claim should be challenged in a particular case, it should be possible to settle the issue by making a further test. If the claim that animal A has learned task X is disputed, we can test animal A again. If it can perform X immediately (or with a reasonable amount of warm-up and practice), then it must be conceded that the criterion had been met and the claim that it had been met was justified. If X is not performed, this does not necessarily entail that the claim previously to have met the criterion was unjustified. Special conditions may have vitiated the test – and if it can be established that they did, the claim to have met the criterion may provisionally be upheld, pending a further test. The situation, then, is complicated, at the level of practice if not that of principle. It is not complicated in principle, because at that level the predictive claim implicit in the assertion that the criterion has been met is perfectly straightforward: the test of *repeatability* is the basic one: in case of doubt, the criterion test must be repeated successfully. This is the essence of scientific method, as avowed by learning theorists. But in the light of this, what sort of criterion can apply to the second-order learning, 'learning-to-learn', which is presently in question?

If the claim that a particular animal has learned to learn should be disputed, what sort of criterion test can be used? Clearly, if one of the tests which has already been passed is simply repeated, the test *on this occasion* will demonstrate no more than that a particular first-order learning has been retained. In order to surmount this difficulty, it will be necessary – if there is to be any chance at all of substantiating the claim that the second-order learning has occurred – to impose as a criterion test some task which the individual animal in question had *not* previously learned to perform. But even this is at best problematical. Suppose the animal learns with suitable rapidity to perform the new task – does this unequivocally entail that the claim for deutero-learning must be conceded? Clearly there are other possibilities. The animal might have learned this task so rapidly, purely by chance. Alternatively, the 'newness' of the criterion test might be impugned. It might be asserted that its novelty was only superficial, that in reality and from the testee's point of view it was so similar to one of the earlier learning tasks that it could not fairly count as a test of *deutero*-learning.

Such criticisms of a criterion-test of 'learning-to-learn' might appear difficult to substantiate. Their logic is close to the argument to limitation used in the 1950s by 'learning theorists' against 'instinct theorists' (see Hebb 1953; Lehrman 1953). It is not identical with those early arguments, however, notably in not resting upon the assumption that one category of behaviour can be defined only as the negation of another. The assumptions in question here are all intrinsic to the 'learning theory' position: the point of the exercise is to expose them to scrutiny, in the belief that conflicts between them will be revealed. The point about the criticisms of a criterion-test of deutero-learning is not so much that they are difficult to substantiate, but that they are difficult to refute. And they cannot merely be ignored or dismissed.

Both their substantiation and their refutation would have to depend, in fact, upon the same process: an extension of the criterion-test to cover not one but a range of learning tasks. In this way the possibility of chance 'learnings' could, in principle, be overcome; so also the possibility of covert similarity to earlier learning tasks provided that a sufficient diversity of learning tasks was included. But this strategy (which is in effect a redefinition of the criterion-test) runs into a different sort of problem.

If the extended criterion-test is passed successfully, presumably the claim for deutero-learning must be conceded – with a proviso which will be mentioned below. But it seems likely that the extension of the criterion to cover a number of learning tasks, especially if some of them are markedly dissimilar to anything learned before, will result in a mixed and equivocal outcome. Instead of the animal learning all the tasks within acceptable time-limits, it is likely to learn some of them properly, but not others. This raises the problem of how many successes, and successes in which of the range of the tasks, is to count as 'passing the criterion'? Unless one is prepared arbitrarily to set a certain percentage of successes as definitive of the extended criterion, all the problems which arose for the single criterion-test arise again now for each component of the extended criterion. In practical terms, the only way around this difficulty seems to be, to take the first horn of the dilemma and set a purely quantitative criterion of percentage success over a range of novel learning-tasks. But it must be emphasized that to do this is to be arbitrary: this dodges rather than solves the problem.

The basic general issue may be represented in this way: When it is claimed that deutero-learning has taken place, i.e. an individual has 'learned to learn', precisely how much predictive generality is involved in the claim? Conversely, how much is it merely retrospective? At one extreme, to say that it is nothing more than retrospective is to trivialize the very notion of deutero-learning. If 'learning to learn' is to have any substance, there must be at least some implicit claim to extrapolate from what has already been learned to what has not yet been learned. It is unnecessary to go to the extreme of demanding that *anything whatever* should be 'learnable' before a claim for deutero-learning could be allowed – the range of possible 'learnings' must obviously be restricted in relation to the species of animal involved, apart from any other consideration – but below this limit, at least some predictive generality must necessarily follow from a deutero-learning claim; and the 'predictive generality' claim, as against the retrospective one, is what makes a criterion-test intrinsically problematical.

The standard assumptions of 'learning theory' cannot, then, accommodate the open-endedness of deutero-learning if this is interpreted substantively, i.e. as involving a predictive claim. If 'deutero-learning', 'learning-to-learn', 'learning set' and similar concepts are intended to bridge the theoretical gap between the 'closed' operationalism of learning theory and the open-ended and adaptive components inherent in most notions of intelligence, it seems impossible that they should do this. The reason is, in essence, that 'learning' in first-order usage presupposes closure in the behaviour-possibilities; while in its second-order usage, as in 'learning to learn', it must approximate to 'intelligence', 'creativity,' etc. in demanding an indefinite openness of response. In short, 'deutero-learning' amounts at the operational level virtually to a contradiction in terms; and as an attempt to fit the assumptions of learning theory to the fact of adaptive variability in behaviour, it must fail. In showing why it must fail, a conflict has been revealed between the predictive or 'open-ended' component inherent in 'deutero-learning' and related concepts, and the methodological (specifically experimental) assumptions of learning theory.

6. 'LEARNING'—INTRA-PARADIGMAL INCOMPLETENESS

The attempt to cope with adaptive variability of behaviour on the basis of learning theory has in fact utilized, not the notions of 'deutero-learning', etc. but much more restricted concepts, notably that of 'habit reversal learning' (Bitterman 1960, 1965). This oblique approach to the problem is no more effective than ordinary learning theory in providing a general explanation of either the causal mechanisms or the adaptive significance of variable behaviour. Its experimental findings are, however, of some relevance, since although the experimental procedure starts with a fixed pattern of behaviour and ends with a fixed pattern, the fact that the intervening processes involve a 'reversal' of pattern entails some contact with the mechanisms of variability. Changing one fixity for another *is* a form of variability; and though Bitterman, like Ewer (1961), apparently does not visualize the possibility of the individual animal retaining or even increasing its capacity for variation, let alone developing a propensity for doing so spontaneously, the capacity for induced or 'forced' variation is undoubtedly a fundamental prerequisite for all variability, hence its assessment in different animal groups is a very substantial contribution towards a general explanatory theory. (Some issues relevant to this topic have already been discussed in Chapter XI. The present section is intended to be supplementary to that earlier discussion.)

Bitterman's findings provide data on the different rates of habit reversal learning in different vertebrate classes – they do not offer an explanation of why habit reversal learning should be of greater adaptive significance in some groups than in others. Although the importance of adaptive variability of behaviour is implicitly presupposed, no attempt is made to relate it to the natural way of life of different animals or to the processes of evolution. The present theory, on the other hand, sets out to argue explicitly on a basis of adaptation and evolution and has been able to specify certain functional requirements necessary for the transition from instinct to intelligence. One of these, the P-factor, would appear to have great relevance for habit reversal learning. Such learnings can occur, no doubt, in animals virtually devoid of P-factor; but they would be greatly facilitated by its operation. (It may be of interest to remark that in a very early draft of the present work, in 1961, I postulated that adaptive variability of behaviour could be achieved either by 'insight' or by 'rapid learning-and-unlearning'. This formulation was abandoned as the crucial importance of an 'abstention factor' became clear to me. Shortly afterwards a friend brought Halstead's work to my notice; after which I decided, on the basis of the apparent similarity of our findings (see Chapter IV), to follow his terminology. Bitterman's work on habit reversal learning did not come to my attention until his paper stimulated the publication of an abstract of my own general theory (Stenhouse 1965 b).)

The phenomenon of habit reversal learning poses a problem which in the absence of adaptive/evolutionary considerations appears impossible of solution. Why should an animal unlearn an already-formed habit? In particular, if the animal's circumstances provide reinforcement for the *un*learning of the habit, whence came the reinforcement necessary for the formation of the habit in the first place? It is easy enough in the laboratory to reverse reinforcement schedules; but common observation in the field suggests that the vast majority of animals have no need to reverse their habits. It becomes implausible to suggest that there could be strong selection pressures in favour of habit reversal learning – its existence in various 'higher' animals, as demonstrated by Bitterman, therefore becomes an enigma.

In the present general theory, by contrast, habit reversal learning *per se* (or intelligence itself) does not need to occur or be selected for in lower forms. The factors if intelligence can occur severally and be under selection independently of each other (to at least a considerable extent) in the earlier phases of evolutionary development, as indicated in Chapters VI and VII. The P-factor, for example, can be seen as under positive selection through:

(a) Exploratory behaviour;
(b) The adaptive value of stability (the absence of overt agonistic behaviour) within social dominance hierarchies, working through reduction in dominance displays in the dominant males of the social group (the 'impassivity display', see Chapter VI pp. 136, also Chapter IX, pp. 219ff);
(c) The value of a 'confidence display' as a part of maternal behaviour (see Chapter VI, pp. 147ff);
(d) The suppression and/or control of reproductive displays by juvenile members of the social group (see Chapter VIII, pp. 205–7);

and in other ways besides. Thus the P-factor and the other factors both of intelligence and instinct can be developed, during evolu-

tion, to a great extent independently. Once they have reached a sufficient level of development they can, as it were, be put into conjunction. They can then act in concert and be mutually reinforcing. At this point very rapid and extensive evolutionary changes are likely to occur: namely, the almost explosive appearance of the grade Psychozoa (Huxley 1959) in the form of the hominids. Thereafter psychosocial evolution can supplement (or in some areas supplant) the processes of general biological evolution, and radical changes in the direction and nature of evolution become possible. A new factor in this is intelligence itself. (The nature of its interactions with the rest of the hominisphere and the biosphere remains at the present time largely unexplored, though a few relevent issues have been touched upon in Chapters V, IX and X.)

Thus the twin demands which perplexed Bitterman (1960, see Chapter XI), for evolutionary continuity *and* for the production of new behavioural capacities, can both be met provided that the new behavioural capacities are not presupposed to be monolithic or to be produced on an all-or-nothing basis. Functional continuity is not necessarily at variance with the resorting of potentiality-factors to produce an entirely new phenotypic method of adaptation – but the perception of significancies among an immense array of *a priori* possible factors must be dependent upon the acceptance and appreciation of the basic category of adaptiveness and its place in the mechanisms of evolution. 'Learning' cannot be regarded as a single unitary phenomenon. It cannot be assumed to involve a single set of causal processes, nor to have a consistent set of functions, through even the 'upper levels' of the evolutionary series. Adherence to it cannot but distort the course of an investigation, even when, as with Bitterman, the basic concept of 'learning' is modified to 'habit reversal learning'. Had Bitterman conceived his investigations in terms not of learning but simply of 'habit reversal' or 'flexibility of behaviour', it is possible that more far-reaching explanatory power might have been achieved.

7. INCREASING THEORETICAL PROVISION FOR VARIABILITY

The history of behaviour theory in this century shows a progressive increase in the variability allowed for in behaviour.

From the fixed and 'forced' movements which were the only

ones allowed under the 'reflexology' regime and in terms of tropisms and taxes (see Chapter II), ethological theory imported the possibility of variability and spontaneity. This variability is confined, in the ethological accounts of instinctive behaviour, to the appetitive phases of a sequence: the consummatory phases of each individual's behaviour substantially exclude variations (except by chance or accident). Not only is variation within the consummatory sequence ruled out, but so also is the non-performance of the consummatory act – provided that the total releaser situation plus 'internal state' is otherwise adequate for it, and provided that no 'higher priority' stimulus supervenes. Hierarchical organization is apparent not only within each instinct (see Tinbergen 1951) but also between instincts – at least in the higher animals. In the rat, for example, exploration has a measure of priority over feeding (Barnett 1958), and escape from predators over all other forms of behaviour. It is clear enough, in general, that overall adaptiveness would demand just this sort of arrangement.

The adaptive value of automatic and fixed-pattern consummatory sequences is also clear enough in general. Wild animals must seize opportunities as they occur for the performance of the consummatory sequences which are such only because they are essential for the perpetuation of the species. Instinctive behaviour may be regarded as built-in opportunism, quick and automatic.

The adaptive value of variability within the behavioural repertoire of the individual is also clear. It would be of value, of course, only in species the individuals of which were long-lived, and where a varied and changing environment provided both the opportunity for and advantage to be gained from variability. Optimal conditions would be achieved, furthermore, where experience-synthesized 'improvements' in consummatory sequences could be disseminated through a social group and from one generation to the next by means of cultural transmission. The present theory seeks to allow for variability within consummatory phases (as well as to allow enhanced adaptiveness in all variable behaviour). One highly significant feature in the 'variability' envisaged in consummatory phases is their non-performance. This is in one sense a total or absolute 'variation'; and it is one which in terms of 'instinct theory' is not possible. The causal factor (which must be regarded as additional to, and separate and distinct from, the causal factors actuating instinctive behaviour *per se*) which at the operational level issues in the non-performance or the temporary withholding of a consummatory sequence has been designated the P- or Power-factor (following Halstead 1947), or Power of Abstention.

It is only through an appreciation of the significance for the evolution of intelligence of the *non*-performance of previously-standard consummatory sequences that the need, indeed the absolute necessity, of a 'power to withhold' can be perceived. The P-factor which confers this 'power to withhold' is contributed to the evolutionary picture of intelligence only by the present theory. As argued in Chapter IV, it or something virtually identical with it had been discovered, earlier and independently and on the basis of quite different evidence, by Halstead (1947). Halstead did not, however, attempt to relate the P-factor to evolutionary developments or, in particular, to the withholding and control of instinctive behaviour in humans. His account is not incompatible with this function being ascribed to the P-factor, and some of his experimental findings do implicitly support the view of its importance for the adaptive variability of behaviour which is advanced in this book (see discussion of P-factor in Chapter IV, pp. 80–2, and Chapter V.)

In historical context, then, the present theory can be seen as continuing the attempt to allow for increased variability of behaviour which began as soon as specific mechanisms of a determinate and determinative nature were adduced as explanatory models. To be acceptable as scientific explanations, the mechanisms proposed have had to be relatively fixed; but this has led to a tension between the theories and the observable behaviour which they were supposed to explain. Pavlov's conditioned reflexes and the elaboration of ethological theory by Lorenz, Tinbergen, and others, can be seen as successive attempts to enlarge the theoretical establishment to accommodate more and more variability. Two classes of behaviour have hitherto been insusceptible, in theory, to variation: the consummatory phases of instinctive sequences; and 'learned' or experience-synthesized patterns. The present theory provides a rationale for the occurrence of variation in consummatory phases. Its relationship with 'learning theory', and in particular with the work done on 'unlearning' by Bitterman and others, has now been outlined. There remains only to ask: is there any *other* area of behaviour in which fixity and rigidity is needlessly imposed by theoretical presuppositions?

There is such an area, which has been touched upon already: the behaviour affected by education.

8. EDUCATION IN PHILOSOPHY FOR ADAPTIVE VARIABILITY

It was implicit in much of Chapter X that a great deal of present education tends towards the reduction in the variability and hence of the potential adaptiveness of behaviour. The mechanisms tending towards rigidification and towards flexibility both need to be explored, first at the conceptual level and then practically. Undoubtedly there are many diverse mechanisms and tendencies which are relevant – but to conclude this chapter I propose to investigate only one set of interrelationships and interactions, a set which is of fairly central importance. This is the set of relationships between philosophy, science, education and intelligence.

If we accept that one feature of the contemporary revolution in philosophy (Ayer et al. 1956) is that it is now to be regarded as an activity to be practiced rather than a body of knowledge to be known, we may suppose that philosophy itself can be regarded, in some respects at least, as a type of behaviour. It may then be asked: is this behaviour fixed or variable? In particular, does learning to do philosophy result in relatively fixed and stereotyped patterns which may be adaptive enough in their way – or does it lead to variability of pattern which opens up the possibility of (though it does not ensure) new types and levels of adaptation?

From the fact that the student of philosophy studies Plato and Aristotle, and counts as 'modern' people from the seventeenth century like Descartes and Locke, it might be assumed that the tendency would be toward fixity and stereotyping of behaviour. So indeed it turns out, in many cases. Many courses do consist in little more than learning about the doctrines of the great figures of the past. But in other courses the doctrines of past thinkers are used as raw material in the understanding and appraisal of which the student can develop his own powers of critico-creative thinking. The emphasis in one course and for one person may lie upon the 'critico-' component; while for others the 'creative' side may be emphasized. And on either side of the dichotomy there may be an *over*-emphasis, criticism and fault-finding being indulged in promiscuously, or loose speculation being labelled 'creative thinking' to cover the absence of real ability. Nevertheless many people do manage to learn philosophy as a balanced activity of critico-creative thinking, and the question arises, does 'creative' then mean what it says? Is there a real tendency to increase the variability of intellectual behaviour?

To sketch an answer to this, it will be useful to revert to one of the criticisms of philosophy which is often linked to the charge that it 'deals only with the past'. This particular indictment is, that teaching courses in philosophy deal with doctrines so extreme, in one way or another, that no sensible person would ever hold them; and that they are in fact rejected as soon as they have been examined. This is held to be an admission of their uselessness. Since they are eventually rejected, why bother with them at all?

So runs the argument. It is based on a misunderstanding of what is being attempted. Extreme possibilities are worth examining, not so much because they may be adopted, as because serious consideration of them activates the individual's discriminatory processes and this in turn leads to clarification of what *can* be believed and adopted. By critical examination of a 'particular position' one comes to a delineation of one's 'personal position' (Stenhouse 1968). By seeing precisely what one does not believe, one comes to see what one does.

If one's personal position on a topic is stipulatively defined as a 'set of beliefs' which may be represented as a set 'A, B, C, D . . .' the question arises: are the members of the set consistent with each other and with the appropriate sector of 'reality'?

To answer this question, it appears we would have to determine the nature of each member or element of the set and the possibilities of its interaction and compatibility with each of the other elements; also the compatibility of each element, and each combination of elements, with reality. If this is attempted on a piecemeal basis, it runs into very serious difficulties. These depend on the size of the set, i.e. the number of elements within it. While the number of elements increases arithmetically, the number of possible interrelationships increases geometrically. In a set of seven elements, for example, there are forty-two dyadic relationships to be investigated (assuming that relationship AB need not be the same as BA – it may be, but cannot be assumed so until it has been investigated). If triadic relationships are also to be considered, the total number of relationships to be investigated becomes: $42 + 210 = 252$.

It is clear that, in a system containing a large number of elements, and especially if complex interactions between large groups of them are to be considered, the number of interrelationships to be investigated rapidly becomes astronomical. The piecemeal investigation of all possible interrelationships thus becomes impracticable.

But the situation is not necessarily hopeless. Ashby (1956:261) discusses the selection procedures open to the physicist looking for *one particular atom* in the whole universe (i.e. 1 in 10^{73}). If each atom is examined separately but rapidly, 1×10^6 of them per second, it will take many millions *of centuries* to find the atom required. If, on the other hand, there were some practicable method available for utilizing the method of dichotomy – finding first in which half of the universe the wanted atom lies, then in which half of that half, and so on, and assuming a much slower rate of sorting, of only one operation per second – the wanted atom could be found in 'just over four minutes'!

Could the method of dichotomy perhaps be used in sorting through a set of beliefs to determine their possibilities of interaction and coherence? Another method designed to be economical, which could be used perhaps in analogue, is what Quine (1950) has called, when used in symbolic logic, the 'Fell Swoop'. This does not guarantee reaching a decision in every case, but where it is applicable it shortens the decision-procedure immensely. It involves a sort of 'devil's advocate' principle, of assuming the opposite (or negation) of what one hopes to prove. If this negative assumption generates an anomaly in the system set up, it must be concluded that the negative assumption should not have been made, hence that whatever positive assumption was originally in question should, or at least could, be made.

It appears that the process, common in philosophy, of looking at the extreme possibilities on any given topic, might be regarded as analogous to a combination of the 'Fell Swoop' and the method of dichotomy. Looking at the extremes is economical in two ways. Usually some of the extremes can be rejected on bare inspection. This not only reduces the number of elements, and thus interrelationships between elements, which have to be dealt with, but also in clarifying that one holds *an* 'in-between' position helps towards determining precisely what the nature of one's in-between position really is, in detail. If there are non-conscious processes that go on inside us all the time – and so many of the observables of human behaviour seem to force us towards the assumption that there are –

then it would seem reasonable to assume that in a normal healthy (i.e. non-schizophrenic) individual the conscious and the non-conscious 'thinkings' together form a single system. Freud is widely accepted as having demonstrated that the conscious part of the system is not fully intelligible on its own. If, then, the examination of extremes stimulates the unconscious parts of the system to take over the exploration of the various elements of a set of possible beliefs and their interconnections with each other and with reality, then this would seem to lead to an increased likelihood of systematization of the beliefs and, eventually, to an enhanced adaptive relationship between beliefs and reality. The non-conscious exploration which seems to occur would take time, but would have this advantage over conscious exploration, that other activities could be carried on in the meantime: sleeping, eating, social intercourse, other work, etc. Also, the so-called 'intuitive leap' seems to be a function of the non-conscious part of the system; and it might be said that the non-conscious 'processes' are not only quicker but more efficient than conscious ones (see examples in Beveridge 1950, Chapter 6; also McClelland 1963.)

It is apparent that if a 'philosophical approach' is to be productive of the sort of outcome indicated above, the individual must develop various habits:

(1) Of 'trusting to' his 'intuition', 'his non-conscious thinkings'. (This is not to say that the products of intuition are to be taken completely on trust, quite the contrary. But they must be entertained before they can be critically investigated.)

(2) Of looking even at the most extreme and perhaps bizarre possibilities. That is, the individual must develop the capacity *not* to 'shy off' from extremes, and must be able to entertain a wide variety of possibilities without being confused by them. This means he must develop the habit:

(3) Of withholding 'closure' either endogenous or exogenous (i.e. socially induced), so as to allow the other habits to operate.

What is being argued is, in effect, that training in philosophy, if of the right sort, can tend towards the development of 'diverger'-rather than 'converger'-type characteristics (see Guilford 1950; Hudson 1966, 1968). If these characteristics are conducive to creativity, as in general they seem to be, then education of the appropriate sort in philosophy could be expected to lead to enhancement not only of the *critical* side of the critico-creative spectrum, but also of the *creative* side. Specifically in relation to the present theory, the development of habit (3) above (with-holding 'closure') can be seen as involving the exercise of the P-factor of intelligence.

The question posed earlier, as to whether philosophical training could result in greater rather than reduced variability of be-haviour, can now be answered. In one sense and at one level it could be said that variability would tend to be reduced: for instead of 'pausing to consider a wide range of possibilities' only sometimes, philosophical habits should lead one to do this most or all of the time (on issues of importance, anyway, and when emergency action is not required). But increased constancy of behaviour in this way is of significance only because it is the prerequisite for adaptive variability at first-order level. And at this level it is adaptiveness which must be primary. Variability is important only because it is necessary, in changing circumstances, as a means towards adaptation. The central issue is the pragmatic one; and if circumstances were unvarying, an achieved optimal response must also be unvarying.

Thus the function of the present theory must be that variability of behaviour is *allowed for* – variability must not be necessitated, for this would conflict with the need to adapt, on occasion, to circumstances which were for a period *un*changing. Basic to the position is the fundamental concept of *adaptation*, whether by means of immediate action to the physical demands of the environment or by means of abstract theoretical argumentation to the need for compatibility between different conceptual schemata, different paradigms, and between conceptual schemata and reality. The present theory can provide the basis, too, for a reflexive account of its own genesis. The history of the develop-ment of a general theory of behaviour involves, as was indicated in Chapter I, an apparent succession of personality-cum-intellectual types among the individuals contributing to the successive elaborations of theory. There seems to be a functional succession analogous to what is involved in an ecological succes-sion. It may be suggested that this should be interpreted as a successive emphasis on different factors of intelligence, as embodied in differing propositions in the several individuals of the series.

9. FUNCTIONAL SUCCESSION OF FACTORS, IN INNOVATION AND DEVELOPMENT

First, the primary creatives or 'revolutionaries' (in Kuhn's 1962 sense) are constituted as such by a predominant endowment of P-factor. This enables them to perform the essential first step of breaking through or away from the previous orthodoxies, to reveal the existence of a problem or an hiatus (see Chapter V). It may also militate against the clear formulation of a new paradigm, however, since P-factor 'negativism' may tend to keep the theoretical system too much in a state of flux (cf. Hudson 1966: 86).

In the second phase, the 'breakthrough' having been made by the high-P individual(s), the relevance and worth of the new idea is most likely to be perceived by those who, possessing sufficient P-factor to be able to escape the domination of the established orthodoxies, have an unusual A-factor ability to appreciate relevancies between hitherto unrelated sectors of knowledge. Seeing the positive implications is only one side of the story – sensitivity to discrepancies (see discussion in Chapter VI, pp. 151ff) is also a function of A-factor, and leads to systematization of the new schema. In short, if the 'external' implications of the new idea are seen to make it worthwhile, any 'internal' inconsistencies and ambiguities will be regarded as worth fixing up. The definitive formulation of a new paradigm, then, and the extension of its range of applicability, is more likely to be carried out by individuals with a predominance of A-factor rather than P-factor.

It is only in the third phase of the development of the new paradigm, subsequently to its worth having been demonstrated (probably most effectively by the high-A individuals), that individuals with a predominance of C-factor come to make a major contribution. Their inclusion in fact signifies that the 'revolutionary' phase is over and is being succeeded by 'normal science'. Predominance of C-factor implies that knowledge supplants critico-creativity to a great extent. 'Learnedness' becomes the criterion of competence, rather than 'insight', 'imagination', 'percipience', or other qualities valued in the early phases. While a reasonable amount of knowledge is always necessary for any advance, the situation can and sometimes does arise when *nothing but* knowledge comes to be valued. This is a phase of petrifaction: previously-established knowledge is passed on uncritically, fossilized doctrines are perpetuated merely 'because they are there'. A tendency to degenerate into this type of scholasticism can be perceived, fleetingly, at different times in the history of most branches of science. It is reinforced by various features of science education: textbooks, examinations, differential selection of personnel, respect for 'authority figures', and so on – in short, the conformist pressures within education in science (see Stenhouse 1965a, 1971). The breakdown of education in science – which would be but the precursor to the breakdown of science itself – is illustrated in an anecdote told by McClelland (1963:184). He tells of:

> . . . a discussion I once had with a graduate student who had read all the literature on a certain subject. He knew practically every research result that had ever been reported on this particular topic, and after he had finished summarizing in a seminar the present state of knowledge in this area, I somewhat unthinkingly asked him what he thought ought to be done next. It was obvious that the question took him by surprise, and he finally stated that he was in no position to judge, since authorities had differed on what the crucial variables were and much of the evidence was conflicting. I pressed him further. I pointed out that he must by now know as much about this field as anyone else and that he ought to be willing to make a decision as to what the most promising line of inquiry was. He still showed some unwillingness to do this and ended by suggesting that perhaps a massive research attack on all fronts at once would pay dividends. I pointed out the research design he had in mind, in which one entertained all hypotheses at once and varied all possible variables simultaneously, was really quite impracticable and that it would take several lifetimes to carry out.

It would be unkind to suggest that the education which produces young men of this type is completely useless, but in fact, in its extreme manifestation, this type of education is not only useless but pernicious. Fortunately many people find it uncongenial, and withdraw themselves from it, either overtly by changing to something else (hence the 'drift from science' and other phenomena), or covertly by withdrawing their belief and acceptance even when they continue practicing the outward forms. (The danger in the latter course is that cynicism may be engendered which can spread through all aspects of social life.)

In the sequence of emphasis upon the different factors of intelligence, during the transition from 'revolution' to 'normality' in science, no mention has yet been made of the D-factor. It would appear that, just as in the transition from instinct to intelligence, neglect of the Sensorimotor Factor has been due not to any intrinsic lack of importance, but simply because it may be common to all the phases in question. High D-factor endowment must be useful to the most conservative of 'normal' scientists just as to the most daring of 'revolutionaries'. A remark by Thomson (1961:55) suggests that part of Rutherford's power of discovery may have been due to a high D-factor endowment: 'Rutherford's genius lay in his uncanny physical insight, combined with a flair for designing very simple experiments which worked. In both these he showed what is fundamentally the same quality, namely the ability to see the essential and discard all the rest.'

The phrase 'uncanny physical insight' suggests a sensorimotor function. Talk of a 'flair' suggests that Rutherford did not (and perhaps could not) explain how he arrived at his ideas, thus making it unlikely that he 'reasoned' his way towards them utilizing the A-factor on the basis of C-factor 'information stores'. Probably his innovation was a matter of immediate and emphathic 'feeling' or 'seeing' of possibilities, a co-operation of A- and D-factors. The 'ability to see *the essential*' suggests A; while to 'discard all the rest' suggests the negativism of P. As suggested in Chapter V, logical and temporal priority in the processes of innovation must go to the P-factor; but once the 'conceptual gap' has been created by the P-factor, it can be filled by a substantive innovation produced by any of a variety of combinations of A, C, and D.

Epilogue

In offering a theory to account for the evolution of intelligence from an instinct-system, the picture of intelligence which has resulted has been progressively elaborated. From the three factors peculiar to the emergence of intelligence from instinct (the C, A, and P-factors), it has been necessary to enlarge to include a fourth, the D-factor, which was under selection from the earliest development of an instinct system. If our intentions are themselves extended to give not merely an account of mechanisms of evolutionary significance for the emergence of intelligence, but also an account of the present nature and basis of intelligent behaviour in humans, then it must be pointed out that even the four-factor theory is inadequate. The inadequacy has been generated by the methodology of the investigation. I have been investigating the nature of what is peculiar to intelligence, of what is entailed as a prerequisite for adaptively variable behaviour. Halstead (1947) also was concentrating upon only a part of the total phenomenon of what we call 'intelligent behaviour' when he isolated the D-factor additionally to C, A, and P. He was concerned with the 'control' aspect, as I have been – we have been investigating, in our different ways, 'what makes behaviour *intelligent*'. But a complete account demands attention to the complementary issue: what makes the phenomenon *behaviour*? Paraphrased to 'What makes for behaviour?', it is clear that the complete account of intelligent behaviour must involve an explanation of 'basic reactivity' and also of the spontaneity of endogenous behaviour. Both of these are accommodated under the rubric of 'instinctive behaviour' as understood within ethological theory. Thus 'intelligent behaviour', understood generically and inclusively rather than specifically and exclusively, incorporates instinctive behaviour.

This conclusion has been implicit in much of this book. In Chapters V and X the complementarity between 'instinct' and 'intelligence' has been to the fore; in Chapters III and IX, on the other hand (and in various other shorter passages), the antagonistic relation between them has received most stress. Both emphases are necessary in a balanced assessment.

A final word may now be said to clarify relationships between intelligence, instinct and learning.

Learning, as was argued in Chapters XI and XII, is itself an instinctive activity. Intelligence on the other hand is not instinctive. It is superordinate to instinct, hence also to learning. In a broad and descriptive sense, 'intelligent behaviour' includes both instinct and learning. Analytically, the capacity for intelligent behaviour which we conveniently call 'intelligence' is, precisely, the capacity for the control and direction of instinct and therefore also of learning. This capacity has been selected for and elaborated in the course of evolution, through the effect it has had on the instinctive-learned complex of behaviour at the phenotypic level.

This picture of intelligence in relation to learning and especially instinct is reminiscent of Freud's simile of the ego-id relationship to that of a man on horseback. Intelligence tries to control and direct instinct and learning just as the ego tries to control the id. Freud wrote (1934:98): 'The functional importance of the ego is manifested in the fact that normally control over the approaches to motility devolves upon it. Thus in its relation to the id it is like a man on horseback, who has to hold in check the superior strength of the horse; with this difference, that the rider seeks to do so with his own strength while the ego uses borrowed forces.'

As Halstead appositely remarked (1947:4): 'It appears that Freud's successors in the psychoanalytic idiom have made more progress in elucidating the "horse" than in advancing our understanding of the "rider".' What was true in 1947 with regard to the psychoanalytic field is still substantially true today over a much wider field concerned with the understanding and explanation of human behaviour. There has been little attempt until very recently to come to terms with the 'upper level' directional systems of human behaviour: those comprising 'intelligence' (which may perhaps be equatable with the Freudian 'ego'). The creativity movement might be regarded as a flank attack on the citadel of the 'upper level' systems. But the attempt to understand intelligence-cum-ego, as it develops in the coming years, will have a vastly more secure basis than it would have had in 1934 or 1947. Presumably there have been advances in our understanding of the id – I cannot speak for the psychoanalytic field – but there have been enormous advances in our understanding of other aspects of human behaviour and behaviour generally. Learning theory and the research findings upon which it is based have both extended in a variety of directions (see Hilgard 1958; Thorpe 1963). Work on intelligence itself has become subtle and sophisticated, as well as having increased in volume (see Guilford 1967). Finally, although instinctive behaviour in humans has received relatively little attention (but see Russell and Russell 1961; Huxley 1964; Morris 1967, 1969), there has been both a revolution and a 'population explosion' in the study of instinctive behaviour in a wide variety of other animals. The theoretical 'revolution' (Kuhn 1962) occurred prior to 1951; the quantitative 'explosion' came subsequently, stimulated largely by Tinbergen's *The Study of Instinct* (1951) and culminating in Lorenz's *The Evolution and Modification of Behaviour* (1965). Other outcomes have been valuable critical syntheses by Thorpe (1963) and Hinde (1966) among others. These many lines of study, especially the two decades of 'normal science' (Kuhn 1962) development in ethology, have provided a basis of research findings and theoretical interpretation from which the very difficult enterprize of exploring intelligence itself may be attempted. Seen in evolutionary perspective, we are now working reflexively, using our most phyletically-recent endowment, intelligence, in the examination of its own nature and origins. Thus we are throwing ourselves into the paradoxical activity of philosophers, who attempt to determine the limits of the human understanding (see Locke 1689, Hume 1739–40) by using the human understanding. This may seem paradoxical, possibly futile and ill-advised, but what else are we to do? Referring to the triad of propositions presented in the Introduction, are we to 'solve' the problem of intelligence by dodging it, by ignoring its existence? Shall we, because the problem is difficult, tacitly deny our own intelligence? For myself, I think that our instinctive drives to know and to understand are too strong: we are not able to avoid making the attempt to understand and explain. But whether the attempt will be successful, and when, are other matters. We shall certainly have to avoid simple-mindedness. The educational background desirable for coming to grips with intelligence will certainly demand changes from much current practice. Besides being 'literate' and 'numerate', behavioural scientists will have to become 'philosophicate': they will have to be acclimatized – as the leaders of 'revolutions' in science have always been – to factual/criterial 'weightlessness' (Stenhouse

1968). They will have to cut lose from many of the paradigms that have served us well so far. They will have to create 'unknowns' where none is recognized now. . . . But I have discussed all this already, in Chapter V. And the book as a whole is an attempt to exemplify the sort of endeavour which will be needed. The exploration of intelligence is bound to be an interdisciplinary venture. I started, in Chapter II, with ethological theory, and worked out in Chapter III an evolutionary picture of the problem posed in Chapter I of the genesis of intelligence. Then in the later chapters a few of the implications of the 'evolutionary intelligence' theory have been explored at the conceptual level in several distinct fields, and have been found to be in accord with a good deal of empirical evidence. I am most sharply conscious of the many ramifications of implication from the theory which remain to be explored. I hope to continue my own explorations; and I hope that others will find the theoretical sketch-map here presented sufficiently enticing to join in the exploration.

Appendix

Further Support for Four-Factor Evolutionary Intelligence

Several further papers have come to hand, since the rest of the book was despatched for publication, that add worthwhile support to the arguments for four-factor evolutionary intelligence and especially the P-factor. If discussion of these additional (and in most cases very recent) sources were to be incorporated in the main text, I should have to embark on a substantial programme of revision and re-writing, and this, in such a dynamic, expanding, and interdisciplinary subject-field, might well prove to be a never-completed task. Instead of attempting this, I have compromised by scattering a number of bare references to recent work through the body of the main text – and by attempting some very brief discussion of some of the more significant issues in this Appendix.

1. It is apparent that the confounding of intelligence with 'learning' and other notions still continues. Campbell (1967), for example, in an otherwise useful book and despite the fact that he says that '. . . there is certainly some distinct feature of human mental activity that is convenient to define as "intelligence" and that can be distinguished from a highly developed capacity to learn' (p. 301), gives an account of intelligence that, because of his preoccupation with 'learning', limits itself to the functions covered by the C, A, and D-factors. The reader is referred to his full discussion (pp. 300–4) – his position seems to be an interesting part-parallel to my own. Even his illustrative figure (p. 303) of the operation of intelligence is similar to my own (Stenhouse 1965; also above, Fig. 8). The most notable difference is the omission of the P-factor.

Some of the limitations of Campbell's position seem to be an outcome of his unstated, and presumably unconscious (cf. Stenhouse 1972) methodological presuppositions. In particular he seems confused between entities being logically distinguishable, and their being ontologically separable or independent. For example, he finds it necessary explicitly to protest: 'But intelligence cannot exist alone, it is a capacity that interacts with knowledge. . . Acquired knowledge is the fuel with which intelligence burns, without which intelligent action is impossible' (p. 302). And later: 'Truly intelligent thinking (*if such a thing is possible*) may be considered to be free from the direct influence of any existing cerebral facilitory activity' (p. 303; my italics). The point is, why make an issue of ontological separability or independence? Surely nobody imagines that intelligence could exist without legs, eyes, etc. . . . It is in this sort of context that the dangers of reification of intelligence become obvious: it would be a corrective, here to talk about 'intelligent *behaviour*'. My own approach is to talk of memory, for example, as part of intelligence, an intrinsic component or factor among the necessary prerequisites for intelligent behaviour (see above, Chapters I, XI, XII) – but then an extensive training in philosophy does generate a sharpened awareness of logical and methodological issues and the pitfalls therein.

Campbell does appear to base his views on a concept of intelligence approximating to 'the capacity for adaptively variable behaviour' (pp. 300–4). Unfortunately, however, he initiates his whole discussion of intelligence, as contrasted with instinct and learning, as follows: 'There is another and very different determinant of behaviour that has not yet been mentioned; it may be termed "intelligent thought". It is in all probability a uniquely human characteristic and so is a novelty of great importance in man's evolution' (p. 22).

This seems likely to run into problems of salatory evolution and the 'monolithic' view of intelligence, see above, p. 287; hence it is not surprising that little advance is made towards an evolutionary account of intelligence. However, Campbell is at least forward-looking, for he goes on: 'Action determined by intelligent thought may draw on learning but is itself not the result of a learned behaviour pattern; it is original.' This might have served to set the stage for the argument of my Chapter III.

2. White (1969) utilizes a pragmatic and holistic methodology, and essentially an ostensive definition of his basic terms, in investigating 'competence' in three- and six-year-old children. One finding of his research team is that, between a group of 'high

competence' and a group of 'low competence' three-year-olds, a highly significant difference is apparent in their powers of 'dual focus': a type of attending behaviour characterized by simultaneous attention (or rapid alternation of attention) to both a central task and an imposed peripheral stimulation. Ability for 'dual focus' seems obviously heavily dependent on the individual's level of P-factor endowment. It would appear that some of Halstead's (1947) tests, notably those designated 'Dynamic Visual Field – Form' and 'D.V.F. – Colour', are in fact virtually 'pure' tests of ability for 'dual focus'. These tests were found by Thurstone and Holzinger to carry substantial 'loadings' for what Halstead finally characterized as the P-factor.

3. Kagan (1970b) provides an excellent discussion of the adaptiveness of what he categorizes as 'inhibition' in young children. The first sign of this, in ontogeny, is in breaking free of the domination of a repeated visual stimulus, at about the age of ten weeks. Another significant change occurs between five and seven years. A number of features of the child's behaviour can be observed to have altered, often in subtle but nonetheless important ways. Kagan says (p. 201) that the child's 'play becomes more planned and his attack upon problems more reflective and pensive'. It appears that the child develops an increased independence both from the immediate stimulus and – especially significant in relation to the theoretical emphasis given in the present book (see p. 63) – from its own previously acquired habits. Kagan may be quoted at greater length on this (p. 201): 'One of the processes common to many of the psychological changes can be described as an increase in reflection: an increased tendency to pause, to consider the differential validity or appropriateness of a response; the ability to select the correct response rather than emit one that happens to sit on top of the hierarchy when an incentive stimulus appears.'

This appears strongly suggestive of P-factor functioning. The generalized descriptions of behaviour are exactly what could be derived from Fig. 8. Some of my own statements of P-factor operation could be transposed with various of Kagan's statements without altering the sense of the several longer passages in either argument. Kagan talks of 'an increased tendency to pause, to consider the differential validity or appropriateness of a response'. I have characterized P-factor operation as providing 'time for (imaginative) testing of action or possible action' leading to 'increased adaptiveness and adaptability of behaviour'. (But see proviso regarding 'time', p. 195).

Kagan also acknowledges evidence pointing to a genetic basis for differences in reflectiveness/impulsiveness hence for the potentiality for 'reflectiveness' (= P-factor operation) itself. This again is in accord with my own arguments.

The findings cited by Kagan and my inferences from them regarding P-factor functioning, taken along with the picture of the ontogenetic development of the four factors emerging from brief and generalized comparison with the work of Piaget (see Chapter VIII), suggest some detailed complications of interaction between various factors of individual and environment. For example, in the outline reinterpretation of the Piagetian sequence of stages of intellectual development given in Chapter VIII, it was suggested that the three major stages are as follows:

(1) Sensorimotor – D-factor – $0-1\frac{1}{2}$ years.
(2) 'Concrete thinking operations' – A- and C-factors – $1\frac{1}{2}-11/12$ years.
(3) 'Formal Thinking Operations' – P-factor added – 14/15 years onward.

As was emphasized (p. 195), this is only a 'thumbnail sketch'; many complexities of detail have to be added; and from consideration of Kagan's paper it can now be suggested what some of the significant additions may be.

'One of the more important developmental benchmarks concerns an important change that occurs between five and seven years of age,' says Kagan (p. 200). 'The child shows a marked increase in his tendency to inhibit irrelevant acts and to select appropriate ones. For example, in learning situations the child stops adopting position habits in solution of problems . . .' If this 'inhibition' is indeed due to the P-factor, it is clear that P-factor is not held in abeyance until the stage of 'Formal Thinking Operations' at the age of 14/15 years. Thus the developmental stages are not stages in the categorical *appearance* (as it were *de novo*) of the factors of intelligence – rather they are to be characterized by *major changes in relative emphasis* between the factors. All factors are always present. They can vary in development, it would seem, both in the relative rates of development of each as capacity or potentiality, and also in the extent to which capacity is actualized in behaviour. The increase in effective P-factor at age 5–7 years seems likely to be due to changes in degree of actualization of capacity, which are probably due in turn to social and other environmental influences – though there may also be an increase in relative potential. The major increase in P-factor capacity at puberty, however, is likely to be an endogenous change associated with the other physiological changes occurring at this time. This is not to say that environmental influences cannot be important. One of the most promising (and most urgently needed) lines of research at the present time is into the modifiable variables affecting the behavioural emanations of P-factor, especially in adolescents. But the increase of P-factor influence which actuates the change to Piaget's 'Formal Thinking Operations' must itself have a genetic basis.

4. From various general considerations regarding the ontogenetic acquisition of language we are led to suspect, says Chomsky (1970:81) 'that we are dealing with a species-specific capacity with a largely innate component', i.e. that the tendency towards the acquisition of language and the mechanisms for doing so are endogenous (in the sense used in this book, following Ewer 1957). 'It seems to me', he goes on, 'that this initial expectation is strongly supported by a deeper study of linguistic competence.'

It seems to *me* that Chomsky is in fact supporting, in a particular instance, viz. the 'learning' of language, my general thesis of Chapter XI that 'learning' itself is an instinctive activity; and further, that various of his conclusions support the position taken in Chapter VII regarding the relationship between the several factors of intelligence and the acquisition of language. On the latter point, he places great emphasis on what he calls the 'creative aspect of language use', meaning by this not what is normally meant by 'creative writing' and the like, but something much more basic, something that is common to *all* uses of language 'creative' or 'uncreative' in that sense, something intrinsic to the very nature of language itself. What he is talking about is 'the ability to produce and interpret new sentences in independence from "stimulus control", i.e. external stimuli or independently identifiable internal states. The normal use of language is "creative" in this sense, as was widely noted in traditional rationalist linguistic theory.' This 'independence of stimulus control' seems to be exactly what I argued for as resulting from, and being dependent upon, the operation of the P-factor of intelligence. The same dependence on P-factor is brought out *within* language. 'Sentences may have very similar underlying structures [i.e., roughly, similar "meaning"] despite great diversity of physical form, and diverse underlying structures despite similarity of surface form. A theory of language acquisition must explain how this knowledge of abstract underlying forms and the principles that manipulate them comes to be acquired and freely used' (p. 84).

This last challenge can be met, Chomsky suggests, only on the assumption that the 'underlying forms and principles' (or 'grammar') are largely innate. Before giving a longer quotation which outlines the argument for this position, let me remark that the relevance of the other factors of intelligence besides the P-factor, viz. the A-, C-, and D-factors, is also implicit in Chomsky's account. While much of what he calls 'abstraction' is actually dependent more on the P- than the A-factor (as was the case with 'abstraction' in Piaget's arguments, see Chapter VIII), the A-factor itself is clearly necessary in the perception of similarities and differences, the C-factor for information storage, and the D-factor for mediating interaction with the environment.

Chomsky's position may best be indicated by quoting in full a key paragraph from his paper:

So far as evidence is available, it seems that very heavy conditions on the form of grammar are universal. Deep structures seem to be very similar from language to language, and the rules that manipulate and interpret them also seem to be drawn from a very narrow class of conceivable formal operations. There is no *a priori* necessity for a language to be organized in this highly specific and most peculiar way. There is no sense of 'simplicity' in which this design for language can be intelligibly described as 'most simple'. Nor is there any content to the claim that this design is somehow 'logical'. Furthermore, it would be quite impossible to argue that this structure is simply an accidental consequence of 'common descent'. Quite apart from questions of historical accuracy, it is enough to point out that this structure must be rediscovered by each child who learns the language. The problem is, precisely, to determine how the child determines that the structure of his language has the specific characteristics that empirical investigation of language leads us to postulate, given the meagre evidence available to him. Notice, incidentally, that the evidence is not only meagre in scope, but very degenerate in quality. Thus the child learns the principles of sentence formation and sentence interpretation on the basis of a corpus of data that consists, in large measure, of sentences that deviate in form from the idealized structures defined by the grammar that he develops (p. 84).

5. The excellent paper by Kagan (1970a, first published 1967) on the need for relativism in psychology strongly supports the methodological position taken in this book. In emphasizing the importance of attention and expectancy, Kagan is led to a view of perception and 'knowing' which is virtually identical with that advanced in Chapter III (see Fig. 8), except, again, that while the importance of P-factor functioning is implicit, it is not explicitly recognized. In discussing alternative explanations of smiling in infants when they are presented with various 'face' drawings, veridical and distorted, Kagan says (p. 143): 'The smile is released following the perceptual recognition of the face, and reflects the assimilation of the stimulus to the infant's schemaa. . . This hypothesis is supported by the fact that the typical latency between the onset of looking at the regular face (in the four-month-old) and the onset of smiling is about three to five seconds.' In terms of evolutionary intelligence, the external stimulus (mediated by D-factor) is compared (A-factor) with a C-factor 'schema', P-factor keeping the comparison 'open' in the meantime, i.e. withholding a response (hence the 'latency' period) until its 'appropriateness' has been determined. Significantly, Kagan talks of the assimilation of the stimulus to the infant's schemaa 'a small but significant act of creation'.

On the elicitation of 'attention', the evidence cited by Kagan appears to be fully compatible with the 'neuronal model' of Sokolov (1960) and the 'discrepancy mechanism' involving the four factors of intelligence which was discussed in Chapter VI, pp. 151ff. On the more important question of motivation, Kagan says (p. 141): 'The joint ideas that man is a pleasure-seeker and that one can designate specific forms of stimulation as sources of pleasure are central postulates in every man's theory of behaviour. Yet we find confusion when we seek a definition of pleasure.' The basic methodological 'confusion' or mistake is in fact one which Kagan does not state. It is one which is well known *as* a mistake, on the one hand, to philosophers: and which evolutionary biologists, on the other, would be unlikely to make. It is the logical mistake of assuming that, because we do in fact feel pleasure in attaining specific goals, we therefore set out to achieve these goals because of the pleasure they will bring. Psychology, it would appear, in posing the problem of motivation in the form indicated by Kagan, has committed itself to the 'hedonistic fallacy' (see Nowell-Smith 1954:135 ff.; Scriven 1966:220). In very simple terms, we do not seek food because of the (future) pleasure we will get from eating – we seek food because at the moment we are hungry. Such a fallacy is unlikely to entrap students of behaviour of the ethological school, however, coming as they do to the problem of motivation via the concepts of adaptation and instinctive mechanisms. Human action derives in direct evolutionary continuity from action in infra-human animals – like our near and remote ancestors, we are motivated towards the actions necessary for our perpetuation by series of mechanisms which operate immediately to lead us into appetitive and then consummatory sequences. We do eventually feel pleasure in the performance of the latter, but in most instances the prospective pleasure is not a significant actuating factor.

This leads to a last remark on the learning/instinct question. 'Learning theory' has achieved its considerable plausibility, not because it offers an adequate explanation alternative to instinct theory, but because it has in fact been conflated with it. Learning experiments have been dependent upon instinctive motivation: only those tasks have been learned which could be tied to patterns flowing from instinctive motivation. The fact that given species could learn some types of action and not others was simply accepted – no attempt was made to explain it. The ubiquity of instinctive bases for the everyday activities of living has allowed a spurious credibility to the illusion that learning itself was responsible for the widespread uniformities observable in behaviour. If learning were conceptually dissociated from instinctive tendencies to behave in some ways rather than others, it would become obvious, as Chomsky hints, that the relative uniformity observed, in relation to the enormous diversity which would theoretically be possible, is itself a major phenomenon in need of explanation. This could be stated as a 'central tendency' within learnings; and as such it cannot be ascribed to learning.

As to methodology, Kagan (1970a:141) says that 'The understanding of pleasure and reinforcement in man is difficult enough without having to worry about infra-human considerations. Let us restrict the argument to the human.' On the contrary, I would argue that many of the difficulties have in the past been generated precisely because the argument was restricted to the human frame of reference. I feel confident that Kagan and other percipient researchers, having quite naturally worked within the 'learning theory' paradigm which they inherited and having become aware of its limitations, will be able to appreciate the enhanced power of understanding which comes with the extension of relativity to include the dimension of evolution.

Human behaviour *cannot* adequately be understood except in terms of its evolution.[1]

6. Our knowledge of primate behaviour has been transformed in the decade during which I have been writing this book. Some of the literature has been available to me as the main part of the text was being written, e.g. the excellent work of George Schaller on gorillas. A great deal that I have only recently heard about has had, through pressure of time and circumstance, to be virtually neglected – much to my sorrow. Other works I have been able, through the fortunate grant of a Sabbatical year's study leave, to read and digest at least to some extent. Chief among these has been Jane van Lawick-Goodall's wonderful *In the Shadow of Man* (1971). In this section I shall try to show, very briefly, some of the support which may be gleaned from the observations on chimpanzees she reports, for the four-factor view of evolutionary intelligence and especially the P-factor.

The dedication of *In the Shadow of Man* concludes with '. . . and to the memory of David Greybeard'. The reason for this is, undoubtedly, that much of the breakthrough in understanding

---

[1] Although there is still, at time of last writing (April 1972), an enormous extent of ignorance of and/or resistance to the evidence for instinctive behaviour in humans – as noted in the Introduction – it should be emphasized that for well over twenty years the more percipient leaders of theoretical understanding have been putting the true situation before the learned public. One need only mention Bowlby (1951, 1952), Tinbergen (1951) and Lorenz (1952) as first-order investigators from the biological sciences, but practising social scientists, e.g. Fletcher (1957), have also argued the same general position. Fletcher indeed, in *Instinct in Man* (1957) and also more recently (1971), has shown at length and in convincing detail that many of the 'old' psychologists and theorists of social behaviour, e.g. Lloyd Morgan, Freud, McDougall, Hobhouse, and others, had a clear understanding of most of the essentials of the position being established at the present time, even though the amount of formally 'validated' empirical evidence available to them was much less than it has since become. From 1960 or thereabouts, along with the proliferation of third-generation ethologists, there has been a substantial increase in the spread of ethological assumptions and approaches into social science, e.g. Goffman (1969), Tiger (1970), and also into 'popular' literature, e.g. Morris (1967, 1969, 1971), Ardrey (1970). Since about 1967, too, there has been a real 'explosion' in writing about and public

which Jane Goodall achieved was dependent upon her crossing the 'communication gap' between human and chimpanzee; and this was dependent upon David Greybeard's high endowment of P-factor. This was what allowed the withholding of the normal fear-response of flight. 'David Greybeard was less afraid of me, from the start, than were any of the other chimps. I was always pleased when I picked out his handsome face and well marked silvery beard in some chimpanzee group, for, with David to calm the others, I had a better chance of approaching to observe them more closely' (p. 42). David's toleration of the human observer also allowed Jane to make two crucial observations – without which financial support for the continuance of the project might not have been forthcoming, and which were of the greatest theoretical interest in their own right – namely, that chimpanzees eat meat and evince by group behaviour a strong desire for it, and that they not only use tools but also in a rudimentary way make them.

The high average intelligence of the chimps is clear in innumerable incidents. So also is the great range of individual variation in intelligence and especially (as will be shown below) in some of its facets or factors. So also is the very substantial adaptive value of high intelligence to the individual and the group.

Seeing the potential in a pair of empty paraffin cans as 'display apparatus' enabled the chimpanzee Mike, previously lowest in the adult male dominance hierarchy, to depose the top ranking Goliath (pp. 109 ff.). This could be ascribed to high endowment of A-factor: a new significance was perceived. Dr van Lawick-Goodall herself says 'Mike's deliberate use of man-made objects was probably an indication of superior intelligence' (p. 110).

---

interest in the biology of the 'population explosion', environment and conservation, pollution, and related matters (see Nicholson 1970, Terry 1971, for references; my own *Crisis in Abundance* 1966 was an early attempt to explain the biology behind the use and conservation of resources, for non-biologist readers). Undoubtedly we are about to witness a revolution in the social and human sciences (Pringle 1970; Palermo 1971; Davies 1971). One of the consequences of this will be the recognition in both theory and, eventually, practice, that instinctive mechanisms are enormously important even in what might appear to be purely intellectual pursuits (see Stenhouse, *Active Philosophy in Education and Science*, in press). As this is recognized, the need will also be appreciated for intrinsic (endogenous) control systems, e.g. the P-factor of intelligence. In short, recognition of instinct will demand changes in other explanatory theories regarding human behaviour: and the P-factor of intelligence is bound to be highlighted.

Demands would probably be made on D-factor, Sensorimotor Efficiency, for the manipulation of the cans, a novel task for a chimp especially on some occasions when three of them were used. High P-factor, and interacting A and C, are also suggested: 'Charging displays usually occur when a chimpanzee becomes emotionaly excited. . . . But it seemed that Mike actually *planned* his charging displays – almost, one might say, in cold blood. Often, when he got up to fetch his cans, he showed no visible signs of frustration or excitement . . .' (p. 111).

It is in the history of Figan, however, that the clearest indications of the social and adaptive value of intelligence, and notably of the P-factor, are to be obtained.

We meet him as an adolescent (p. 95), apparently in 1964: 'It was during these months that we first realized what an exceptionally gifted chimp was young Figan . . . One day, some time after the group had been fed, Figan suddenly spotted a banana that had been overlooked – but Goliath was resting directly underneath it. After no more than a quick glance from the fruit to Goliath, Figan moved away and sat on the other side of the tent so that he could no longer see the fruit. Fifteen minutes later, when Goliath got up and left, Figan, without a moment's hesitation, went over and collected the banana. Quite obviously he had sized up the whole situation: if he had climbed for the fruit earlier Goliath, almost certainly, would have snatched it away. If he had remained close to the banana he would probably have looked at it from time to time: chimps are very quick to notice and interpret the eye movements of their fellows, and Goliath would possibly, therefore, have seen the fruit himself. And so Figan had not only refrained from instantly gratifying his desire, but had also gone away so that he could not give the game away by looking at the banana.'

Other incidents demonstrate the great intelligence, and in particular the truly enormous power of withholding of a desired response until conditions were appropriate of some individuals. 'Figan was timed on one occasion waiting half an hour for higher-rank chimps to go away before he opened a food-box' (p. 134).

On the question of sex differences in, as well as selection for, P-factor endowment in the young (see above, especially p. 205): 'Male juveniles show much caution when interacting with their elders. It is probably their increasing respect for the adult males that leads to a reduction of, or complete abstinence from, the mounting of pink females [i.e. females in heat] . . .' (p. 158).

And '. . . even Figan, with all his skill in avoiding punishment, was often attacked as an adolescent; and others of his age seemed to be attacked more frequently' (p. 163).

As 'negative' evidence of the importance of P-factor, the behaviour of the orphan Merlin may be cited. He showed maladaptive perseveration (p. 208; cf. Ellen and Wilson 1963; Lawicka and Konorski 1961; Thorpe 1965; Bogen and Bogen 1969). He appeared unable to control the drive to approach adult males, often approached displaying individuals, and as a result got hurt. In fact most of his behavioural tendencies seem to have been exaggerated, both submission and aggression (p. 206). This suggests the implication of P-factor in the *mediation* processes between major instincts. Perhaps even in 'instinctive' (lower) animals the 'combination' of two drives, to produce overt behaviour different from that autochthonous to either drive on its own, is dependent on the initial or stronger drive being held in check so that the allochthonous outflow channel can be used. Such a mechanism for breaking away from an all-or-nothing type of responsiveness would provide an avenue of selection for rudimentary P-factor even within a pure instinct system.

Finally, I suggest that the following passages regarding some of Figan's adult activities show vividly the power of P-factor withholding of response, the 'impassivity display' (see above, p. 136): 'Figan was particularly in awe of Humphrey [second ranking male], but, at the same time, he began to show a good deal less respect for Mike [top ranking male]. Mike began to get very worried about this young upstart. When Mike displayed, the other chimps still rushed out of his way – all but Figan who sat quite calmly, his back to Mike. This happened again and again, and Mike became increasingly uneasy. When Figan was near he displayed more and more frequently, and most of his displays were directed towards the young male. Indeed he swayed the very branches on which Figan sat, his back still firmly turned towards Mike. Yet Mike, it seemed, dared not actually attack Figan' (p. 238).

'Hugo and I suspect that eventually Figan will become the top-ranking male – possibly after Humphrey has had an innings. For Figan is not only more intelligent than Evered [a peer and possible rival] – he also has the backing of a large family [offspring of Flo, the dominant female, a chimpanzee sex-goddess-cum-dynasty-mother]. The proximity of [brother] Faben will probably also give him that feeling of confidence that David Greybeard once gave to Goliath' (p. 239). (See note, p. 358.)

Thus we can begin to see the complex interplay of selection pressures favouring intelligence in general and also the more specific manifestations of its several factors, operating through the individual but also through the blood line of the family (which in the case of the chimpanzees is identifiable only matrilineally). Beyond this there may be natural selection on a 'social group' basis. Since culture is essentially a group function, the corollary of cultural selection, as Darlington (1969) implies, is group selection – though this is never likely to be simple and straightforward.

7. The book *Social Groups of Monkeys, Apes and Men* by Chance and Jolly (1970) provides a valuable conspectus and discussion of work published up to that time (and mainly in the decade 1960–9). Although the greater part of the work considered consisted of field studies, the background of the two authors appears to be mainly related to studies on animals in captivity; and their methodological presuppositions furnish an interesting and instructive contrast with those of some other students of primite behaviour. A brief examination of some of these methodological issues, besides being worthwhile in itself, will lead on to a reinterpretation of various observations which will both illuminate the function of the P-factor of intelligence and resolve some anomalies in the argument given by Chance and Jolly.

First, their methodology.

They criticize Goodall (1963, 1965) on the grounds that no assessment was made of 'observer influence' (p. 49). This is one of the standard criticisms levelled against field as opposed to captivity/laboratory investigations (see Discussion sections in Primates volume of *Symp. zool. Soc. Lond.*, vol. 10, 1963, especially remarks by Zuckerman). It is easy to make, and difficult to refute without going at length into questions of methodology. But what is its real import? Is it, like Vernon's criticism of the absence of follow-up validation in Getzels and Jackson (1962), an implicit demand for something that is unattainable? (see above, Chap. V, sections 4, 5, 6). Are they suggesting that field workers should be able to observe without observing, provide 'control' data over against which any bias in observed data could be detected and allowed for? Stated thus explicitly, it is clear that they are asking the impossible. Hence their criticism is no criticism. This is not of course to say that no criticism can be made of field studies, but the rather sweeping condemnation implicit in their remarks on Goodall's work (see p. 49) is based on simplistic assumptions which ramify throughout their own arguments. They fall into the 'Bogus Precision' fallacy (see above p. 100); and associated with and behind this are simplistic epistemological assumptions about the 'objectivity' of observational knowledge. They talk about 'established knowledge' (p. 49), about the *discovery* of emergent patterns' of behaviour (p. 13, my emphasis), and assert that it is '. . . incorrect to attempt functional explanations before completing a description of, preferably, the whole of a species' behaviour . . .' (p. 15). Later, they advocate the use of techniques of quantitative analysis to build up a (presumably 'true') picture 'from indications derived from other *less rigorous* studies' (p. 164; again my emphasis added).

It would be inappropriate in the present context to attempt a proper dissection of the logical and epistemological substrate of their position. I shall confine myself to remarking that the necessary and inescapable participation of the *subjective* element in all knowledge, all observation, which has been emphasized by philosophers of science such as Toulmin (1961) and Kuhn (1962) as well as by 'pure' philosophers from Kant to Wittgenstein, is totally overlooked. Putting what is part of the same issue a different way: Chance and Jolly illustrate the danger of importing the methodology of 'normal' science (Kuhn 1962) prematurely into a science where a single paradigm has not yet been established (Palermo 1971). In particular, 'observer influence' is always relative to the conclusions to be drawn from a study. The use of feeding stations to facilitate observations of feeding behaviour in chimpanzees, for example, is not likely to make much difference to the motor patterns of feeding activity – though it may well distort the normal relative incidence of frequency of the several patterns. Conclusions about the latter would therefore be subject to provisos which would not apply to the former.

The foregoing considerations can be brought into focus as follows: I suggest that Chance and Jolly, along with perhaps a majority of ethologists at the present time, have distorted their methodological approach by a misinterpretation of Tinbergen's (1951) advice to find out what a species' behaviour *is* before attempting to analyse and explain it. Hence their emphasis, noted above, on 'description', 'discovery', etc. But it is necessary in this context to make and keep always in mind the sharpest possible distinction between:

(a) Formulating a description; and
(b) Experiencing the natural behaviour of the species, in the field.

It is clear that *formulating a description* involves the use of subjective categorizations, not only those of the individual observer but also those of his culture and society as embodied in the language (including mathematical conventions, etc.) that is used. Language must eventually be used, of course, and descriptions formulated – but the whole point of Tinbergen's advice is that the categorizations embodied in the language at the commencement of the study of a particular species may not be entirely appropriate to that species' behaviour. The categories may need to be altered, or new ones made. The only way in which appropriate, sound, and revealing categories are likely to eventuate is if the investigator allows the 'objective reality' in the form of the natural behaviour of the species to impinge very extensively upon his own subjectivity. Hence the need for extensive unstructured experiencing in the field. Note that here and in (b) above I talk of 'experiencing', not 'observing'. The latter term usually carries, as part of its connotation, implications of particularity: we expect to know what we expect to observe; whereas 'experience' presupposes little more than that there will be something-or-other to experience.

Tinbergen himself, in discussing the boom in ethological research which has occurred subsequently to, and largely stimulated by, the publication of *The Study of Instinct* in 1951, is perturbed at the uncritical adoption of ready-made categorizations exhibited by what I have called the 'third-generation ethologists'. He says: 'Having myself always spent long periods of exploratory watching of natural events . . . I find this tendency of prematurely plunging into quantification and experimentation, which I observe in many younger workers, really disturbing' (1969: vi).

It is precisely because Jane Goodall allowed chimpanzee behaviour to determine her categorizations of it (this carrying through into the descriptions she gives in her 1971 publication) that I find her book more revealing, of more use for the elucidation of 'intelligence', than the publications of some other workers, e.g. Chance and Jolly 1970. Her descriptions are less loaded with

M

theoretical presuppositions than theirs, and this facilitates the re-interpretation which is necessary in assessing the relationship between the behaviour of the animals and the new conceptual schema presented in this book. Chance and Jolly, in covering a much more extensive field, inevitably 'encode' behaviour before they can incorpoiate it in their 'map'. This forces upon the proponent of an alternative conceptual schema the necessity of 'de-coding' their argument before 're-encoding' it in the new terms. This task I will shortly begin, but before doing so, I feel that one last methodological issue must be touched upon.

The majority of ethologists who have produced books on the subject have mentioned the enhanced understanding of human behaviour which can result from understanding the behaviour of infra-human animals especially that of the primates. At the same time, everyone emphasizes the need to avoid the fallacy of anthropomorphism. This generates a paradox. If there are similarities between human behaviour and that of the great apes – and on general evolutionary grounds we should expect close similarity – then we should expect a certain amount of insight to be transferable in the ape-to-human comparison. The converse transfer, the human-to-ape comparison, is however, supposed to be forbidden – yet, since we are incomparably better acquainted with human behaviour than we are with ape behaviour, it is precisely this comparison which *prima facie* would appear likely to be most enlightening! Williams (1967, see also Williams 1965) in a sensible discussion of this problem, quotes Lorenz: 'The refraint from humanizing animals would imply the loss of a superlatively important source of knowledge. . . . With monkeys we must expect to find many characteristics which are inherited from common ancestry by them and by us. The similarity is not only functional, but historical, and it would be an actual fallacy not to humanize!' (Williams 1967: 54)

There are dangers either way, as Williams himself emphasizes: 'Subjective humanizing can of course be as pathetic as scientific pedantry.' The cure is the same in every case – to use one's wits, to be aware of all the alternative pitfalls, and to push forward, trying to avoid the traps but undeterred by them.

To return now to Chance and Jolly (1970).

The key to their position is the distinction, in relation to the social behaviour of monkeys and apes, between 'acentric' and 'centripetal' societies (see pp. 178 ff.). Species having acentric social organization (Langur, Patas monkey, Gibbon, Hamadryas baboon) are not organized upon the basis of continuous 'leader-ship' (whether by one individual or several). They do have mechanisms of social dominance which result in hierarchical structure to some extent; but, notably under external threat by a predator, the social group can break up both spatially and functionally at times of stress. At these times the attention of the individual is directed solely towards the environment, into which escape is made. In contrast to these acentric societies, the centri-petal societies (in which are included most monkey and baboon spp. so far studied, also the 'more hominoid' of the apes) are organized, functionally and to a great extent also spatially, round a dominant individual or individuals. The reaction of a centripetal society to the threatening presence of a predator is built round aggression: the dominant male(s) tends to move aggressively towards the predator. Subsequently, the group as a whole can either retreat, covered by the threat of the dominant male(s); or attack, led by the aggressive 'alpha(s)'. These dif-ferences in strategy and organization are associated with differences in habitat, feeding, etc., the 'total behavioural com-plex' being the adaptive unit (cf. Dobzhansky 1956).

Another significant feature of the behaviour of acentric species is the occurrence of what is called 'protean' behaviour (p. 155): erratic components in escape movement, comprising as it were 'unstrung' components from other behaviour patterns. This is felt to have the double function of distracting and making more difficult the predator's attack, and of inhibiting the predator's tendency to attack.

The two types of social organization, with their associated characteristic behaviour, can be summarized briefly as follows:

Acentric	*Centripetal*
'Protean' behaviour	'Organized behaviour
'Individualistic'	Group-orientated
Attention to environment	Social attention

Three points can be made, relative to this analysis, which indicate selection for the P-factor of intelligence (or a functional homologue or precursor of it) in the centripetal species. It is significant that the three species widely accepted on a variety of grounds as being the most intelligent – namely chimpanzee, gorilla, and orang-utan – are in fact classified by Chance and Jolly in the type of society now to be argued as providing the strongest selection pressures for intelligence and especially the P-factor.

(a) Although controlled behaviour is obviously demanded in all societies and all species, it is implicit in Chance and Jolly's argument that there are greater demands on control in centripetal societies. Even assuming that nothing more is involved in this than the selection, from moment to moment, of the appropriate instinctual pattern, this still involves a modicum of withholding of other possible responses in order that a 'chosen' one may be given priority (the 'mediation between instincts' mechanism, see above, p. 350), and thus entails P-factor selection.

(b) In a group-orientated centripetal species the individual can no more afford to neglect the environment than can an individual of acentric species. The social group to which the 'centripetal' individual has to pay proportionately more attention is therefore to be regarded as adding to and extending the 'quantity of relevant environment'. Thus greater demands could be expected to be made upon behaviour; and since the social environment is likely on the whole to be more changeable than the environment in the general sense, the quantitive demands upon all factors of intelligence are likely to be greater.

(c) Additional to this general premium on intelligence, in centripetal species the imposition of frequent or continuous *split attention* – both the environment as such, and also the other members of the social group especially the dominant male(s), having to be kept under attention – must be regarded as generating substantial selection pressures for P-factor (see above, p. 341; Kagan 1970; also Halstead's 1947 tests for central *v.* peripheral attention – which can thus be seen to have very substantial evolutionary and adaptive significance).

Since the phenomenon of social dominance is basic to the centripetal type of society, the mechanisms by which it is achieved and ordered become of prime importance. It is in this area that re-interpretation can result, I suggest, in considerable clarification. Chance and Jolly state that 'the dominant animal may be said to dominate the attention of others at most, if not all, times, and usually without taking any specific actions to achieve this' (p. 172). Later they remark that 'in many species the most active part of this relationship is played by the subordinate rather than by the dominant animal itself' (p. 173). An objection to this can be put in two different ways: is the suggestion that the 'passive display' of the female Grayling Butterfly should count as 'behaviour', indeed as 'display' (see Tinbergen 1951 for references), to be rejected? The idea of 'passive' or 'inactive display' has been accepted, up till now by most ethologists – are Chance and Jolly really suggesting that this notation should be abandoned? Or, to put the same point in more general form, is 'behaviour' now to include only observable physical movements, which would previously have been called observable *changes* in 'behaviour' in the old sense of 'behaviour'? Their position on this is not altogether clear.

Getting down to specifics (and bearing in mind that the social behaviour of gorillas and especially chimpanzees is mainly in question), Chance and Jolly say that: '. . . the cohesion of centripetal societies is maintained by frequent threats [by the dominant male(s)] spaced out in time, which, after a delay when the initial high level of arousal has dissipated somewhat, brings about approach to the dominant animals, which are not only the focus of persistent attention but maintain this state by well-timed threats' (p. 197).

Neglecting several possible quibbles over details, the major point to be made is that of the gross inconsistency between this passage and the previous ones quoted (and several others could be cited in addition) to the effect that there is little or no overt aggressive behaviour, in many chimp and gorilla groups, on the part of the dominant male(s). The relative absence of such aggression is attested also by Schaller (1963, 1967) and by van Lawick-Goodall (1971), yet the dominants are supposed to maintain their position by frequent and 'well-timed threats'.

Chance and Jolly can in fact be rescued from inconsistency only by recognizing the existence of something approximating to the 'impassivity display' (see above, p. 136), a second-order display which, like the display of the Grayling Butterfly, does not involve overt movement or any specific gesture or expression. Depending on context, an absence of display may be a display. It seems hardly necessary to elaborate on this, at this stage. Neither should it be necessary to pile up further evidence of an 'impassivity display', associated with and to a considerable extent dependent upon P-factor endowment, in gorillas and chimpanzees.

I suggest, in conclusion, that far from being antagonistic to the theoretical structuring accomplished by Chance and Jolly (1970) and others, the four-factor theory of evolutionary intelligence which has been argued in this book actually provides a means of resolving various anomalies within other theories, and also a bridge for extending them across inter-disciplinary chasms to achieve a synthesis of wider scope and sounder foundation than has been possible for several decades (see Palermo 1971). I am sharply conscious that my own efforts, so far, have cut only a few trails through the jungle. They seem, however, to make new, useful, and to my mind exciting connections – and perhaps others will think it worthwhile to join in turning the trails into highways.

With regard to the chimpanzee, Figan, Dr van Lawick-Goodall wrote (5 June 1973): '. . . at this very moment, Figan is more or less alpha male—at a very young age, as he can't be more than twenty at the most.'

This offers dramatic confirmation, I suggest, of the social value of intelligence and especially the P-factor. I am very grateful to Dr van Lawick-Goodall for allowing me to cite her observations.

References

Grateful acknowledgement is made to the authors and publishers of works cited in the research incorporated in this volume.

ANDERSON, A. W. 1960. 'A note on high intelligence and low academic performance in a university', *Educand* (W. Australia University Press, Perth, W.A.) 4: 1.
ARDREY, R. 1967. *The Territorial Imperative*, London: Collins.
    1970. *The Social Contract*, London: Collins.

ARONSON, L. R. 1951. 'Orientation and jumping behaviour in the gobiid fish *Bathygobius soporator*', *Amer. Mus. Novitates* **1486**: 1–22.

ASHBY, W. R. 1960. *Design for a Brain*, London: Chapman & Hall.

1964. *An Introduction to Cybernetics*, London: Methuen.

AUSUBEL, D. P. 1963. *The Psychology of Meaningful Verbal Learning*, N.Y.: Grune and Stratton.

AYER, A. J. *et al.* 1956. *The Revolution in Philosophy*, London: Macmillan.

BARNETT, S. A. 1958a. 'The nature and significance of exploratory behaviour', *Proc. roy. Physical Soc. Edinburgh* **27**: 41–45.

1958b. 'Exploratory behaviour', *Brit. J. Psychol.* **49**: 289–310.

1963. *A Study in Behaviour*, London: Methuen.

1967. *'Instinct' and 'Intelligence'*, London: McGibbon & Kee.

BARNETT, S. A. (ed.) 1958c. *A Century of Darwin*, London: Heinemann.

BARRACLOUGH, G. 1962. 'Universal History' in *Approaches to History*, H. P. R. Finberg (ed.) London: Routledge & Kegan Paul.

BARZUN, J. 1961. *The House of Intellect*, London: Secker & Warburg.

BEACH, F. A. 1948. *Hormones and Behaviour*, N.Y.: Hoeber.

BEREITER, C. and ENGELMANN, S. 1966. *Teaching Disadvantaged Children in the Preschool*, Engelwood Cliffs, N.J.: Prentice-Hall.

BERLYNE, D. E. 1960. *Conflict, Arousal and Curiosity*, N.Y.: McGraw-Hill.

BEST, J. B. and RUBINSTEIN, I. 1962. 'Environmental familiarity and feeding in a Planarian', *Science* **135**: 916–8.

BEVERIDGE, W. I. B. 1950. *The Art of Scientific Investigation* (3rd ed., 1957), London: Heinemann.

BIDDLE, B. J. and THOMAS, E. 1966. *Role Theory: concepts and research*, N.Y.: Wiley.

BINET A. and SIMON T. 1905. *Méthodes nouvelles pour le diagnostic du niveau intellectual des anormaux'*, *Année psychol.* **11**: 191–244.

BITTERMANN, M. E. 1960. 'Toward a comparative psychology of learning' *Amer. Psychol.* **15**: 704–12.

1965. 'The evolution of intelligence', *Sci. American* **212**: 92.

BLEST, A. D. 1961. 'The Concept of Ritualization' in *Current Problems in Animal Behaviour*, W. H. Thorpe and O. L. Zangwill (eds), Cambridge: Cambridge University Press.

BLISS, E. L. 1962. *Roots of Behaviour*, N.Y.: Harper.

BOGEN, J. E. and BOGEN, G. M. 1969. 'The other side of the brain. III. The Corpus Collosum and Creativity', *Bull. Los Angeles Neurol. Soc.* **3** (4): 191–220.

BOLWIG, N. 1960. 'Some thoughts on man, primates and their evolution', *Acta 6th Congr. Int. Sci. Anthrop. Ethnol*, Paris.

1961. 'An intelligent tool-using baboon', *S. Afric. J. Sci* **57**: 147–152 (213).

BOURNE, G. H. 1972. *The Ape People*, London: Hart-Davis.

BOWLBY, J. 1951. *Maternal Care and Mental Health*, London: H.M.S.O.

1952. *Child Care and the Growth of Love*, London: Penguin.

360 THE EVOLUTION OF INTELLIGENCE

BOWLBY, J. 1969. *Attachment and Loss. I. Attachment*, London: Penguin, 1971.

BRACE, C. L. 1967. *The Stages of Human Evolution*, Englewood Cliffs, N.J.: Prentice-Hall.

BRISTOWE, W. S. 1958. *The World of Spiders*, London: Collins.

BRUNER, J. S. 1961. *The Process of Education*, Cambridge, Mass.: Harvard University Press.

BRUNER, J. S. 1962. In Gruber *et al.*, *Contemporary Approaches to Creative Thinking*,

BURT, C. 1962. 'The psychology of creative ability', *Brit. J. Ed. Psych.* 32: 292–8.

BUTCHER, H. J. 1968. *Human Intelligence: its nature and assessment*, London: Methuen.

CAMPBELL, B. 1967. *Human Evolution: an introduction to man's adaptations*, London: Heinemann.

CANNON, W. B. 1932. *The Wisdom of the Body*, N.Y.: Norton.

CARNEGIE CORPORATION OF NEW YORK QUARTERLY 9 (3), 1961.

CHANCE, M. R. A. and JOLLY, C. J. 1970. *Social Groups of Monkeys, Apes, and Men*, London: Cape.

CHOMSKY, N. 1970 'Recent contributions to the theory of innate ideas' in Hudson, 1970 (first pub. 1967 *Synthese* 17: 2–11).

COLBERT, E. H. 1955. *Evolution of the Vertebrates*, N.Y.: Wiley.

COMFORT, A. 1970, in a review of Huxley, J. S. 1970. *Memories*, London: Allen & Unwin, in *The Australian* (Canberra). 29/8/70.

COOPER, B. and FOY, J. M. 1967. 'Examinations in higher education – a review', *J. Biol. Educ.*, 1: 139–51.

CRAVIOTO, J., De LICARDIE, E. R., and BIRCH, H. G. 1966. 'Nutrition, growth and neurointegrative development: an experimental and ecologic study', *Pediatrics* 38: 319–372.

CROOK, J. H. 1968. 'The Nature and Function of Territorial Agression' in *Man and Agression*, Montagu, M. F. Ashley (ed.).

(ed.) 1970. *Social Behaviour in Birds and Mammals*, London: Academic Press.

CRUTCHFIELD, R. S. 1962. 'Conformity and Creative Thinking' in *Contemporary Approaches to Creative Thinking*, Gruber *et al.* (eds).

DARCHEN, R. 1952. '*Sur l'activité exploratrice de* Blatella germanica' *Z. Tierpsychol.* 9: 362–72.

DARLINGTON, C. D. 1969. *The Evolution of Man and Society*, London: Allen & Unwin.

DART, R. A. 1963. 'Carnivorous propensities of baboons', *Symp. Zool. Soc. Lond.* 10: 89–102.

DART, R. A. and CRAIG, D. 1959. *Adventures with the Missing Link*, London: Hamilton.

DARWIN, C. 1873. *The Expression of the Emotions in Man and Animals*, London: Murray.

DAVIES, J. KERI, 1971. 'Man unexplored,' *Biol. and Human Affairs* **36** (2): 33–7.

DELACOUR J. and MAYR E. 1945. 'The Family Anatidae', *Wilson Bull.* **57**: 3–55.

De VORE, I. (ed.) 1965. *Primate Behaviour*, N.Y.: Holt, Rinehart.

DIMOND, S. J. 1970. *The Social Behaviour of Animals*, London: Batsford.

DOBZHANSKY, Th. 1956. 'What is an adaptive trait?' *Amer. Nat.* **90**: 337–347.

1967. *Evolution, Genetics, and Man*, N.Y.: Wiley.

DOBZHANSKY, Th., and ASHLEY MONTAGU, M. F. 1947. 'Natural selection and the mental capacities of mankind', *Science* **105**: 587–590.

ECCLES, J. C. 1953. *The Neurophysiological Basis of Mind*, London: Oxford University Press.

EIBL-EIBESFELDT, I. 1955. '*Zur Biologie des Iltis* (Putorius putorius L.)' *Verhandle. Deut. Zool. Ges. Erlangen*, 304–23.

1963. '*Angeborenes und Erworbenes im Verhalten einiger Säuger*', *Z. Tierpsychol*, **20**: 705–54.

EISELEY, L. C. 1958. *Darwins' Century*, N.Y.: Doubleday.

ELLEN, P. and WILSON, A. S. 1963. 'Perseveration in rat following hippocampal lesions', *Exp. Neurol.* **8**: 310–17.

ELTON, C. S. 1958. *The Ecology of Invasions by Animals and Plants*, London: Methuen.

ETKIN, W. 1954. 'Social behaviour and the evolution of man's mental faculties', *Amer. Nat.* **88**: 129–42.

(ed.) 1964. *Social Behaviour and organization among Vertebrates*, Chicago: Chicago University Press.

EWER, R. F. 1957. 'Ethological concepts', *Science* **126**: 599–603.

1961. 'Further observations on suckling behaviour in kittens, together with some general considerations of the interrelations of innate and acquired responses', *Behaviour* **17**: 247–260.

FIELD, T. W. and CROPLEY, A. J. 1969. 'Cognitive Style and Science Achievement', *J. Res. Sci. Teaching* 6 (1).

FLETCHER, R. 1957. *Instinct in Man*, (2nd ed. 1968), London: Allen & Unwin.

1971a. *The Making of Sociology: a study of sociological theory. I. Beginnings and Foundations*, London: Michael Joseph.

1971b. *The Making of Sociology. II. Developments*, London: Michael Joseph.

FOREHAND, G. A. and LIBBY, W. L. 1962. 'Effects of educational programs and perceived organizational climate upon changes in innovative administrative behaviour' in *Innovative Behaviour*, Chicago: University of Chicago Centre for Programs in Government Administration.

FREEDMAN, D. G. 1967. 'A Biological View of Man's Social Behaviour,' in *Social Behaviour from Fish to Man*, W. Erkin (ed.), Chicago: Phoenix.

FREUD, S. 1934. *The Future of an Illusion*, London: Hogarth Press.

FURTH, H. G. 1969. *Piaget and Knowledge*, Englewood Cliffs, N.J.: Prentice-Hall.

GARSTANG, W. 1922. 'The theory of recapitulation: a critical restatement of the biogenetic law', *J. Linn. Soc. Lond.* **35**: 81–101.

GERARD, R. W. 1961. 'The fixation of experience' in *Brain Mechanisms and Learning*, J. F. Delafresnaye, (ed.) C.I.O.M.S. Symposium, London: Oxford, University Press.

GETZELS, J. W. and JACKSON, P. W. 1962. *Creativity and Intelligence*, N.Y.: Wiley.

GIBB, J. A. 1954. 'Feeding ecology of tits, etc.', *Ibis* **96**: 513–43.

GOFFMAN, E. 1969. *Where the Action Is*, London: Allen Lane.

GOODALL, J. 1963. 'Feeding behaviour of wild chimpanzees', *Symp. Zool. Soc. Lond.* **10**: 49–56.

1965. 'Chimpanzees of the Gombe stream reserve' in *Primate Behaviour*, De Vore (ed.), I.

1968. 'The behaviour of Free-living Chimpanzees in the Gombe Stream Reserve,' *Anim. Behav. Monog.* **1** (3): 161–311.

GRAVES, W. L., Freeman M. G., and Thompson, J. D. 1968. 'Culturally related reproductive factors in mental retardation'. Paper read at Conference on Sociocultural Aspects of Mental Retardation, Peabody College, Nashville, Tenn.

GREEN, B. F. 1966. 'Current Trends in Problem Solving' in *Problem Solving: Research, Method, and Theory*, B. Kleinmuntz (ed.), N.Y.: Wiley.

GREENE, J. C. 1959. *The Death of Adam*, Iowa: Iowa State University Press.

GREGORY, R. L. 1961. 'The Brain as an Engineering Problem' in *Current Problems in Animal Behaviour*, W. H. Thorpe and O. L. Zangwill (eds), Cambridge: Cambridge University Press.

GRUBER, H. E., TERRELL G. and WERTHEIMER, M. 1962. *Contemporary Approaches to Creative Thinking*, N.Y.: Atherton Press.

GUILFORD, J. P. 1950. 'Creativity', *American Psychologist* **5**: 444.

1967. *The Nature of Human Intelligence*, N.Y.: McGraw-Hill.

HALL, K. R. L. 1963. 'Observational learning in monkeys and apes', *Brit. J. Psychol.* **54**: 201–26.

1965. 'Social organization of the old world monkeys and apes', *Symp. Zool. Soc. Lond.* **14**: 265–90.

HALSTEAD, W. C. 1947. *Brain and Intelligence*, Chicago: University of Chicago Press.

1951. 'Biological Intelligence' *J. Pers.* **20**: 118–20.

HARLOW, H. F. 1958. 'The Evolution of Learning' in *Behaviour and Evolution*, A. Roe and G. G. Simpson (eds), New Haven, Conn.: Yale University Press.

HASSENSTEIN, B. 1959. '*Optokinetische Wirksamkeit bewegter periodischer Muster*', *Zeits. f. Naturforschung*, II, 6.

HAYES, C. 1952. *The Ape in our House*, London: Gollancz.

HEBB, D. O. 1949. *The Organization of Behaviour*, N.Y.: Wiley (1961 Science Editions).

1953. 'Heredity and environment in mammalian behaviour', *Brit. J. Anim. Behav.* 1: 43–7.

1965. 'The evolution of mind', *Proc. roy. Soc. B* 161: 367–83.

HILGARD, E. R. 1958. *Theories of Learning*, (2nd ed.), London: Methuen.

HILL, K. (ed.) 1964. *The Management of Scientists*, Boston: Beacon Press.

HINDE, R. A. 1955. 'Appetitive behaviour and consummatory act, etc.', *Behaviour* 5: 189–224.

1956a. 'Ethological models and the concept of drive', *Brit. J. Philos. Sci.* 6: 321–31.

1956b. 'The biological significance of the territories of birds', *Ibis* 98: 340–69.

1966. *Animal Behaviour*, N. Y.: McGraw-Hill.

HINDE, R. A., and FISHER, J. 1951. 'Further observations on the opening of milk bottles by birds', *Brit. Birds* 44: 393–6.

HINDE, R. A. and TINBERGEN N. 1958. 'The Comparative Study of Species-specific Behaviour' in *Behaviour and Evolution*, New Haven, Conn.: Yale University Press.

HOCKETT, C. F. 1967. 'The Origin of Speech' in *Human Variation and Origins*, W. S. Laughlin and R. H. Osborne (eds), San Francisco: Freeman (originally in *Scientific American*, Sept. 1960).

HOFFMAN, B. 1962. *The Tyranny of Testing*, N.Y.: Crowell-Collier.

HOGAN, J. A. 1965. 'An experimental study of conflict and fear, etc.', *Behaviour* 25: 45.

HOYLE, F. 1963. *Of Men and Galaxies*, London: Heinemann.

HOWARD, L. 1952. *Birds as Individuals*, London: Collins.

HUDSON, L. 1962. 'Intelligence, divergence and potential originality', *Nature* 196: 601–2.

1963. 'Personality and scientific aptitude', *Nature* 198: 913–14.

1966. *Contrary Imaginations*, London: Methuen.

1968. *Frames of Mind*, London: Methuen.

HUDSON, L. (ed.) 1970. *The Ecology of Human Intelligence*, London: Penguin.

HUNT, J. MCV. 1961. Intelligence and Experience, N.Y.: Ronald.

HUXLEY, A. L. 1946. *The Perennial Philosophy*, London: Collins, 1961.

1954. *The Doors of Perception*, London: Chatto & Windus.

HUXLEY, J. S. 1942. *Evolution: the modern synthesis* (2nd edn., 1963), London: Allen & Unwin.

1953. *Evolution in Action*, London: Chatto and Windus; Pelican Books 1963.

1959. 'Clades and Grades' in *Function and Taxonomic Importance*, A. J. Cain (ed.) London: Systematics Assn. Pub. 3.

1964. 'Psychometabolism: general and Lorenzian' in *Perspectives in Biology and Medicine* 7 (4): 399–432.

364 THE EVOLUTION OF INTELLIGENCE

HUXLEY, T. H. 1871. *Man's Place in Nature*, London: Williams & Norgate.
JERISON, H. L. 1963. 'Interpreting evolution of the brain,' *Hum. Biol.* 35: 263.
JENSEN, A. R. 1969. 'How much can we boost I.Q. and scholastic achievement?', *Harvard Educ. Review* 39: 1–123.
KAGAN, J. 1970a. 'On the Need for Relativism' in Hudson 1970.
1970b. 'Biological Aspects of Inhibition Systems' in Hudson 1970.
KAWAMURA, S. 1959. 'The process of sub-culture propagation among Japanese macaques', *J. Primat.* 2: 43–60.
KELLOGG, W. N. 1961. *Porpoises and Sonar*, Chicago: Chicago University Press.
KERKUT, G. A. 1959. *The Implications of Evolution*, London: Pergamon.
KOESTLER, A. 1964. *The Sleepwalkers: a history of man's changing vision of the universe*, London: Penguin.
KORTLANDT, A. and KOOIJ, M. 1963. 'Protohominid behaviour in primates,' *Symp. Zool. Soc. London* 10: 61–87.
KÖHLER, W. 1925. *The mentality of apes*, London: Routledge & Kegan Paul; Harmondsworth: Penguin 1957.
KOHTS, N. 1935. 'Infant Ape and Human Child', *Sci. Mem. Mus. Darwin, Moscow* 3: 1–586 (cited in Morris and Morris 1966).
KRECHEVSKY, I. 1932. '"Hypothesis" versus "chance" in the pre-solution period in sensory discrimination-learning', *Univ. Calif. Publ. Psychol.* 6: 27–44.
KUHN, T. S. 1962. *The Structure of Scientific Revolutions*, Chicago: University of Chicago Press.
KUO, Z-Y. 1967. *The Dynamics of Behaviour Development*, N.Y.: Random House.
LACK, D. 1954. *The Natural Regulation of Animal Numbers*, Oxford: Oxford University Press.
LANYON, W. E. and TAVOLGA, W. N. (eds) 1960. 'Animal Sounds and Communication', *A.I.B.S. Publ. No.* 7, Washington D.C.
LASHLEY, K. S. 1938. 'Experimental analysis of instinctive behaviour', *Psychol. Rev.* 45: 445–71.
1949. 'Persistent problems in the evolution of mind', *Quart. Rev. Biol.* 24: 28–42.
LAUGHLIN, W. S. and OSBORNE, R. H. (eds) 1967. *Human Variation and Origins*, San Francisco: Freeman.
LAWICK-GOODALL, JANE VAN. 1967. *My Friends the Wild Chimpanzees*, Washington D.C.: National Geographical Society.
1971. *In the Shadow of Man*, London: Collins.
LAWICKA, W. and KONORSKI, J. 1961. 'The effects of prefrontal lobectomies on the delayed responses in cats', *Acta Biol. Experimentalis* 21: 141–56.
LEACH, E. R. 1968. *A Runaway World?*, London: Macmillan.
1970. *Levi-Strauss*, London: Collins.

LEDERBERG, J. 1958. *J. Cell. Comp. Physiol.* (Suppl. 1) **52**: 398.

LEHRMAN, D. S. 1953. 'A critique of Konrad Lorenz's theory of instinctive behaviour', *Quart. Rev. Biol.* **28**: 337–63.

LEWONTIN, R. C. 1965. 'Selection in and of Populations' in *Ideas in Modern Biology* J. A. Moore (ed.), N.Y.: Natural History Press.

LILLEY, J. C. 1962. *Man and Dophin*, London: Gollancz.

LOEB, J. 1918. *Forced movements, Tropisms, and Animal Conduct*, Philadelphia: Lippinocott.

LORENZ, K. 1935. *'Der Kumpan in der Umwelt des Vogels'*, *J. Ornithol.* **83**: 137–214, 289–413.

1950. 'The comparative method in studying innate behaviour patterns in *Physiological Mechanisms in Animal Behaviour*, London: Cambridge University Press (Symp Soc. Expl. Biol.).

1952. *King Solomon's Ring*, London: Methuen.

1965. *Evolution and Modification of Behaviour*, Chicago: University of Chicago Press.

1966. *On Aggression*, London: Methuen.

MCCLELLAND, D. C. 1963a. 'On the Psychodynamics of Creative Physical Scientists' in *Contemporary Approaches to Creative Thinking* Gruber *et al.* (eds).

1963b. 'The Calculated Risk: an aspect of scientific performance' in *Scientific Creativity: its recognition and development*, Taylor and Barrow (eds).

MACRAE, D. G. 1958. 'Darwinism and the Social Sciences' in *A Century of Darwin*, S. A. Barnett (ed.), London: Heinemann.

MAINARDI, D. and PASQUALI, n. 1968. 'Cultural transmission in the house mouse', *Atti Soc. Ital. Sci. Nat. Mus. Civ. Milane.* **107**: 147–52.

MALLESON, N. 1961. 'Instinct and history', *Behavioural Science* **6**: 117–26.

MARLER, P. 1959. 'Development in the Study of Animal Communication' in *Darwin's Biological Work* P. R. Bell (ed.) Cambridge: Cambridge University Press.

1961. 'The logical analysis of animal communication', *J. Theoret. Biol.* **1**: 295–317.

1961. 'The filtering of external stimuli during instinctive behaviour' in *Current Problems in Animal Behaviour*, Thorpe and Zangwill (eds), Cambridge: Cambridge University Press.

MAYR, E. 1942. *Systematics and the Origin of Species*, N.Y.: Columbia University Press.

1958. 'Behavior and Systematics' in *Behaviour and Evolution*, Roe and Simpson (eds).

1963. *Animal Species and Evolution*, Cambridge, Mass.: Bellknap Press.

MCDOUGALL, W. 1926. *Outline of Abnormal Psychology*, N.Y.: Charles Scribner's Sons.

MACKINNON, D. W. 1962. 'The Nature and Nurture of Creative Talent', *Amer. Psychol* **17**: 484–95.

MCREYNOLDS, P. 1962. 'Exploratory behaviour: a theoretical interpretation', *Psychol. Rpts.* **11**: 311–18.

MEAD, A. P. 1960. 'A quantitative method for the analysis of exploratory behaviour in the rat', *Anim. Behav.* **8**: 19–31.

MEAD, MARGARET, 1956. *New Lives for Old: cultural transformation, Manus 1928–1953*, London: Gollancz.

MEDAWAR, P. B. 1960. *The Future of Man*, London: Methuen.

MEYER-HOLZAPFEL, M. 1955. '*Das Spiel bei Säugetieren*', *Handbuch der Zoologie* **8** (10): 1–36.

MONTAGU, M. F. ASHLEY. 1956. *Toynbee and History*, Boston: Porter Sargent.

1968. *Man and Aggression* N.Y.: Oxford University Press.

MORRIS, D. 1957. 'Typical intensity' and its relationship to the problem of ritualization', *Behaviour* **11**: 1–13.

1967. *The Naked Ape*, London: Cape.

1969. *The Human Zoo*, London: Cape.

1971. *Intimate Behaviour*, London: Cape.

MORRIS, R. and MORRIS, D. 1966. *Men and Apes*, London: Hutchinson.

MUNN, N. L. 1955. *The Evolution and Growth of Human Behavior*, London: Harrap.

MURRY, J. M. 1954, *D. H. Lawrence: Son of Woman*, London: Cape (first published 1931).

NICHOLSON M. 1970. *The Environmental Revolution*, London: (Penguin 1972).

NISSEN, H. W. 1958. 'Axes of Behavioral Comparison' in *Behaviour and Evolution*, Roe and Simpson (eds).

NOWELL-SMITH, P. H. 1954. *Ethics*, London: Penguin.

NYBERG, D. 1971. American Naturalist, **105**: 183–5.

OAKLEY, K. P. 1961. *Man the Toolmaker*, London: British Museum (Nat. Hist.).

1962. 'On man's use of fire, with comments on tool-making and hunting' in Washburn, S. L. (ed.) *Social Life of Early Man*, London: Methuen.

PACKARD, V. 1957. *The Hidden Persuaders*, London: Penguin, 1960.

1959. *The Status Seekers*, London: Pelican 1961.

PALERMO, D. S. 1971. 'Is a scientific revolution taking place in psychology?' *Science Studies* **1**: 135–56.

PARKINSON, C. N. 1958. *Parkinson's Law*, London: Murray.

1960. *The Law and the Profits*, London: Murray.

PASSMORE, J. 1958. *A Hundred Years of Philosophy* (2nd ed 1963), London: Duckworth.

1967. 'On Teaching to be Critical' in *The Concept of Education*, R. S. Peters (ed.).

PASSMORE, J. 1969. *Philosophy in the Last Decade*, Sydney: Sydney University Press.

PASTORE, N. 1954. 'Discrimination learning in the canary', *J. comp. physiol. Psychol.* **47**: 288–9, 389–90.

PAVLOV, K. P. 1927. *Conditioned Reflexes: an investigation of the activity of the cerebral cortex*, (Trans. G. V. Anrep), London.

PAYNE, R. 1962. *Lawrence of Arabia: a triumph*, N.Y.: Pyramid Books.

PETERS, R. S. 1967. *The Concept of Education*, London: Routledge & Kegan Paul.

PETTERSSON, H. L. R. 1956. 'Greenfinches and *Daphne mezereum*', *Bird Study* **3**: 147–8.

PIAGET, J. 1967. *Biologie et connaissance*, Paris: Gallimard (see Furth 1969).

PITTENDRIGH, C. S. 1958. 'Adaptation, natural selection and behaviour' in *Behaviour and Evolution*, Roe and Simpson (eds).

POTTER, S. 1948. *Gamemanship*, London: Hart-Davis.

PRECHTL, A. F. R. 1956. '*Neurophysiologische Mechanismen des form-starren Verhaltens*', *Behaviour* **9**: 243–319.

PRINGLE, J. W. S. 1951. 'On the parallel between learning and evolution', *Behaviour* **3**: 174–215.

— 1970. 'Biology as a human science', *J. Inst. Biol.* **17**: 204–7.

QUINE, W. V. O. 1950. *Methods of Logic*, N.Y.: Holt Rinehart.

REED, S. C. 1965. 'The evolution of human intelligence', *Amer. Sci.* **53**: 317–26.

RENSCH, B. 1956. 'Increase of learning capability by increase of brain size', *Amer. Nat.* **90**: 81–95.

— 1966. *Evolution Above the Species Level*, N.Y.: Wiley Science Editions (Original German publication 1954).

RÉVÉSZ, G. 1924. 'Experiments on animal space perception', *Brit. J. Psychol.* **14**: 386–414.

RIESMAN, D. 1953. *The Lonely Crowd*, N.Y.: Doubleday.

RIESS, P. E. 1954. 'The effect of altered environment and of age on the mother-young relationship among animals', *Ann. N.Y. Acad. Sci.* **57**: 606–10.

ROE. A. 1952. *The Making of a Scientist*, N.Y.: Dodd-Mead.

— 1953. 'A Psychological study of eminent psychologists and anthropologists and a comparison with biological and physical scientists', *Psychol. Monogr.* **67** (352).

— 1964 'The Psychology of Scientists' in *The Management of Scientists*, K. Hill (ed.).

ROE, A. and SIMPSON, G. G. 1958. *Behavior and Evolution*, New Haven: Yale University Press.

RUDNER, R. S. 1966. *Philosophy of Social Science*, Englewood Cliffs, N.J.: Prentice-Hall.

RUSSELL, Clair and W. M. S. RUSSELL, 1957. 'An approach to human ethology', *Behavioural Science* 2: 169–200.

1961. *Human Behaviour*, London: Deutsch.

RYLE, G. 1949. *The Concept of Mind*, London: Hutchinson.

1967 in *The Concept of Education*, R. S. Peters (ed.), London: Routledge & Kegan Paul.

SCHALLER, G. 1963. *The Mountain Gorilla*, Chicago: University of Chicago Press (1967. *The Year of the Gorilla*, Harmondsworth: Penguin).

SCHEFFLER, I. 1960. *The Language of Education*, Springfield, Ill.: Thomas.

SCHRIER, A. M. (ed.) 1965. *Behavior of Non-human Primates*, Vols 1 & 2, London: Academic Press.

SCRIVEN, M. 1966. *Primary Philosophy*, N.Y.: McGraw-Hill.

SHERRINGTON, C. S. 1906. *The Integrative Action of the Nervous System*, London: Cambridge University Press.

SHILLITO, E. E. 1963. 'Exploratory behavior in the Short-tailed Vole *Microtus agrestis*', *Behaviour* 21: 145–54.

SHOUKSMITH, G. 1970. *Intelligence, Creativity, and Cognitive Style*, London: Batsford.

SHWAYDER, D. S. 1965. *The Stratification of Behaviour*, London: Routledge & Kegan Paul.

SIMPSON, G. G. 1953. *The Major Features of Evolution* N.Y.: Columbia University Press.

1967. *The Meaning of Evolution*, New Haven: Yale University Press.

SNOW, C. P. 1960. *The Two Cultures and the Scientific Revolution*, London: Cambridge University Press.

SOKOLOV, E. N. 1960. 'Neuronal models and the orienting reflex' in *The Central Nervous System and Behavior*, M. A. B. Brazier (ed.) Washington.

SOROKIN, P. 1956. 'Toynbee's Philosophy of History' in Montagu, M. F. Ashley *Toynbee and History*, Boston: Porter Sargent.

SOUTHWICK, C. H. 1963. *Primate Social Behavior*, Princeton, N. J.: Van Nostrand.

SPURWAY, H. and HALDANE, J. B. S. 1953. 'The comparative ethology of vertebrate breathing', *Behaviour* 6: 8–24.

STEBBING, L. S. 1933. *A Modern Introduction to Logic*, London: Methuen.

STENHOUSE, D. 1962. 'A new habit of the Redpoll *Carduelis flammea* in New Zealand', *Ibis* 104: 250–2.

1965a. 'Teleonomic Teaching and the Supply of Biologists', *Nature* 206: 867–868.

1965b. ' A general theory for the evolution of intelligent behavior', *Nature* 208: 815.

1966. *Crisis in Abundance*, London: Heinemann.

1968a. 'Multiple and mutual grading plan,' *National Education* 49: 423–6 (Wellington, N.Z.)

STENHOUSE, D. 1968b. 'O'Connor's paradox and the teaching of educational philosophy', *Brit. J. Educ. Studies* **16** (3): 243–57.

1969a. 'Good persons, good teachers and language-games', *Educ. Phil. & Theory*: **1**: 41–50.

1969b. 'Examination "selection pressures" in biology', *J. Biol. Educ.* **3**: 233–47.

1971. 'Scientific creativity: 'normal' or 'revolutionary', *Aust. J. Educ.* **15**: 171–84.

1972. *Unstated Assumptions in Education. . . a cross-cultural investigation*, Wellington, N.Z.: Hicks, Smith.

(in press) Active Philosophy in Education and Science, London: Allen & Unwin.

STERN, W. 1914. *The Psychological Methods of Testing Intelligence*, Baltimore: Warwick and York.

TAYLOR, C. W. (ed.) 1956. *Research Conference on the Identification of Creative Scientific Talent*, Salt Lake City: University of Utah Press.

TAYLOR, C. W. and BARRON, F. (eds) 1963. *Scientific Creativity: its recognition and development*, N.Y.: Wiley.

TAYLOR, C. W. and HOLLAND, J. L. 1962. 'Development and application of tests of creativity', *Rev. Educ. Res.* **32**: 91–102.

TAYLOR, C. W. and HOLLAND, J. L. 1964. 'Predictors of creative performance' in *Creativity: progress and potential*, C. W. Taylor (ed.), N.Y.: McGraw-Hill.

TERMAN, L. M. and ODEN, M. H. 1947. *The Gifted Child Grows Up*, Stanford, Calif.: Stanford University Press.

TERRY, M. 1971. *Teaching for Survival: a handbook for environmental education*, N.Y.: Ballantine.

THOMPSON, W. R. 1957. 'Influence of prenatal maternal anxiety on emotionality in young rats', *Science* **125**: 698–9.

THOMPSON, W. R., WATSON, J. and CHARLESWORTH, W. R. 1962. 'The effects of prenatal maternal stress on offspring behaviour in rats', *Psychol. Monogr.*

THOMSON, G. 1961. *The Inspiration of Science*, London: Oxford University Press.

THORNDIKE, E. L. 1921. 'Intelligence and its measurement: a symposium', *J. Ed. Psychol.* **12**: 124–7.

1926. *The Measurement of Intelligence*, New York: Teachers' College.

THORNTON, J. B. 1968. 'The drift from science to arts', *Aust. J. Sci.* **31**: 206–9.

THORPE, W. H. 1951. 'The learning abilities of birds' *Ibis* **93**: 1–52.

1961. 'Sensitive Periods in the Learning of Animals and Men: a study of imprinting with special reference to the induction of cyclic behaviour' in *Current Problems of Animal Behaviour*, eds. Thorpe and Zangwill.

N*

THORPE, W. H. 1963. *Learning and Instinct in Animals*, Cambridge, Massachusetts: Harvard University Press.

1965. 'The Ontogeny of Behaviour' in *Ideas in Modern Biology*, J. A. Moore, (ed.). N.Y.: *Proc. XVI Int. Cong. Zool.*

THORPE, W. H. and ZANGWILL, O. L. 1961. *Current Problems in Animal Behaviour*, Cambridge: University Press.

TIGER, L. 1969. *Men in Groups*, London: Nelson.

TINBERGEN, N. 1942. 'An objectivistic study of the innate behaviour of animals', *Biblioth. biother.* 1: 39–98.

and Van Iersel, J. J. A. 1947. '"Displacement" reactions in the Three-spined stickleback', *Behaviour* 1: 56–63.

TINBERGEN, N. 1951. *The Study of Instinct*, London: Oxford University Press.

1952. '"Derived" activities; their causation, biological significance, origin and emancipation during evolution', *Quart. Rev. Biol.* 27: 1–32.

1953a. *The Herring-gull's World*, London: Collins.

1953b. *Social Behaviour in Animals*, London: Methuen.

1957. 'The functions of territory', *Bird Study* 4: 14–27.

1964. 'Behaviour and Natural Selection' in *Ideas in Modern Biology*, Moore, J. A. (ed.), N.Y.: Natural History Press.

1968. 'On war and peace in animals and man', *Science* 160: 1411–18.

1969. Preface to re-issue of *The Study of Instinct*, London: Oxford University Press.

TINBERGEN, N. and KRUYT, W. 1938. '*Über die Orienterung des Bienenwolfes* (*Philanthus triangulum* Fabr.)' *Z. vergul. Physiol.* 25: 292–334.

TORRANCE, E. P. 1963. *Education and the Creative Potential, Minneapolis*: University of Minnesota Press.

1964. 'Education and Creativity', in *Creativity: Progress and Potential*, C. W. Taylor (ed.) N.Y.: McGraw-Hill.

TOULMIN, S. E. 1961. *Foresight and Understanding*, London: Hutchinson.

TOULMIN, S. E. and GOODFIELD, J. 1965. *The Discovery of Time*, London: Penguin Books.

TOYNBEE, A. J. 1933–47. 'A Study of History' abridged D. C. Somervell London: Oxford University Press, 1947.

TREVOR-ROPER, H. R. 1957. *Encounter* 8 (6) 14–28.

van BEUSEKOM, G. 1948. 'Some experiments on the optical orientation in *Philanthus triangulum* Fabr.' *Behaviour* 1: 195–225.

van HOOFF, J. 1962. 'Facial expressions in the higher primates', *Symp. zool. Soc. Lond.* 8: 97–125.

VERNON, P. E. 1964. 'Creativity and Intelligence', *Educ. Res.* 6: 163–9.

1969. *Intelligence and Cultural Environment*, London: Methuen.

VIAUD, G. 1960. *Intelligence, its evolution and forms*, London: Hutchinson.

VON HOLST, E. and VON ST. PAUL, U. V. 1963. 'On the functional organization of drives,' *Animal Behaviour* 11 (1): 1–20.

VON UEXKULL, J. 1921. *Umwelt und innenwelt der Tiere*, Berlin.

WADDINGTON, C. H. 1957. *The Strategy of the Genes*, London: Allen & Unwin.

1958. 'Theories of Evolution' in *A Century of Darwin*, S. A. Barnett (ed.), London: Heinemann.

WALLACH, M. A. and KOGAN, N. 1965a. *Modes of Thinking in Young Children*, N.Y.: Holt, Rinehart & Winston.

1965b. 'A new look at the creativity-intelligence distinction', *J. Personal.* 33: 348–69.

WALTER, G. 1953. *The Living Brain*, London: Duckworth.

WASHBURN, S. L. 1962. *Social Life of Early Man*, London: Methuen.

WASHBURN, S. L. and DE VORE, I. 1962. 'Social Behaviour of Baboons and Early Man' in *Social Life of Early Man*, Washburn, S. L. (ed.).

WEINER, J. S. 1971. *Man's Natural History*, London: Weidenfeld & Nicholson.

WESTOLL, T. S. 1962. 'Some Crucial Stages in the Transition from Devonian Fish to Man' in *The Evolution of Living Organisms*, G. D. Leeper (ed.), Melbourne: Melbourne University Press.

WHITE, B. L. 1969. 'Child Development Research: an edifice without a foundation', *Merrill-Palmer Quarterly* 15: 49–79.

WHITE, J. P. 1967. 'Indoctrination' in *The Concept of Education*, R. S. Peters (ed.), London: Routledge & Kegan Paul.

WHITEHEAD, A. N. 1933. *Adventures of Ideas*, London: Cambridge University Press; Penguin: 1948.

WHYTE, W. H. 1963. *The Organization Man*, London: Cape.

WILLIAMS, L. 1965. *Samba and the Monkey Mind*, London: Bodley-Head.

1967. *Man and Monkey*, London: Deutsch.

WIRZ, R. 1950. '*Studien über die Cerebralisation: zur Quantitativen Bestimmung der Rangordnung bei Saugetiere*', *Acta Anatomica* 9: 134.

WISDOM, J. 1952. *Other Minds*, Oxford: Blackwell.

WISEMAN, S. (ed.) 1967. *Intelligence and Ability*, London: Penguin.

WITTGENSTEIN, L. 1953. *Philosophical Investigations*, Oxford: Blackwell.

WRIGHT, S. 1938. 'Size of population and breeding structure in relation to evolution', *Science* 87: 430.

YERKES, R. M. and YERKES, A. W. 1929. *The Great Apes*, New Haven, Conn.: Yale University Press.

YOUNG, J. Z. 1963. *Symp. zool. Soc. Lond.* 10: 207–8.

ZIEDINS, R. 1956. 'Conditions of observation and states of observers' *Phil. Rev.* 65: 299–323.

ZUCKERMANN, S. 1932. *The Social Life of Monkeys and Apes*, London: Routledge & Kegan Paul.

Analytic Index

NOTE: In a work which argues for theoretical re-orientation statements, on particular issues can be intelligible only within the framework of the argument as a whole. Since the standard meanings of many technical terms are shown as demanding changes in the meanings of other terms, no one passage can legitimately be considered in isolation from the rest of the book. This might seem to make an index redundant and perhaps misleading. It does so indeed, in the author's opinion. Nevertheless, one is offered. It can serve a purpose, it is hoped, in facilitating recall of particulars *subsequent* to the book's being read as a whole.

Relating to this, references are usually not just to the page nominated, but to all the argument around and following that page. Thus most entries in this index should properly be in the form: *pp 999 ff.*—but for the sake of brevity, only single page numbers are given. These normally indicate the start of the relevant discussion.

Darchen, R. 154
Darlington, C. D. 229, 351
Dart, R. A. 167
Darwin, C. 56, 124
Davies, J. K. 348
Delacour, J. 22
Definitions:
  of intelligence 24, 29
  of instinct 29
  of learning 317
Development, ontogenetic 189
Dewey, J. 238
Discrimination, 134
Displacement 51
Divergent thinking 93, 330
Dobzhansky, Th. 31, 58, 232, 355
'Dual focus' ('Split attention') 340

Eccles, J. C. 135
Education: 238, 259
  pragmatic 261
  examinations in 274
Eibl-Eibesfeldt, I. 298
Eiseley, L. C. 231
Ellen, F. 88, 159, 350
Elton, C. S. 235
Emancipation 51
Ethology 35
Etkin, W. 25, 58, 130
Evolution:
  Darwinian 233
  future human 177
  of intelligence 287
  Lamarckian 233
  of increased variability 324
Ewer, R. F. 52, 60, 202, 301, 322, 343
Exploratory behaviour 64
  mechanisms of 154

'Facts' 22, 259, 290
Fighting 47
Fletcher, R. 34, 222, 347
Four-factor intelligence:
  support for 85
  sequential appearance of factors:
    in pylogeny and ontogeny 195
    in scientists 332
    individual differences in 277
Freud, S. 336
Furth, H. G. 186, 202

Galileo 107, 261
Garstang, W. 186
Gaugin 121
Gerard, R. W. 71
Getzels, J. W. 66, 91, 214, 244, 351
Gibb, J. A. 125, 232
Goffman, E. 347
Goodall, J.—see Lawick-Goodall, Jane van
Green, B. F. 294
Gregory, R. L. 79, 88, 159, 312
Gruber, H. E. 98
Guilford, J. P. 21, 93, 109, 114, 277, 330

Halstead, W. C. 73, 88, 115, 192, 323, 356
Harlow, H. F. 22, 57, 68, 134, 165, 318
Hayes, C. 130
Hebb, D. O. 19, 22, 30, 62, 85, 130, 297, 320
Hierarchical model 41
Hilgard, E. R. 288, 337
Hinde, R. A. 21, 34, 43, 47, 58, 125, 151, 290, 317
Hobhouse, L. T. 69
Hockett, C. F. 176
Hoffman, B. 248
Hogan, J. A. 151
Hominoidea, C-, A-, and D-factors in 131
Howard, L. 52
Hoyle, F. 258
Hudson, L. 66, 93, 108, 121, 245, 270, 278, 283, 330
Huxley, A. L. 26, 66
Huxley, J. S. 19, 34, 63, 111, 179, 201, 229, 324, 337
Huxley, T. H. 124

Ideas, new 210
Impassivity display (P-factor):
  in social dominance 136, 357
  in parent-offspring relations 147
  in politics 220
Instinctual mechanisms:
  in teaching 269
  in advertising 217
  in politics 220
  in imitation 284